Complete Course in Astrobiology

Edited by
Gerda Horneck and Petra Rettberg

1807–2007 Knowledge for Generations

Each generation has its unique needs and aspirations. When Charles Wiley first opened his small printing shop in lower Manhattan in 1807, it was a generation of boundless potential searching for an identity. And we were there, helping to define a new American literary tradition. Over half a century later, in the midst of the Second Industrial Revolution, it was a generation focused on building the future. Once again, we were there, supplying the critical scientific, technical, and engineering knowledge that helped frame the world. Throughout the 20th Century, and into the new millennium, nations began to reach out beyond their own borders and a new international community was born. Wiley was there, expanding its operations around the world to enable a global exchange of ideas, opinions, and know-how.

For 200 years, Wiley has been an integral part of each generation's journey, enabling the flow of information and understanding necessary to meet their needs and fulfill their aspirations. Today, bold new technologies are changing the way we live and learn. Wiley will be there, providing you the must-have knowledge you need to imagine new worlds, new possibilities, and new opportunities.

Generations come and go, but you can always count on Wiley to provide you the knowledge you need, when and where you need it!

William J. Pesce
President and Chief Executive Officer

Peter Booth Wiley
Chairman of the Board

Complete Course in Astrobiology

Edited by
Gerda Horneck and Petra Rettberg

WILEY-VCH Verlag GmbH & Co. KGaA

The Editors

Dr. Gerda Horneck
DLR
Inst. of Aerospace Medicine
51170 Köln

Dr. Petra Rettberg
DLR
Inst. of Aerospace Medicine
51147 Köln

Cover Picture

Picture courtesy of ESO
(The European Southern Observatory)

Library of Congress Card No.:
applied for

British Library Cataloguing-in-Publication Data
A catalogue record for this book is available from the British Library.

Bibliographic information published by the Deutsche Nationalbibliothek
Die Deutsche Nationalbibliothek lists this publication in the Deutsche Nationalbibliografie; detailed bibliographic data are available in the Internet at <http://dnb.d-nb.de>.

Typesetting Dörr + Schiller GmbH, Stuttgart
Printing betz-Druck GmbH, Darmstadt
Binding Litges & Dopf Buchbinderei GmbH, Heppenheim

Printed in the Federal Republic of Germany
Printed on acid-free paper

ISBN: 978-3-527-40660-9

Table of Contents

Complete Course in Astrobiology. Edited by Gerda Horneck and Petra Rettberg

ISBN: 978-3-527-40660-9

Preface

Astrobiology is a relatively new research area that addresses questions that have intrigued humans for a long time: "How did life originate?" "Are we alone in the Universe?" "What is the future of life on Earth and in the Universe?" These questions are jointly tackled by scientists converging from widely different fields, reaching from astrophysics to molecular biology and from planetology to ecology, among others.

Whereas classical biological research has concentrated on the only example of "life" so far known – life on Earth – astrobiology extends the boundaries of biological investigations beyond the Earth to other planets, comets, meteorites, and space at large. Focal points are the different steps of the evolutionary pathways through cosmic history that may be related to the origin, evolution, and distribution of life. In the interstellar medium, as well as in comets and meteorites, complex organics are detected in huge reservoirs that eventually may provide the chemical ingredients for life. More and more data on the existence of planetary systems in our Galaxy are being acquired that support the assumption that habitable zones are frequent and are not restricted to our own Solar System. From the extraordinary ability of life to adapt to environmental extremes, the boundary conditions for the habitability of other bodies within our Solar System and beyond can be assessed. The final goal of astrobiology is to reveal the origin, evolution, and distribution of life on Earth and throughout the Universe in the context of cosmic evolution, and thereby to build the foundations for the construction and testing of meaningful axioms to support a theory of life.

The multidisciplinary character of astrobiology is a challenge on the one hand because it complies with modern science approaches; on the other hand, the full expertise in astrobiology is not always available at a single university. To overcome this problem, experts from seven different European universities or research centers, specialized in leading fields of astrobiology, gathered in an astrobiology lecture course network with live tele-teaching and an interactive question-and-answer period. This book is based mainly on this multidisciplinary lecture series in astrobiology, and each chapter corresponds to a 90-minute lecture. The main fields of astrobiology are covered in a very competent and instructive manner.

Complete Course in Astrobiology. Edited by Gerda Horneck and Petra Rettberg

ISBN: 978-3-527-40660-9

The book starts with a general introduction to the fascinating world of astrobiology. The next chapters provide insights into the different steps of cosmic evolution, from the Big Bang through the formation of galaxies and stellar systems, with emphasis on the evolution of matter required for life: the elements and molecules of life. The history of life on Earth is covered in the next chapters, including the latest results about the RNA world and concepts of a "window for life" as inferred from life's strategies to adapt to factually every location on Earth. This leads to a definition of habitability that is applied to the planets and moons of our Solar System, especially our neighbor planets Venus and Mars and the satellites of the giant planets, Titan and Europa. With the advent of space exploration, space and the bodies of our Solar System are now within our reach; therefore, the technology required for astrobiology missions is also covered, exemplified by astrobiology experiments in low Earth orbit and astrobiology missions to Mars. The book concludes with a chapter on the legal and scientific issues of planetary protection required for each space mission within our Solar System.

This book is intended as a textbook in astrobiology for students and teachers from various fields of science that are interested in astrobiology. In each chapter, a list of questions for students is included. The CD is based on the original lectures that were given at the astrobiology lecture course network. The lectures can be followed on the Web streaming network of the European Space Agency (ESA) (streamiss.spaceflight.esa.int) under "Astrobiology Lecture Course Network (a.y. 2005–2006)."

The editors and authors are grateful to the ESA for providing the platform and support for realizing the Astrobiology Lecture Course Network that this book is based on. Special thanks go to Daniel Sacotte, Director of Human Spaceflight, Microgravity and Exploration Programmes at ESA for providing continuous support, and to Dieter Isakeit, Massimo Sabbatini, and their staff at the Erasmus User Centre & Communication Office of ESA for their competent and efficient performance in providing the required video production and infrastructure tools. Thanks go to former and present coworkers and students of the authors for their contributions to the research in the different chapters. Special thanks go to the following colleagues or organizations: Frances Westall for collaboration on the postulate for habitability, which Chapter 6 is based on, and for providing Figure 5.1; Birgit Huber for providing Figure 5.11; Léna Leroy for drawing Figure 3.2; Audrey Noblet for drawing Figure 3.8; Rainer Facius for providing figures on cosmic radiation in Chapter 11; and Gerhard Kminek for valuable comments on the ESA space missions and planetary protection for human missions in Chapter 13. Support for the laboratory work of Helga Stan Lotter (Chapter 5) by the Austrian FWF grants P16260 and P18256 is gratefully acknowledged. The PPARC provided financial support for the research of Monica Grady (Chapter 8). We appreciate the encouraging support from Christoph von Friedeburg and Nina Stadthaus from the Physics Department of Wiley-VCH, who gave us valuable advice during the preparation of the book. During the editing process, we had continuous support by Folk Horneck, Lisa Steimel, and Thomas Urlings. Their commitment in reviewing and proofreading the manuscripts and in reworking electronic versions of the figures is highly appreciated.

29 June 2006

Gerda Horneck
Petra Rettberg
Cologne, Germany

List of Contributors

Pietro Baglioni
ESA-ESTEC
Directorate of Human Spaceflight
Microgravity and Exploration
Keplerlaan 1
2201 AZ Noordwijk
The Netherlands
e-mail: Pietro.Baglioni@esa.int

André Brack
Centre de Biophysique Moleculaire
CNRS
Rue Charles Sadron
45071 Orleans Cedex 2
France
e-mail: brack@cnrs-orleans.fr

Charles S. Cockell
Planetary and Space Sciences Research Institute
Open University
Milton Keynes, MK7 6AA
United Kingdom
e-mail: c.s.cockell@open.ac.uk

Hervé Cottin
Laboratoire Interuniversitaire des Systèmes Atmosphériques
Universités Paris 12 – Paris 7
CNRS UMR 7583
61 Avenue du Général de Gaulle
94010 Créteil Cedex
France
e-mail: cottin@lisa.univ-paris12.fr

André Debus
Planetary Protection Advisor
CNES
18 Avenue Edouard Belin
31401 Toulouse Cedex 9
France
e-mail: andre.debus@cnes.fr

Stefanos Fasoulos
Institute for Aerospace Engineering
Space Systems and Space Utilisation
Dresden University of Technology
01062 Dresden
Germany
e-mail: fasoulas@tfd.mw.tu-dresden.de

Complete Course in Astrobiology. Edited by Gerda Horneck and Petra Rettberg

ISBN: 978-3-527-40660-9

Monica Grady
Planetary and Space Sciences Research Institute
The Open University
Walton Hall
Milton Keynes MK7 6AA
United Kingdom
e-mail: M. M.Grady@open.ac.uk

Gerda Horneck
DLR
Institute of Aerospace Medicine
Linder Höhe
51170 Cologne
Germany
e-mail: gerda.horneck@dlr.de

Gerhard Kminek
ESA-ESTEC
Directorate of Human Spaceflight
Microgravity and Exploration
Keplerlaan 1
2201 AZ Noordwijk
The Netherlands
e-mail: Gerhard.kminek@esa.int

Harry Lehto
University of Turku
20014 Turku
Finland
and
NORDITA
Blegdamsrej 17
2100 Copenhagen
Denmark
e-mail: hlehto@utu.fi

Kirsi Lehto
Department of Plant Physiology and Molecular Biology
University of Turku
20014 Turku
Finland
and
NORDITA
Blegdamsrej 17
2100 Copenhagen
Denmark
e-mail: klehto@utu.fi

Peter Mani
Tecrisk GmbH
Postfach 298
3047 Bremgarten
Switzerland
e-mail: peter@tecrisk.com

Daniel Prieur
Université de Bretagne Occidentale
Institut Universitaire Européen de la Mer
Technopole Brest Iroise
Place Nicolas Copernic
29280 Plouzane
France
e-mail: Daniel.Prieur@univ-brest.fr

François Raulin
Laboratoire Interuniversitaire des Systèmes Atmosphériques
UMR CNRS 7583
Universités Paris 12 & Paris 7
61 Avenue du General de Gaulle
94010 Créteil Cedex
France
e-mail: raulin@lisa.univ-paris12.fr

Massimo Sabbatini
ESA-ESTEC
Erasmus User Centre &
Communication Office
Keplerlaan 1
2201 AZ Noordwijk
The Netherlands
e-mail: massimo.sabbatini@esa.int

Tino Schmiel
Institute for Aerospace Engineering
Dresden University of Technology
01062 Dresden
Germany
e-mail: tino.schmiel@tu-dresden.de

Christophe Sotin
Laboratoire de Planétologie et
Géodynamique de Nantes (LPGN)
Faculté des Sciences
University of Nantes
B. P. 92208
44322 Nantes
France
e-mail: Christophe.Sotin@univ-nantes.fr

J. Andrew Spry
Biotechnology and Planetary Protection
Group
Jet Propulsion Laboratory
Oak Grove Drive
Pasadena, CA 91109
USA
e-mail: James.a.Spry@jpl.nasa.gov

Helga Stan-Lotter
Division of Molecular Biology
Department of Microbiology
University of Salzburg
Hellbrunnerstr. 34
5020 Salzburg
Austria
e-mail: helga.stan-lotter@sbg.ac.at

Jorge Vago
ESA
Directorate of Human Spaceflight
Microgravity and Exploration
Keplerlaan 1
2201 AZ Noordwijk
The Netherlands
e-mail: Jorge.Vago@esa.int

1
Astrobiology: From the Origin of Life on Earth to Life in the Universe

André Brack

This chapter covers the different theories about the steps toward the origin and evolution of life on Earth, and the major requirements for these processes and for life at large are discussed. Conclusions are drawn on the likelihood of life originating and persisting on other places of our Solar System, such as the terrestrial planets and the moons of the giant planets, or beyond in the Universe.

1.1
General Aspects of Astrobiology

1.1.1
Historical Milestones

Humans in every civilization have always been intrigued by their origin and the origin of life itself. For thousands of years, the comforting theory of spontaneous generation seemed to provide an answer to this enduring question. In ancient China, people thought that aphids were spontaneously generated from bamboos. Sacred documents from India mention the spontaneous formation of flies from dirt and sweat. Babylonian inscriptions indicate that mud from canals was able to generate worms.

For the Greek philosophers, life was inherent to matter. It was eternal and appeared spontaneously whenever the conditions were favorable. These ideas were clearly stated by Thales, Democritus, Epicurus, Lucretius, and even by Plato. Aristotle gathered the different claims into a real theory. This theory safely crossed the Middle Ages and the Renaissance. Famous thinkers such as Newton, Descartes, and Bacon supported the idea of spontaneous generation.

The first experimental approach to the question was published in the middle of the 17th century, when the Flemish physician Van Helmont reported the gener-

Complete Course in Astrobiology. Edited by Gerda Horneck and Petra Rettberg

ISBN: 978-3-527-40660-9

ation of mice from wheat grains and a sweat-stained shirt. He was quite amazed to observe that they were identical to those obtained by procreation. A controversy arose in 1668, when Redi, a Toscan physician, published a set of experiments demonstrating that maggots did not appear when putrefying meat was protected from flies by a thin muslin covering.

Six years after Redi's treatise, the Dutch scientist Anton Van Leeuwenhoek observed microorganisms for the first time through a microscope that he made himself. From then on, microorganisms were found everywhere and the supporters of spontaneous generation took refuge in the microbial world. However, Van Leeuwenhoek was already convinced that the presence of microbes in his solutions was the result of contamination by ambient air. In 1718, his disciple Louis Joblot demonstrated that the microorganisms observed in solutions were, indeed, brought in from the ambient air, but he could not convince the naturalists.

Even Buffon, in the middle of the 18th century, thought that nature was full of the germs of life able to scatter during putrefaction and to gather again, later on, to reconstitute microbes. His Welsh friend John Needham undertook many experiments to support this view. He heated organic substances in water in a sealed flask in order to sterilize the solutions. After a while, all solutions showed a profusion of microbes. The Italian priest Lazzaro Spallanzani argued that the sterilization was incomplete. He heated the solutions to a higher temperature and killed all the microbes, but he could not kill the idea of microbial spontaneous generation.

The controversy reached its apotheosis one century later when Felix Pouchet published his treatise in 1860. He documented the theory of spontaneous generation in the light of experiments that, in fact, were the results of contamination by ambient air. Pasteur gave the finishing blow to spontaneous generation in June 1864 when he designed a rigorous experimental set up for sterilization. By using flasks with long necks that had several bends and were filled with sterilized broth or urine, he showed that no life appeared in the infusions as long as the flask remained intact.

The beautiful demonstration of Pasteur opened the fascinating question of the historical origin of life. Because life can originate only from preexisting life, it has a history and therefore an origin, which must be understood and explained by chemists.

Charles Darwin first formulated the modern approach to the chemical origin of life. In February 1871, he wrote in a private letter to Hooker:

> *If (and oh, what a big if) we could conceive in some warm little pond, with all sorts of ammonia and phosphoric salts, light, heat, electricity, etc., present that a protein compound was chemically formed, ready to undergo still more complex changes, at the present day such matter would be instantly devoured or adsorbed, which would not have been the case before living creatures were formed.*

For 50 years, the idea lay dormant. In 1924, the young Russian biochemist Aleksander Oparin pointed out that life must have arisen in the process of the evolution of matter thanks to the nature of the atmosphere, which was considered to be reducing. In 1928, the British biologist J. B. S. Haldane, independently of Oparin, speculated on the early conditions suitable for the emergence of life. Subjecting a mixture of water, carbon dioxide, and ammonia to UV light should produce a variety of organic substances, including sugars and some of the materials from which proteins are built up. Before the emergence of life they must have accumulated in water to form a hot, dilute "primordial soup." Almost 20 years after Haldane's publication, J. D. Bernal conjectured that clay mineral surfaces were involved in the origin of life. In 1953, Stanley Miller, a young student of Harold Urey, reported the formation of four amino acids – glycine, alanine, aspartic acid and glutamic acid – when he subjected a mixture of methane, ammonia, hydrogen, and water to electric discharges. Miller's publication really opened the field of experimental prebiotic chemistry (see Chapter 3).

1.1.2
Searching for Emerging Life

Defining life is a difficult task, and the intriguing and long lasting question "What is life?" has not yet received a commonly accepted answer, even for what could be defined as *minimal life*, the simplest possible form of life. On the occasion of a Workshop on Life, held in Modena, Italy, in 2003, each member of the International Society for the Study of the Origins of Life was asked to give a definition of life. The 78 different answers occupy 40 pages in the proceedings of the workshop.

Perhaps the most general working definition is that adopted in October 1992 by the NASA Exobiology Program: "Life is a self-sustained chemical system capable of undergoing Darwinian evolution." Implicit in this definition is the fact that the system uses external matter and energy provided by the environment. In other words, primitive life can be defined, *a minima*, as an open chemical system capable of self-reproduction, i.e., making more of itself by itself, and capable of evolving. The concept of evolution implies that the chemical system normally transfers its information fairly faithfully but makes a few random errors. These may potentially lead to higher complexity/efficiency and possibly to better adaptation to changes in the existing environmental constraints.

Schematically, the premises of an emerging life can be compared to parts of "chemical robots." By chance, some parts self-assembled to generate robots capable of assembling other parts to form identical robots. Sometimes, a minor error in the building generated more efficient robots, which became the dominant species.

In a first approach, present life, based on carbon chemistry in water, is generally used as a reference to provide guidelines for the study of the origins of life and for the search for extraterrestrial life. It is generally assumed that the primitive robots emerged in liquid water and that the parts were already organic molecules. The early molecules that contain carbon and hydrogen atoms associated with oxygen, nitrogen, and sulfur atoms are often called the CHONS, where C stands for carbon, H for hydrogen, O for oxygen, N for nitrogen, and S for sulfur.

1.1.3
The Role of Water

Liquid water played a major role in the appearance and evolution of life by favoring the diffusion and exchange of organic molecules. Liquid water has many peculiarities. Water molecules establish hydrogen bonds with molecules containing hydrophilic groups. In water, organic molecules containing both hydrophilic and hydrophobic groups self-organize in response to these properties. This duality generates interesting prebiotic situations, such as the stereo-selective aggregation of short peptide sequences of alternating hydrophobic– hydrophilic residues into thermostable β-sheet structures endowed with chemical activity, as shown below.

In addition to H-bonding capability, water exhibits a large dipole moment (1.85 debye) as compared to alcohols (<1.70 debye). This large dipole moment favors the dissociation of ionizable groups such as -NH_2- and -COOH-generating ionic groups, which can form additional H bonds with water molecules, thus improving their solubility.

With a high dielectric constant ε of 80, water is an outstanding dielectric compound. When organic groups with opposite charges Q and Q', separated by a distance r, are formed, their recombination is unfavorable because the force of attraction for reassociation is given by

$$F = Q \times Q'/\varepsilon \times r^2. \quad (1.1).$$

This is also true for metal ions, which have probably been associated with organic molecules since the beginning of life.

Liquid water was probably active in prebiotic chemistry as a clay producer and heat dissipator. Further, liquid water is a powerful hydrolytic chemical agent. As such, it allows pathways for chemical combinations that would have few chances to occur in an organic solvent. Liquid water is therefore generally considered as a prerequisite for the emergence of life on Earth.

1.1.4
The Physicochemical Features of Carbon-based Life

Life is autocatalytic in essence and is able to evolve. To evolve, i.e., to increase its diversity, the molecules bearing the hereditary memory must be able to be extended and diversified by combinatorial dispersive reactions. This can best be achieved with a scaffolding of polyvalent atoms. From a chemical viewpoint, carbon chemistry is by far the most productive in this respect. Another clue in favor of carbon is provided by radio astronomers: about 90 carbon-containing molecules have been identified in the interstellar medium while only 9 silicon-based molecules have been detected (see Chapter 2). To generalize, a carbon-based life is not just an anthropocentric view, it appears as a highly plausible prerequisite.

Carbon atoms exhibit two remarkable features relative to life, one-handedness and stable isotope distribution. One-handedness, also called homochirality (from the Greek *kheiros*, the hand), of the proteins synthesized via the genetic code is a characteristic of all living systems (see Chapter 3). Each central carbon atom of the amino acid molecule occupies the center of a tetrahedron. Except for the amino acid glycine, the four substituents of the central carbon atom are different. The carbon atom is therefore asymmetrical and it is not superimposable onto its mirror image. Each amino acid exists in two mirror-image forms called enantiomers (from the Greek *enantios*, opposed). All protein amino acids have the same handedness: they are homochiral. They are left-handed and hence known as L-amino acids. Their right-handed mirror images are known as D-amino acids. L-amino acids engaged in a protein chain generate right-handed, single-strand α-helices and asymmetrical, multi-strand β-sheet structures. Nucleotides, the building blocks of the nucleic acids DNA and RNA, are also homochiral. Their nomenclature is more complex because each nucleotide possesses four asymmetrical carbon atoms. In this case, the geometry of a selected carbon atom was chosen. Following this convention, nucleotides are right-handed. Right-handed nucleotides generally generate right-handed helical nucleic acids.

Pasteur was probably the first to realize that biological asymmetry could best distinguish between inanimate matter and life. Life that would simultaneously use both right- and left-handed forms of the same biological molecules appears very unlikely for geometrical reasons. For example, enzyme β-pleated sheets cannot form when both L- and D-amino acids are present in the same molecule. Because the catalytic activity of an enzyme is intimately dependent upon the geometry of the chain, the absence of β-pleated sheets would impede, or at least considerably reduce, the activity spectrum of the enzymes. The use of one-handed biomonomers also sharpens the sequence information of the biopolymers. For a polymer made of n units, the number of sequence combinations will be divided by 2^n when the system uses only homochiral (one-handed) monomers. Taking into account the fact that enzyme chains are generally made up of hundreds of monomers, and that nucleic acids contain several million nucleotides, the tremendous gain in simplicity offered by the use of monomers restricted to one handedness is self-evident.

Finally, if the biopolymers to be replicated were to contain L- and D-units located at specific sites, the replication process would have to not only be able to position the right monomers at the right place but also select the right enantiomer from among two species, which differ only by the geometry of the asymmetrical carbon atoms. For example, the bacterium *Bacillus brevis* is able to synthesize the peptide gramicidin A, which is constructed on a strict alternation of left- and right-handed amino acids. However, the biosynthesis of this peptide involves a set of complex and sophisticated enzymes that are homochiral.

Life on Earth uses homochiral left-handed amino acids and right-handed sugars. A mirror-image life, using right-handed amino acids and left-handed sugars, is perfectly conceivable and might have developed on another planet. Thus, homochirality is generally considered a crucial signature for life.

The uptake of carbon dioxide by living systems can produce biomolecules enriched in ^{12}C carbon isotopes at the expense of ^{13}C isotopes. For example, on Earth, over 1600 samples of fossil kerogen (a complex organic macromolecule produced from the debris of biological matter) have been compared with carbonates in the same sedimentary rocks. The organic matter is enriched in ^{12}C by about 25‰. This offset is now taken to be one of the most powerful indications that life on Earth was active nearly 3.9 billion years ago, because the sample suite encompasses specimens right across the geologic timescale.

1.1.5
Clays as Possible Primitive Robots

Any scientist who has observed the crystallization of minerals initiated by the addition of seeds to a supersaturated solution is tempted to associate life with mineral crystals. Jean Schneider, for example, suggested that complex dislocation networks encountered in crystals may, in some cases, follow the criteria of living units and lead to a crystalline physiology. He also discussed the places of possible occurrence in nature of this kind of physiology, such as terrestrial and extraterrestrial rocks, interplanetary dust, white dwarfs, and neutron stars.

According to Cairns-Smith, there is no compelling reason to necessarily relate the last common ancestor made of organic molecules with first life. Although the easy accessibility of numerous organic building blocks of life has been demonstrated experimentally, the dominant use of these molecules in living organisms can be seen as a result of evolution rather than a prerequisite for its initiation. Cairns-Smith proposed that the first living systems, and the chemical evolution preceding it, might have been based on a chemistry different from that which we know. The structurally and functionally complex genetic system of modern life arose subsequently in a living organism using a less efficient primary system with a much higher probability of spontaneous assembly. As genetic candidates, he advocated crystalline inorganic materials presenting suitable properties, such as the ability to store and replicate information in the form of defaults, dislocations, and substitutions. Clay minerals, such as kaolinite, are particularly attractive because they form at ambient temperatures from aqueous weathering of silicate rock.

The following "genetic takeover" scenario was proposed by Cairns-Smith as a mineral origin of life. Certain clays having properties that favor their synthesis proliferated, and their replication defects, likewise, became more common. In certain clay lineages, the development of crude photochemical machinery favored the synthesis of some non-clay species, such as polyphosphates and small organic compounds. Natural selection favored these lineages of clays because the organic compounds they produce catalyzed the clay formation. Multiple-step pathways of high specificity, including chiral stereoselection, arose through specific adsorption, followed by the appearance of polymers of specified sequence, at first serving only structural roles. Base-paired polynucleotides replicated, giving rise to secondary and minor genetic material. This secondary material proved to be useful in the alignment of amino acids for polymerization. The ability to produce specific enzymes came concomitantly with the ability to produce sequence-specified polypeptides and proteins. More-efficient pathways of organic synthesis ensued, and finally the clay machinery was dispensed with in favor of a polynucleotide-based replication–translation system. Although each step of the hypothetical sequence of events was developed in detail, the scenario has not been supported by experimental facts.

1.2 Reconstructing Life in a Test Tube

It is now a generally accepted conception that life emerged in water and that the first self-reproducing molecules and their precursors were probably organic molecules built up with carbon atom skeletons. Organic molecules were formed from gaseous molecules containing carbon atoms (methane, carbon dioxide, carbon monoxide), nitrogen atoms (nitrogen, ammonia), or sulfur atoms (hydrogen sulfide, sulfur oxide). The energy for these reactions came from electric discharges, cosmic and UV radiation, or heat.

1.2.1 The Quest for Organic Molecules

1.2.1.1 Terrestrial Production

In 1953, Stanley Miller exposed a mixture of methane, ammonia, hydrogen, and water to electric discharges to mimic the effects of lightning. Among the compounds formed, he identified 4 of the 20 naturally occurring amino acids, the building blocks of proteins. Since this historic experiment, 17 natural amino acids have been obtained via the intermediate formation of simple precursors, such as hydrogen cyanide and formaldehyde. It has been shown that spark discharge synthesis of amino acids occurs efficiently when a reducing gas mixture containing significant amounts of hydrogen is used (see Chapter 3). However, the true composition of the primitive Earth atmosphere remains unknown. Today, geochemists favor a non-reducing atmosphere dominated by carbon dioxide. Under such conditions, the production of amino acids appears to be very limited. Strongly reducing environments capable of reducing carbon dioxide were necessary for the synthesis of amino acids.

Deep-sea hydrothermal systems may represent suitable reducing environments for the synthesis of prebiotic organic molecules. The ejected gases contain carbon dioxide, carbon monoxide, sulfur dioxide, nitrogen, and hydrogen sulfide. For instance, high concentrations of hydrogen (more than 40% of the total gas) and methane have been detected in the fluids of the Rainbow ultramafic hydrothermal system of the Mid-Atlantic Ridge. The production of hydrogen, a highly reducing agent favoring prebiotic syntheses, is associated with the hydrous alteration of olivine into serpentine and magnetite, a reaction known as "serpentinization." Indeed, hydrocarbons containing 16 to 29 carbon atoms have been detected in these hydrothermal fluids. Amino acids have been obtained, although in low yields, under conditions simulating these hydrothermal vents.

According to Wächtershäuser (1994), primordial organic molecules formed near the hydrothermal systems; the energy source required to reduce the carbon dioxide might have been provided by the oxidative formation of pyrite (FeS_2), from iron sulfide (FeS) and hydrogen sulfide (H_2S). Pyrite has positive surface charges and bonds the products of carbon dioxide reduction, giving rise to a two-dimensional reaction system, a "surface metabolism." Laboratory work has provided support for this promising new "metabolism-first" approach. Iron sulfide, hydrogen sulfide, and carbon dioxide react under anaerobic conditions to produce hydrogen and a series of thiols, including methanethiol. Methanethiol and acetic acid have also been obtained from carbon monoxide, hydrogen sulfide, iron and nickel sulfides, and catalytic amounts of selenium. Under specific conditions, thioesters are formed that might have been the metabolic driving force of a "thioester world," according to de Duve (1998). Thioesters are sulfur-bearing organic compounds that presumably would have been present in a sulfur-rich, volcanic environment on the early Earth.

Hydrothermal vents are often disqualified as efficient reactors for the synthesis of bioorganic molecules because of their high temperature. However, the products that are synthesized in hot vents are rapidly quenched in the surrounding cold water, which may preserve those organics formed.

1.2.1.2 Delivery of Extraterrestrial Organic Molecules

Comets and meteorites may have delivered important amounts of organic molecules to the primitive Earth (see Chapter 3). Nucleic acid bases, purines, and pyrimidines have been found in the Murchison meteorite. One sugar (dihydroxyacetone), sugar alcohols (erythritol, ribitol), and sugar acids (ribonic acid, gluconic acid) have been detected in the Murchison meteorite, but ribose, the sugar moiety of ribonucleotides, themselves the building blocks of RNAs (see Chapter 4), has not been detected.

Eight proteinaceous amino acids have been identified in one such meteorite, among more than 70 amino acids found therein. Cronin and Pizzarello (1997) found an excess of about 9 % of L-enantiomers for some non-protein amino acids detected in the Murchison meteorite. The presence of L-enantiomeric excesses in these meteorites points towards an extraterrestrial process of asymmetric synthesis of amino acids asymmetry that is preserved inside the meteorite. These excesses may help us to understand the emergence of biological asymmetry or one-handedness. The excess of one-handed amino acids found in the meteorites may result from the processing of the organic mantles of the interstellar grains from which the meteorite was originally formed. That processing could occur, for example, by the effects of circularly polarized synchrotron radiation from a neutron star, a remnant of a supernova. On the other hand, strong infrared circular polarization, resulting from dust scattering in reflection nebulae in the Orion OMC-1 star formation region, has been observed by Bailey. Circular polarization at shorter wavelengths might have been important in inducing this chiral asymmetry in interstellar organic molecules that could be subsequently delivered to the early Earth.

Dust collection in the Greenland and Antarctic ice sheets and its analysis by Maurette show that the Earth captures interplanetary dust as micrometeorites at a rate of about 20 000 tons per year. About 99% of this mass is carried by micrometeorites in the 50–500 μm size range. This value is much higher than the most reliable estimate of the normal meteorite flux, which is about 10 tons per year. A high percentage of micrometeorites in the 50–100 μm size range have been observed to be unmelted, indicating that a significant fraction traversed the terrestrial atmosphere without drastic heating. In this size range, the carbonaceous micrometeorites represent 20% of the incoming micrometeorites, and they contain 2.5% carbon on average. This flux of incoming micrometeorites might have brought to the Earth about 2.5×10^{22} g carbon over the period corresponding to the late heavy bombardment. As inferred from the lunar craters, during this period planetesimals, asteroids, and comets impacted the Earth until about 3.85 billion years ago (see Chapter 2). For comparison, this delivery represents more carbon than is in the present surficial biomass, which amounts to about 10^{18} g. In addition to organic compounds, these grains contain a high proportion of metallic sulfides, oxides, and clay minerals that belong to various classes of catalysts. In addition to the carbonaceous matter, micrometeorites might also have delivered a rich variety of catalysts. These may have functioned as tiny chemical reactors when reaching oceanic water.

On 15 January 2006, the *Stardust* reentry probe brought cometary grains to the Earth for the first time. The preliminary analyses of these grains from comet Wild-2 confirmed that micrometeorites are probably witnesses of a chemical continuum – via the cometary grains – from the interstellar medium, where they form, to terrestrial oceans. Comets are the richest planetary objects in organic compounds known so far. Ground-based observations have detected hydrogen cyanide and formaldehyde in the coma of comets.

In 1986, onboard analyses performed by the two Russian missions *Vega 1* and *Vega 2*, as well as observations obtained by the European mission *Giotto* and the two Japanese missions *Suisei* and *Sakigake*, demonstrated that Halley's Comet shows substantial amounts of organic material. On average, dust particles ejected from Halley's nucleus contain 14% organic carbon by mass. About 30% of cometary grains were dominated by the light elements C, H, O, and N, and 35% are close in composition to the carbon-rich meteorites. The presence of organic molecules, such as purines, pyrimidines, and formaldehyde polymers, has been inferred from the fragments analyzed by the *Giotto* PICCA and *Vega* PUMA mass spectrometers. However, there was no direct identification of the complex organic molecules probably present in the cosmic dust grains and in the cometary nucleus.

Many chemical species of interest for exobiology were detected in Comet Hyakutake in 1996, including ammonia, methane, acetylene (ethyne), acetonitrile (methyl cyanide), and hydrogen isocyanide. The study of the Hale-Bopp Comet in 1997 led to the detection of methane, acetylene, formic acid, acetonitrile, hydrogen isocyanide, isocyanic acid, cyanoacetylene, formamide, and thioformaldehyde.

It is possible, therefore, that cometary grains might have been an important source of organic molecules delivered to the primitive Earth. Comets orbit on unstable trajectories and sometimes collide with planets. The collision of Comet

Shoemaker-Levy 9 with Jupiter in July 1994 is a recent example of such an event. Such collisions were probably more frequent 4 billion years ago, when the comets orbiting around the Sun were more numerous than today. By impacting the Earth, comets probably delivered a substantial fraction of the terrestrial water (about 35% according to the estimation of Owen based on the relative contents in hydrogen and deuterium) in addition to organic molecules. The chemistry that is active at the surface of a comet is still poorly understood. The European mission *Rosetta*, launched in 2004, will analyze the nucleus of the comet 67 P/Churyumov-Gerasimenko. The spacecraft will first study the environment of the comet during a flyby over several months, and then a probe will land to analyze the surface and the subsurface ice sampled by drilling.

1.2.2 Space Experiments

Ultraviolet (UV) irradiation of dust grains in the interstellar medium may result in the formation of complex organic molecules. The interstellar dust particles are assumed to be composed of silicate grains surrounded by ices of different molecules, including carbon-containing molecules (see Chapter 3). In a laboratory experiment, ices of H_2O, CO_2, CO, CH_3OH, and NH_3 were deposited at 12 K under a pressure of 10^{-5} Pa (10^{-7} mbar) and were then irradiated by electromagnetic radiation representative of that in the interstellar medium. The solid layer that developed on the solid surface was analyzed by enantioselective gas chromatography coupled with mass spectrometry (GC-MS). After the analytical steps of extraction, hydrolysis, and derivatization, 16 amino acids were identified in the simulated ice mantle of interstellar dust particles. These experiments confirmed the preliminary amino acid formation obtained first by Greenberg. The chiral amino acids were identified as being totally racemic (consisting of equal amounts of D- and L-enantiomers). Parallel experiments performed with ^{13}C-containing substitutes definitely excluded contamination by biological amino acids. The results strongly suggest that amino acids are readily formed in interstellar space.

Before reaching the Earth, organic molecules are exposed to UV radiation, both in interstellar space and in the Solar System. Amino acids have been exposed in Earth orbit to study their survival in space. The UV flux of wavelengths <206 nm in the diffuse interstellar medium is about 10^8 photons cm^{-2} s^{-1}. In Earth orbit, the corresponding solar flux is in the range of 10^{16} photons cm^{-2} s^{-1}. This means that an irradiation over one week in Earth orbit corresponds to that over 275 000 years in the interstellar medium. Thin films of amino acids, like those detected in the Murchison meteorite, have been exposed to space conditions in Earth orbit within the BIOPAN facility of the European Space Agency (ESA) onboard the unmanned Russian satellites *Foton 8* and *Foton 11* (see Chapter 11). Aspartic acid and glutamic acid were partially destroyed during exposure to solar UV. However, decomposition was prevented when the amino acids were embedded in clays.

Amino acids and peptides have also been subjected to solar radiation outside the MIR station for 97 days. After three months of exposure, about 50% of the amino acids were destroyed in the absence of mineral shielding. Peptides exhibited a noticeable sensitivity to space vacuum, and sublimation effects were detected. Decarboxylation was found to be the main effect of photolysis. No polymerization occurred and no racemization (the conversion of L- or D-amino acids into a racemic mixture) was observed. Some samples were embedded into mineral material (montmorillonite clay, basalt, or Allende meteorite) when exposed to space. Among the different minerals, the meteoritic powder offered the best protection, whereas the clay montmorillonite was the least efficient. Different thicknesses of meteorite powder films were used to estimate the shielding threshold. Significant protection from solar UV radiation was observed when the thickness of the meteorite mineral was 5 μm or greater.

1.2.3
Attempts to Recreate Life in a Test Tube

By analogy with contemporary living systems, it was long considered that primitive life emerged as cellular entities, requiring boundary molecules able to isolate the system from the aqueous environment (membrane). Further, catalytic molecules would be needed to provide the basic chemical work of the cell (enzymes), as well as information-retaining molecules that allow the storage and transfer of the information needed for replication (RNA).

Fatty acids are known to form vesicles when the hydrocarbon chains contain more than 10 carbon atoms. Such vesicle-forming fatty acids have been identified in the Murchison meteorite by Deamer (1985). However, the membranes obtained with these simple amphiphiles are not stable over a broad range of conditions. Stable neutral lipids can be obtained by condensing fatty acids with glycerol or with glycerol phosphate, thus mimicking the stable contemporary phospholipid. Primitive membranes also could initially have been formed by simple isoprene derivatives.

Most of the catalytic chemical reactions in a living cell are achieved by proteinaceous enzymes made of 20 different homochiral L-amino acids. Amino acids were most likely available on the primitive Earth as complex mixtures, but the formation of mini-proteins from the monomers in water is not energetically favored. Because the peptide bond of proteins is thermodynamically unstable in water, it requires an energy source, such as heat or condensing chemicals, to link two amino acids together in an aqueous milieu. Oligoglycines up to the octamer have been obtained from glycine in a flow reactor simulating hydrothermal circulation. These compounds were formed by repeatedly circulating solutions of glycine and $CuCl_2$ from a high-pressure, high-temperature chamber to a high-pressure, low-temperature chamber, simulating conditions in a hydrothermal vent. Clays and salts may also be used to condense free amino acids in water. When subjecting mixtures of glycine and kaolinite to wet–dry cycles and 25–94 °C temperature fluctuations, the formation of oligopeptides up to five glycines long has been observed.

Bulk thermal condensation of amino acids has been described by Fox, who has shown that dry mixtures of amino acids polymerize when heated to 130 °C to give "proteinoids." In the presence of polyphosphates, the temperature can be decreased to 60 °C. High molecular weights were obtained when an excess of acidic or basic amino acids was present. In aqueous solutions, the proteinoids aggregated spontaneously to form microspheres of 1–2 μm, presenting an interface resembling the lipid bilayers of living cells. The microspheres increased slowly in size from dissolved proteinoids and were sometimes able to bud and to divide. These microspheres were described as catalyzing the decomposition of glucose and were able to work as esterases and peroxidases. The main advantage of proteinoids is their organization into particles, but they also represent a dramatic increase in complexity.

The number of condensing agents capable of assembling amino acids into peptides in water is limited, especially when looking for prebiotically plausible compounds. Carbodiimides, having a general formula R-N=C=N-R', power the ligation of both glutamic and aspartic acids on hydroxyapatite. The simplest carbodiimide, H-N=C=N-H, can be considered a transposed form of cyanamide (NH_2-CN), which is present in the interstellar medium. In water, cyanamide forms a dimer, dicyandiamide, which is as active as carbodiimides in forming peptides. However, the reactions are very slow and do not proceed beyond the tetrapeptide.

Oligomers of glutamic acid greater than 45 amino acids in length were formed on hydroxyapatite and illite after 50 feedings with glutamic acid and the condensing agent carbonyldiimidazole (Im-CO-Im). Amino acid adenylate anhydrides have been reported to condense readily in the presence of montmorillonite. According to de Duve (1998), the first peptides may have appeared via thioesters, which generate short peptides in the presence of mineral surfaces. Among the most effective activated amino acid derivatives for the formation of oligopeptides in aqueous solution are the *N*-carboxyanhydrides.

Chemical reactions capable of selectively condensing protein amino acids more readily than non-protein ones have been described. Helical and sheet structures may be modeled with the aid of only two different amino acids, the first one being hydrophobic and the second hydrophilic. Polypeptides with alternating hydrophobic and hydrophilic residues adopt a water-soluble β-sheet geometry because of hydrophobic side-chain clustering. Because of the formation of β-sheets, alternating sequences display a good resistance toward chemical degradation. Aggregation of alternating sequences into β-sheets is possible only with homochiral (all L or all D) polypeptides, as demonstrated by this author at the Centre de Biophysique Moléculaire in Orléans, France. Short peptides have also been found to exhibit catalytic properties.

In contemporary living systems, the hereditary memory is stored in nucleic acids built up with bases (purine and pyrimidine), sugars (ribose for RNA, deoxyribose for DNA), and phosphate groups. The accumulation of significant quantities of natural RNA nucleotides does not appear to be a plausible chemical event on the primitive Earth. Purines are easily obtained from hydrogen cyanide or by subjecting reduced gas mixtures to electric discharges. No successful pyrimidine synthesis from electrical discharges has been published so far, whereas hydrogen cyanide affords only very small amounts of these bases. Condensation of formaldehyde leads to ribose as well as a large number of other sugars. Although the synthesis of purine nucleosides (the combination of purine and ribose) and of nucleotides has been achieved by heating the components in the solid state, the yields are very low and the reactions are not regioselective. Interestingly, very effective montmorillonite-catalyzed condensation of nucleotides into oligomers up to 55 monomers long has been reported by Ferris from nucleoside phosphorimidazolides.

The synthesis of oligonucleotides is much more efficient in the presence of a preformed, pyrimidine-rich polynucleotide acting as a template. Non-enzymatic replication has been demonstrated by Orgel. The preformed chains align the nucleotides by base-pairing to form helical structures that bring the reacting groups into close proximity. However, the prebiotic synthesis of the first oligonucleotide chains remains an unsolved challenge.

1.2.4
A Primitive Life Simpler than a Cell?

Thomas Cech found that some RNAs, the ribozymes, have catalytic properties (see Chapter 4). For example, they increase the rate of hydrolysis of oligoribonucleotides. They also act as polymerization templates, because chains up to 30 monomers long can be obtained starting from a pentanucleotide. Since this primary discovery, the catalytic spectrum of these ribozymes has been considerably enlarged by the directed test tube molecular evolution experiments initiated in the laboratories of Gerald Joyce and Jack Szostak. Since RNA was shown to be able to act simultaneously as genetic material and as a catalyst, RNA has been considered the first living system on the primitive Earth (the "RNA world"). This is because it can simultaneously be the genotype and the phenotype and can fulfill the basic cycle of life consisting of self-replication, mutation, and selection. Strong evidence for this proposal has been obtained from the discovery that modern protein synthesis in the ribosome is catalyzed by RNA. One should, however, remember that RNAs synthesis under prebiotic conditions remains an unsolved challenge. It seems unlikely that life started with RNA molecules, because these molecules are not simple enough. The RNA world appears to have been an episode in the evolution of life before the appearance of cellular microbes rather than the spontaneous birth of life.

RNA analogues and surrogates have been studied. The initial proposal was that the first RNA was a flexible, achiral derivative in which ribose was replaced by glycerol, but these derivatives did not polymerize under prebiotic conditions. The ease of forming pyrophosphate (double phosphate) bonds prompted investigation of linking nucleotides by pyrophosphate groups. This proposal was tested using the reactions of the diphosphorimidazolides of deoxynucleotides. Their reaction in the presence of magnesium or manganese ions resulted in the formation of 10–20 mers of the oligomers.

Considering the ease of formation of hexose-2,4,6-triphosphates from glycolaldehyde phosphate in a process analogous to the formose reaction, Eschenmoser and coworkers chemically synthesized polynucleotides containing hexopyranose ribose (pyranosyl-RNA or p-RNA) in place of the usual "natural" pentofuranose ribose found in RNA. p-RNAs form Watson-Crick-paired double helices that are more stable than RNA. Furthermore, the helices have only a weak twist, which should make it easier to separate strands during replication. Replication experiments have had marked success in terms of sequence copying but have failed to demonstrate template-catalysis turnover numbers greater than 1. The chemical synthesis of threo furanosyl nucleic acid (TNA), an RNA analogue built on the furanosyl form of the tetrose sugar threose, was also reported by the Eschenmoser group. TNA strands are much more stable in aqueous solution than RNA and are resistant to hydrolysis. They form complementary duplexes between complementary strands. Moreover, of even greater potential importance, they form complementary strands with RNA. This raises the possibility that TNAs could have served as templates for the formation of complementary RNAs by template-directed synthesis. TNA is a more promising precursor to RNA than p-RNA, because tetroses have the potential to be synthesized from glycolaldehyde phosphate and two other carbon precursors, which may have been present in quantities greater than those of ribose on the primitive Earth.

Peptide nucleic acids (PNA), first synthesized by Nielsen and coworkers, consist of a peptide-like backbone to which nucleic acid bases are attached. PNAs form very stable double-helical structures with complementary strands of PNA (PNA-PNA), DNA (PNA-DNA), RNA (PNA-RNA), and even stable PNA_2-DNA triple helices. Information can be transferred from PNA to RNA, and vice versa, in template-directed reactions. Although PNA hydrolyzes rather rapidly, thus considerably restricting the chances of PNA ever having accumulated in the primitive terrestrial oceans, the PNA-PNA double helix illustrates that genetic information can be stored in a broad range of double-helical structures.

Chemists are also tempted to consider that primitive self-replicating systems must have used simpler informational molecules than biological nucleic acids or their analogues. Because self-replication is, by definition, autocatalysis, they are searching for simple autocatalytic molecules capable of mutation and selection. Reza Ghadiri tested peptides as possible templates, and non-biological organic molecules have been screened by Günter von Kiedrowski. In most cases, the rate of the autocatalytic growth was not linear. The initial rate of autocatalytic synthesis was found to be proportional to the square root of the template concentration: the reaction order in these autocatalytic self-replicating systems was found to be 1/2 rather than 1, a limiting factor as compared to most autocatalytic reactions known so far. Autocatalytic reactions are particularly attractive because they might amplify small enantiomeric excesses, eventually extraterrestrial, to homochirality.

1.3
The Search for Traces of Primitive Life

1.3.1
Microfossils

The first descriptions of fossil microorganisms in the oldest, well-preserved sediments (3.5–3.3 billion years old) from Australia and South Africa were made by

William Schopf and Elso Barghoorn. Although their observations have been called into question, Westall and others have more recently established that life was abundant in these rocks. From a theoretical point of view, the earliest forms of life would most likely have had a chemolithoautotrophic metabolism, using inorganic materials as a source of both carbon and energy. Evidence for the presence of chemolithotrophs in the hydrothermal deposits from Barberton and the Pilbara comes from the N and C isotopic data. Considering the high temperatures on the early Earth, it is probable that the earliest microorganisms were thermophilic. Structures resembling oxygenic photosynthetic microorganisms such as cyanobacteria have been described, and carbon isotope data have been interpreted as evidence for their presence. However, these interpretations are strongly disputed. Nevertheless, the fact that the microorganisms inhabiting the early Archean environments of Barberton (South Africa) and Pilbara (Australia) formed mats in the photic zone suggests that some of them may already have developed anoxygenic photosynthesis.

1.3.2
Oldest Sedimentary Rocks

The oldest sedimentary rocks occur in southwest Greenland and have been dated at 3.8–3.85 billion years by Manfred Schidlowski and Stephen Mozsis. The old sediments testify to the presence of permanent liquid water on the surface of the Earth and to the presence of carbon dioxide in the atmosphere. They contain organic carbon. Minik Rosing found that the isotopic signatures of the organic carbon found in Greenland meta-sediments provide indirect evidence that life may be 3.85 billion years old. Taking the age of the Earth as about 4.6 billion years, this means that life must have begun quite early in Earth's history. Although there are serious reservations concerning these studies (contamination by more recent fossilized endoliths, ^{12}C enrichment produced by thermal decomposition of siderite and metamorphic processes), stepped combustion analysis by Mark von Zuilen has verified the existence of the ^{12}C enrichment in these sediments.

1.4
The Search for Life in the Solar System

1.4.1
Planet Mars and the SNC Meteorites

The mapping of Mars by the spacecrafts *Mariner 9*, *Viking 1* and *Viking 2*, *Mars Global Surveyor*, and *Mars Express* revealed channels and canyons resembling dry riverbeds (see Chapter 8). The gamma-ray spectrometer instrument onboard the Mars orbiter *Odyssey* detected hydrogen, which indicates the presence of water ice in the upper meter of soil in a large region surrounding the planet's south pole, where ice is expected to be stable. The amount of hydrogen detected indicates 20–

50% ice by mass in the lower layer beneath the topmost surface. The ice-rich layer is about 60 cm beneath the surface at latitude 60 °S and approaches 30 cm of the surface at latitude 75 °S. The ancient presence of liquid water on the surface of Mars was confirmed by the two American Mars Exploration Rovers (MER A and B), *Spirit* and *Opportunity*, and by the presence of sulfates and clays revealed by the infrared spectrometer OMEGA onboard the European Mars orbiter *Mars Express*. However, Martian oceans were probably restricted to the very early stages of Martian history. Mars therefore possessed an atmosphere capable of decelerating carbonaceous micrometeorites, and chemical evolution may have been possible on the planet.

The *Viking 1* and *Viking 2* lander missions were designed to address the question of extant (rather than extinct) life on Mars (see Chapter 5). Three experiments were selected to detect metabolic activity, such as photosynthesis, nutrition, and respiration of potential microbial soil communities. The results were ambiguous, because although "positive" results were obtained, no organic carbon was found in the Martian soil by gas chromatography–mass spectrometry (GC-MS). It was concluded that the most plausible explanation for these results was the presence at the Martian surface of highly reactive oxidants, such as hydrogen peroxide, which would have been photochemically produced in the atmosphere. Direct photolytic processes were suggested to be responsible for the degradation of organics at the Martian surface. Because the *Viking* landers could not sample soils below 6 cm, the depth of this apparently organic-free and oxidizing layer is still unknown. More information is expected from the upcoming lander missions to Mars: the U. S. *Mars Science Laboratory* (MSL) to be launched in 2009 and the European *ExoMars* mission to be launched in 2013 (see Chapter 12).

There are Martian rocks on Earth represented by a small group of meteorites of igneous (volcanic) origin, known as the SNC meteorites (after their type specimens Shergotty, Nakhla, and Chassigny), that had comparatively young crystallization ages, equal to or less than 1.3 billion years (see Chapter 8). One of these meteorites, designated EETA79001, was found in Antarctica in 1979. It had gas inclusions trapped within a glassy component. Both compositionally and isotopically, this gas matched the makeup of the Martian atmosphere, as measured by the *Viking* mass spectrometer. The data provide a very strong argument that at least that particular SNC meteorite came from Mars, representing the product of a high-energy impact that ejected material into space. There are now 34 SNC meteorites known in total.

Two of the SNC meteorites, EETA79001 and ALH84001, supply new and highly interesting information. A subsample of EETA79001, excavated from deep within the meteorite, has been subjected to stepped combustion. The CO_2 release from 200–400 °C suggested the presence of organic molecules. The carbon is enriched in ^{12}C, and the carbon isotope difference between the organic matter and the carbonates in Martian meteorites is greater than that observed on Earth. This could be indicative of biosynthesis, although some as yet unknown abiotic processes could perhaps explain this enrichment. David McKay has reported the presence of other features in ALH84001 that may represent a signature of relic biogenic activity on Mars, but this biological interpretation is strongly questioned.

Because Mars had a presumably "warm" and wet past climate, it should have sedimentary rocks deposited by running and/or still water on its surface as well as layers of regolith generated by impacts. Such consolidated sedimentary rocks therefore ought to be found among the Martian meteorites. However, no such sedimentary material has been found in any SNC meteorite. It is possible that they did survive the effects of the escape acceleration from the Martian surface but did not survive terrestrial atmospheric entry because of decrepitation of the cementing mineral. The "STONE" experiment, flown by ESA, was designed to study precisely such physical and chemical modifications to sedimentary rocks during atmospheric entry from space (see Chapter 11). A piece of basalt (representing a standard meteoritic material), a piece of dolomite (sedimentary rock), and an artificial Martian regolith material (80% crushed basalt and 20% gypsum) were embedded into the ablative heat shield of the satellite *Foton 12*, which was launched on 9 September 1999 and landed on 24 September 1999. Such an experiment had never been performed before, and the samples, after their return, were analyzed for their chemistry, mineralogy, and isotopic compositions by a European consortium. Atmospheric entry modifications are made visible by reference to the untreated samples. The results suggest that some Martian sedimentary rocks might, in part, survive terrestrial atmospheric entry from space.

Even if convincing evidence for ancient life in ALH84001 has not been established, the two SNC meteorites (EETA79001 and ALH84001) do show the presence of organic molecules. This suggests that the ingredients required for the emergence of primitive life were present on the surface of Mars. Therefore, it is tempting to consider that microorganisms may have developed on Mars and lived at the surface until liquid water disappeared. Because Mars probably had no plate tectonics, and because liquid water seems to have disappeared from Mars' surface very early, the Martian subsurface perhaps keeps a frozen record of very early forms of life.

Currently, a very intensive exploration of Mars is planned, and U. S., European, Russian, and Japanese missions to Mars are taking place or are in the planning. Exobiology interests are included, especially in the analysis of samples from sites where the environmental conditions may have been favorable for the preservation of evidence of possible prebiotic or biotic processes. The European Space Agency (ESA) convened an Exobiology Science Team chaired by this author to design an integrated suite of instruments to search for evidence of life on Mars. Priority was given to the *in situ* organic and isotopic analysis of samples obtained by subsurface drilling. The basic recommendations of the Exobiology Science Team are presently serving as guidelines for the elaboration of the Pasteur exobiology payload of the ESA *ExoMars* mission scheduled for 2013 (see Chapter 12).

1.4.2
Jupiter's Moon Europa

The Jovian moon Europa appears as one of the most enigmatic of the Galilean satellites. With a mean density of about 3.0 g cm^{-3}, the Jovian satellite should be dominated by rocks. Ground-based spectroscopy, combined with gravity data, suggests that the satellite has an icy crust, which is kilometers thick, and a rocky interior. The images obtained by the *Voyager* spacecraft in 1979 showed very few impact craters on Europa's surface, indicating recent, and probably continuing, resurfacing by cryovolcanic and tectonic processes. Recent images of Europa's surface, taken by the *Galileo* spacecraft over 14 months (from December 1997 to December 1999), show surface features – iceberg-like rafted blocks, cracks, ridges, and dark bands – that are consistent with the presence of liquid water beneath the icy crust. Data from *Galileo*'s near-infrared mapping spectrometer show hydrated salts that could be evaporites. The most convincing argument for the presence of an ocean of liquid water comes from *Galileo*'s magnetometer. The instrument detected an induced magnetic field within Jupiter's strong magnetic field. The strength and response of the induced field require a near-surface, global conducting layer, most likely a layer of salty water (see Chapter 10).

If liquid water is present within Jupiter's moon Europa, it is quite possible that it includes organic matter derived from thermal vents. Terrestrial-like prebiotic organic chemistry and primitive life may therefore have developed in Europa's ocean. If Europa maintained tidal and/or hydrothermal activity in its subsurface until now, it is possible that microbial activity is still present. Thus, the possibility of extraterrestrial life present in a subsurface ocean of Europa must seriously be considered. The most likely sites for extant life would be at hydrothermal vents below the most recently resurfaced area. To study this directly would require making a borehole through the ice in order to deploy a robotic submersible. On the other hand, biological processes in and around hydrothermal vents could produce biomarkers that would be pushed up as traces in cryovolcanic eruptions and thereby be available at the surface for *in situ* analysis or sample return. Mineral nutrients delivered through cryovolcanic eruption would make the same locations the best candidates for autotrophic life.

1.4.3 Saturn's Moon Titan

Titan's atmosphere was revealed mainly by the *Voyager 1* mission in 1980, which yielded the bulk composition (90% molecular nitrogen and about 1–8% methane). Also, a great number of trace constituents were observed in the form of hydrocarbons, nitriles, and oxygen-containing compounds, mostly CO and CO_2. Titan is the only other object in our Solar System bearing any resemblance to our own planet in terms of atmospheric pressure, which amounts to $1.5\ 10^5$ Pa (1.5 bar), and carbon/nitrogen chemistry. It therefore represents a natural laboratory for studying the formation of complex organic molecules on a planetary scale and over geological times (see Chapter 9). The European Space Agency's Infrared Space Observatory (ISO) has detected tiny amounts of water vapor in the higher atmosphere, but Titan's surface temperature (94 K) is much too low to allow the presence of liquid water. Although liquid water is totally absent at the surface, the satellite provides a unique milieu to study, *in situ*, the products of the fundamental physical and chemical interactions driving a planetary organic chemistry. Herewith, Titan serves as a reference laboratory to study, by default, the role of liquid water in exobiology.

The NASA–ESA *Cassini–Huygens* spacecraft launched in October 1997 arrived in the vicinity of Saturn in 2004 and performed several flybys of Titan, making spectroscopic, imaging, radar, and other measurements. On 14 January 2005, an instrumented descent probe managed by European scientists penetrated the atmosphere and systematically studied the organic chemistry in Titan's geofluid. For 150 min, *in situ* measurements provided analyses of the organics present in the air, in the aerosols, and at the surface. The GC-MS of the *Huygens* probe measured the chemical composition and the isotopic abundances, from an altitude of 140 km down to the surface. The main findings are as follows:

- nitrogen and methane are the main constituents of the atmosphere;
- the isotopic ratio $^{12}C/^{13}C$ suggests a permanent supply of methane in the atmosphere;
- the surface is "wetted" by liquid methane and rich in organics (cyanogen, ethane);
- the presence of ^{40}Ar suggests the existence of internal geological activity.

1.5 The Search for Life Beyond the Solar System

1.5.1 The Search for Rocky Earthlike Exoplanets

Apart from abundant hydrogen and helium, 114 interstellar and circumstellar gaseous molecules have been identified in the interstellar medium (see Chapter 2). It is commonly agreed that the catalog of interstellar molecules represents only a fraction of the total spectrum of molecules present in space, the spectral detection being biased by the fact that only those molecules possessing a strong electric dipole can be observed. Among these molecules, 83 contain carbon, whereas only 7 contain silicon. Silicon has been proposed as an alternative to carbon as the basis of life. However, silicon chemistry is apparently less inventive and does not seem to be able to generate any life as sophisticated as the terrestrial carbon-based one. May these molecules survive the violent accretion phase generating a planetary system?

The origin and distribution of the molecules from the interstellar medium to the planets, asteroids, and comets of the Solar System are presently at the center of a debate based on isotope ratios. Some molecules might have survived in cold regions of the outer Solar System, whereas others would have been totally reprocessed during accretion. Whatever the case, the interstellar medium tells us that organic chemistry is universal.

What about liquid water? New planets have been discovered beyond the Solar System. On 6 October 1995, the discovery of an extrasolar planet, i.e., a planet outside our Solar System, was announced by Michel Mayor and Didier Queloz. The planet orbits an 8 billion year old star called 51 Pegasus, 42 light years away within the Milky Way. The suspected planet takes just four days to orbit 51 Pegasus. It has a surface temperature around 1 000 °C and a mass about 0.5 the mass of Jupiter. One year later, seven other extrasolar planets were identified. Among them, 47 Ursa Major has a planet with a surface temperature estimated to be around that of Mars (–90 to –20 °C), and the 70 Virginis planet has a surface temperature estimated at about 70–160 °C. The latter is the first known extrasolar planet whose temperature might allow the presence of liquid water. As of May 2006, 193 exoplanets had been observed.

1.5.2 Detecting Extrasolar Life

Extrasolar life, i.e., life on a planet of a solar system other than our own, will not be accessible to space missions in the foreseeable future. The formidable challenge of detecting distant life must therefore be tackled by astronomers and radio astronomers. The simultaneous detection of water, carbon dioxide, and ozone (an easily detectable telltale signature of oxygen) in the atmosphere would constitute the most convincing biomarker but not absolute proof. Other anomalies in the atmospheres of telluric exoplanets (i.e., rocky Earthlike planets), such as the presence of methane, might be the signature of an extrasolar life. European astrophysicists are proposing the construction of a flotilla of four free-flying spacecrafts, each containing an infrared telescope of 3 m in diameter, to search for signs of life on terrestrial-like planets. The mission, called *Darwin-IRSI*, is presently under study at ESA. Finally, the detection of an unambiguous electromagnetic signal (via the Search for Extraterrestrial Intelligence program) would obviously be an exciting event.

1.6 Conclusions

On Earth, life probably appeared about 4 billion years ago, when some assemblages of organic molecules in a liquid water medium began to transfer their chemical information and to evolve by making a few accidental transfer errors. The number of molecules required for those first assemblages is still unknown. The problem is that on Earth, those molecules have been erased. If life started on Earth with the self-organization of a relatively small number of molecules, its emergence must have been fast; therefore, the chance of the appearance of life on any appropriate celestial bodies might be real. On the other hand, if the process required thousands of different molecules, the event risks being unique and restricted to the Earth.

The discovery of a second independent genesis of life on a body presenting environmental conditions similar to those prevailing on the primitive Earth would strongly support the idea of a rather simple genesis of terrestrial life. More than just a societal wish, the discovery of a second genesis of life is a scientific need for the study of the origin of life. It will demonstrate that life is not a magic, one-shot process but a rather common phenomenon. Many scientists are convinced that microbial life is not restricted to the Earth, but such conviction now needs to be supported by facts.

1.7 Further Reading

1.7.1 Books and Articles in Books

Brack, A. (Ed.) *The Molecular Origins of Life: Assembling Pieces of the Puzzle*, Cambridge University Press, Cambridge, 1998.

Gilmour, I., Sephton, M. A. (Eds.) *An Introduction to Astrobiology*, The Open University and Cambridge University Press, Cambridge, 2003.

Clancy, P., Brack, A., Horneck, G. *Looking for Life. Searching the Solar System*, Cambridge University Press, Cambridge, 2005.

De Duve, C. Possible starts for primitive life. Clues from present-day biology: the thioester world, in: A. Brack (Ed.) *The Molecular Origins of Life: Assembling Pieces of the Puzzle*, Cambridge University Press, Cambridge, pp. 219–236, 1998.

Fox, S. W. Dose, K. *Molecular Evolution and the Origin of Life* (Revised Edition), Marcel Dekker, New York, 1977.

Gargaud, M., Barbier, B., Martin, H., J. Reisse, J. (Eds.) *Lectures in Astrobiology*, Springer-Verlag, Berlin, Heidelberg, 2005.

Horneck. G., Baumstark-Khan, C. (Eds.) *Astrobiology – The Quest for the Conditions of Life*, Springer, Berlin, Heidelberg, New York, 2002.

Jakosky, B. *The Search for Life on Other Planets*, Cambridge University Press, 1998.

Jones, B. W. (Ed.) *Life in the Solar System and beyond*, Springer, Berlin, Heidelberg, New York 2004.

Maurette, M. *Micrometeorites and the Mysteries of our Origins*, Springer, Berlin Heidelberg, New York, 2006.

Rauchfuss, H. *Chemische Evolution*, Springer, Berlin, Heidelberg, New York, 2005

Schulze-Makuch, D., Irwin L. N. *Life in the Universe*, Springer, Berlin, Heidelberg, New York, 2004.

Westall, F., Southam, G. The early record of life, in: K. Benn, J.-C. Mareschal, K. C. Condie (Eds.) *Archean Geodynamics and Environments*, American Geophysical Union, Washington DC, pp. 283–304, 2006.

1.7.2 Articles in Journals

Boillot, F., Chabin, A., Buré, C., Venet, M., Belsky, A., Bertrand-Urbaniak, M., Delmas, A., Brack, A., Barbier, B. The Perseus exobiology mission on MIR: behaviour of amino acids and peptides in Earth orbit, *Origins Life Evol. Biosphere* **2002**, *32*, *359–385*.

Brack, A. From amino acids to prebiotic active peptides: a chemical reconstitution, *Pure Appl. Chem.* **1993**, *65*, 1143–1151.

Brack, A., Baglioni, P., Borruat, G., Brandstätter, F., Demets, R., Edwards, H. G. M., Genge, M., Kurat, G., Miller, M. F., Newton, E. M., Pillinger, C. T., Roten, C.-A., Wäsch, E. Do meteoroids of sedimentary origin survive terrestrial atmospheric entry? The ESA artificial meteorite experiment *STONE*, *Planet. Space Sci.* **2002**, *50*, *763–772*.

Cronin, J. R., Pizzarello, S. Enantiomeric excesses in meteoritic amino acids, *Science* **1997**, *275, 951–955.*

Deamer, D. W. Boundary structures are formed by organic components of the Murchison carbonaceous chondrite, *Nature* **1985**, *317*, 792–794.

Eschenmoser, A. Chemical etiology of nucleic acid structure, *Science* **1999**, *284, 2118–2124.*

Ferris, J. P., Hill, A. R. Jr, Liu, R., Orgel, L. E. Synthesis of long prebiotic oligomers on mineral surfaces, *Nature* **1996**, *381*, 59–61.

Luther, A., Brandsch, R., von Kiedrowski, G. Surface-promoted replication and exponential amplification of DNA analogues, *Nature* **1998**, *396*, 245–248.

Maurette, M., Duprat, J., Engrand, C., Gounelle, M., Kurat, G., Matrajt, G., Toppani, A. Accretion of neon, organics, CO_2, nitrogen and water rom large interplanetary dust particles on the early, *Earth Planet. Space Sci.* **2000**, *48*, 1117–1137.

Mayor, M. D., Queloz, D. A. Jupiter-mass companion to a solar-type star, *Nature* **1995**, *378*, 355–359.

Miller, S. L. The production of amino acids under possible primitive Earth conditions, *Science* **1953**, *117*, 528–529.

Munoz Caro, G. M., Meierhenrich, U. J., Schutte, W. A., Barbier, B., Arcones Segovia, A., Rosenbauer, H., Thiemann, W. H.-P., Brack, A., Greenberg, J. M. Amino acids from ultraviolet irradiation of interstellar ice analogues, *Nature* **2002**, *416, 403–405.*

Niemann, H.B et al. The abundances of constituents of Titan's atmosphere from the GC-MS instrument on the Huygens probe, *Nature* **2005**, *438*, 779–784.

Ogata, Y., Imai, E.-I., Honda, H., Hatori, K., Matsuno, K. Hydrothermal circulation of seawater through hot vents and contribution of interface chemistry to prebiotic synthesis, *Origins Life Evol. Biosphere* **2000**, *30*, 527–537.

Ourisson, G., Nakatani, Y. The terpenoid theory of the origin of cellular life: the evolution of terpenoids to cholesterol, *Chem. Biol.* **1994**, *1*, 11–23.

Paecht-Horowitz, M., Berger, J., Katchalsky, A. Prebiotic synthesis of polypeptides by heterogeneous polycondensation of amino-acid adenylates, *Nature* **1970**, *228*, 636–639.

Pizzarello, S., Cronin, J,R. Non-racemic amino-acids in the Murray and Murchison meteorites, *Geochim. Cosmochim. Acta* **2000**, *64(2)*, 329–338.

Schmidt, J. G., Nielsen, P. E., Orgel, L. E. Information transfer from peptide nucleic acids to RNA by template-directed syntheses, *Nucleic Acids Res.* **1997**, *25*, 4797–4802.

Schöning, K.-U., Scholz, P., Guntha, S., Wu, X., Krishnamurthy, R., Eschenmoser, A. Chemical etiology of nucleic acid structure: the α-threofuranosyl-(3'2') oligonucleotide system, *Science* **2000**, *290*, 1347–1351.

Schopf, J. W. Microfossils of the Early Archean Apex Chert: new evidence of the antiquity of life, *Science* **1993**, *260, 640–646.*

Severin, K. S., Leem, D. H., Martinez, J. A., Ghadiri, M. R. Peptide self-replication via template-directed ligation, *Chem. Eur. J.***1997**, *3, 1017–1024.*

Terfort, A., von Kiedrowski, G. Self-replication by condensation of 3-aminobenzamidines and 2-formyl-phenoxyacetic acids, *Angew. Chem. Int. Ed. Engl.* **1992**, *31*, 654–656.

Wächtershäuser, G. Life in a ligand sphere, *Proc. Natl. Acad. Sci. USA* **1994**, *91*, 4283–4287.

Westall, F. Life on the early Earth: a sedimentary view, *Science* **2005**, *308*, 366–337.

Wintner, E. A., Conn, M. M., Rebek, J. Studies in molecular replication, *Acc. Chem. Res.* **1994**, *27*, 198–203.

1.7.3
Web Sites

Jet Propulsion Laboratory / NASA: *Mars Meteorites*, **2003** http://www.jpl.nasa.gov/snc/

Schneider, J. **2003**. http://www.obspm.fr/encycl/f-encycl.html.

2
From the Big Bang to the Molecules of Life

Harry J. Lehto

This chapter deals with the evolution of the Universe that predates both chemical evolution and the evolution of life. First, the origin of the elements critical for life is reviewed. The origin of hydrogen and helium is explained from cosmological models. Hydrogen is required when water is formed. Helium, which appears quite useless for life, turns out to be critical in the stellar synthesis of carbon leading to heavier elements. The second constituent of water, oxygen, is also formed in the hot cores of stars. Once these elements are expelled from stars back into the cool interstellar space, they get mixed into the interstellar clouds resulting in a vast spectrum of molecules. The observed properties of these molecules and of the larger dust particles are reviewed. They play an important role in the formation of planets in the accretion disks of metal-rich populations of stars. This material, which is reprocessed by the newly formed stars and other planets in the planetary system, "rains" on the young planet in the form of meteorites, comets, and asteroids, creating a supply of raw materials for the prebiotic evolution and the eventual biological evolution of life. These processes are considered in the light of the formation of the Earth.

2.1
Building Blocks of Life

Life as we know it is made of compartments, the cells. Within all cells similar molecules are found, no matter which cell is considered.

- Water (H_2O) is the basis for the cytoplasm of all cells: all the essential biological reactions take place in water. Water is also essential for transport of nutrients and waste. The elements in water are quite obviously hydrogen (H) and oxygen (O) (see Chapter 1).
- Nucleic acids (DNA and RNA) are essential to present life forms in providing instructions on how to build proteins and

Complete Course in Astrobiology. Edited by Gerda Horneck and Petra Rettberg

ISBN: 978-3-527-40660-9

in transferring this information from one generation of cells to the next. Nucleic acids are long polymers that consist of nucleotide bases – cytosine, guanine, thymine, and adenine in DNA and cytosine, guanine, uracil, and adenine in RNA – joined by a backbone formed by a sugar – deoxyribose in DNA or ribose in RNA containing five carbon atoms – and a phosphate group, connecting the adjacent sugars. The elements found in the nucleic acids are carbon (C), H, nitrogen (N), O, and phosphorus (P) (see Chapters 3 and 4).

- Amino acids are the building blocks of proteins. They are also quite complex molecules that are exclusively made of C, H, N, O, and sulfur (S) (see Chapter 3).
- Lipids are the fundamental molecules in cellular membranes. Normally, membranes are bi-layered with water-attracting (hydrophilic) layers on the outside and water-repelling (hydrophobic) layers in the "internal" layers. The hydrophobic layers are hydrocarbon chains, and the hydrophilic ends usually contain phosphate groups. The main elements in lipids are C and H, but O, N, and P are also used in the hydrophilic ends.
- Carbohydrates, comprising sugars, starches, collagens, or cellulose, are important for all life forms. Carbohydrates are energy-rich compounds that function as energy-storage and structural compounds in cells. On the first level, they are produced by autotrophic life forms by reducing CO_2, either by using energy derived from geochemical compounds or by binding solar energy. The sugars deoxyribose and ribose play an essential role in forming the backbone of the DNA and RNA. The essential elements in carbohydrates are C, H, and O.

During the dawn of life the prebiotic formation of these complex molecules took place through simpler precursor molecules. These are most likely H_2O, formaldehyde (H_2CO), hydrogen cyanide (HCN), small sugars $(CH_2O)_n$, ammonia (NH_3), methane (CH_4), and other hydrocarbons $-(CH_2)_n-$ (see Chapter 3).

Six elements, namely C, H, O, N, S, and P, make up 98% of all living tissue. They are the most important elements of known life. The remaining 2% of living tissue is made of other elements. The exact list of trace elements differs somewhat among authors. There is common agreement that the most important trace elements are sodium (Na), chlorine (Cl), potassium (K), fluorine (F), calcium (Ca), iodine (I), magnesium (Mg) and iron (Fe). Ca and F are needed for external skeletons; K, Na, Cl, and I for nervous systems; and Mg and Fe for chlorophyll and hemoglobin, respectively. K, Na, and Cl are needed for maintaining the correct osmotic pressure of cells.

Some authors suggest a list of less important trace elements, such as boron (B), aluminum (Al), silicon (Si), chromium (Cr), manganese (Mn), copper (Cu), zinc

(Zn), selenium (Se), strontium (Sr), molybdenum (Mo), silver (Ag), tin (Sn), lead (Pb), nickel (Ni), bromine (Br), and vanadium (V). Other authors would remove some elements or add some other ones. The need for specific trace elements depends on the species and its environment.

In summary, one may conclude that approximately 25–30 elements are needed for life. The remaining 80 elements are not essential and often turn out to be detrimental to life.

2.2 Big Bang: Formation of H and He

At the beginning, the Universe was small, dense, and hot and there were no elements. What happened during the first 10^{-43} seconds is beyond the realm of physics. At 10^{-43} seconds the temperature was 10^{19} GeV (for convenience, 1 eV k^{-1} ~11 605 K, so 1 GeV corresponds to 10^{13} K; k is the Boltzmann constant). Hence, 10^{19} GeV corresponds to a temperature of 10^{32} K, a number so high that it is hard to imagine.

When considering the evolution of the early Universe, the single most important parameter is temperature. The physics of the early Universe is easiest to understand in terms of temperature because equations of state of its different components are most conveniently calculated as a function of temperature. The beginning of the Universe took place some 13.7 ± 0.15 billion years ago. The Universe started to expand immediately (Fig. 2.1).

The Universe was dominated by radiation. Elementary particles changed continuously to radiation and vice versa. At 10 GeV, when 0.01 μs had elapsed, particles called bosons started to decouple from radiation; protons, neutrons, and mesons began to form. This continued until temperature fell to about 100 MeV (10^{12} K), when the temperature of the Universe became too low for the formation of these particles. At this point all protons, neutrons, and mesons were formed. In the meantime, about 100 μs had elapsed. Later, the protons would become the nuclei of hydrogen atoms, and neutrons and protons would combine to form alpha particles or helium nuclei. The Universe continued expanding and its temperature fell.

Within the next second the temperature fell enough for the formation of electron-positron pairs. Most of these pairs annihilated back to radiation, but because of an antimatter-matter symmetry, a surplus of electrons was left.

For astrobiology, the next relevant moment in the evolution of the Universe was the time of nucleosynthesis. It started at about 1 s from the beginning, when the number of photons in the Universe, with energies smaller than 2.2 MeV, corresponding to the binding energy of a deuteron (proton + neutron) fell below the number of neutrons. The nucleosynthesis continued for about 5 min, until the temperature fell to about 70 keV (or 8×10^8 K). The number of neutrons during this era was such that there was 1 neutron (n_n) for 7 protons (n_p). This fixed the mass ratio of helium (He) to hydrogen (H) at:

$$\frac{2\left(\frac{n_n}{n_p}\right)}{\left(1+\frac{n_n}{n_p}\right)} = \frac{1}{4} \qquad (2.1).$$

At that time, almost all neutrons became bound to He. In addition, some light nuclei lithium (Li), B, and beryllium (Be), as well as ^{3}He and deuterium (D), were formed around this time. Heavier nuclei did not form because there are no stable five-nucleon nuclei and because heavier nuclei would have required the presence of He nuclei and even higher temperatures. Within the next few minutes, the He nuclei were formed, and the last remaining neutrons decayed with a half lifetime of 615 s (or a mean lifetime of 886 s).

When the temperature of the Universe fell to $T = 0.8$eV or 9 284 K ($t = 47\,000$ years) it was time for photons to decouple from matter. The Universe became

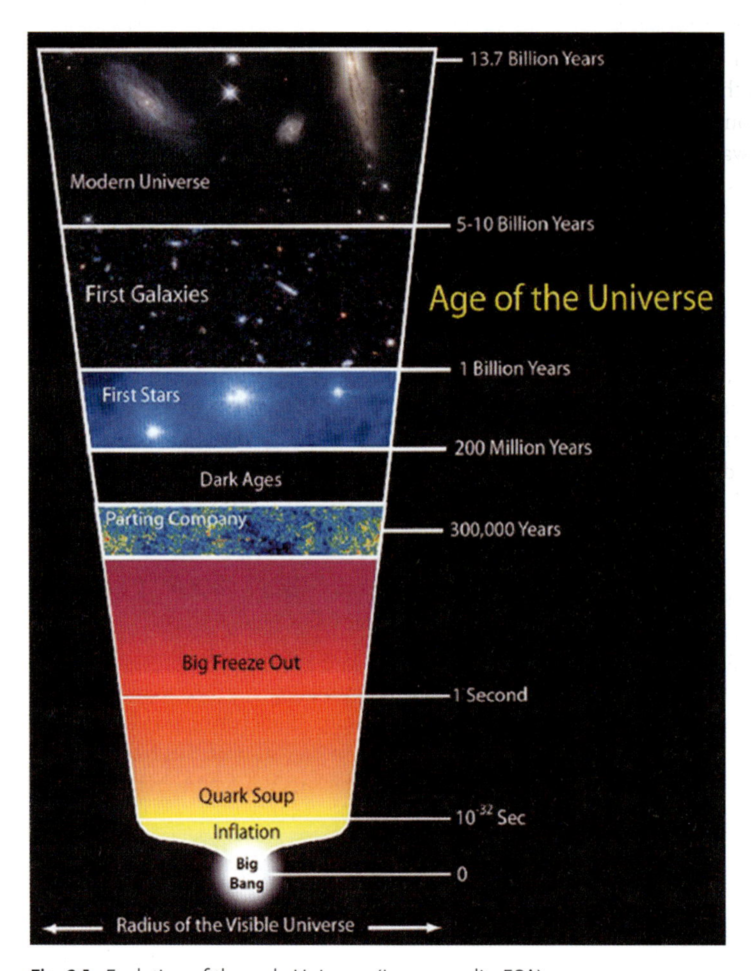

Fig. 2.1 Evolution of the early Universe (Image credit: ESA).

matter dominated. Relatively soon, at $T = 0.3$ eV or 3 000 K, or 380 000 years from the Big Bang, the average background photon encountered its last scattering.

The distant faint glow of the temperature (and matter) fluctuations of that moment can be seen as the cosmic microwave background. Around the time of the last scattering, the temperature had fallen enough so that atoms could also form. Protons combined with electrons into H, and two electrons and an alpha particle formed He atoms. The scale factor (a sort of a measure of size) of the Universe at the time was about 0.09% of the present value.

The production of light elements in the Big Bang was first calculated in 1948 by Ralph Alpher, Hans Bethe and George Gamow. They predicted a microwave background radiation of 5 K. This basic idea was then confirmed in 1964 by the Nobel-winning, serendipitous observations of background radiation by Penzias and Wilson. They measured the temperature to be 3 K. More recently, in 1990, the observations from the NASA satellite *Cosmic Background Explorer* (COBE) showed that the radiation spectrum of the cosmic microwave background follows very accurately a blackbody spectrum.

In 2003, the *Wilkinson Microwave Anisotropy Probe* (WMAP) team led by Charles Bennett from John Hopkins University provided the best fit to the temperature of the microwave background, with a temperature of $T = 2.725 \pm 0.002$ K. These observations were further analyzed by subtracting from each point the average spectrum of the sky and by removing the dipole component, which is at a level of about 3 mK and is caused by our movement with respect to the microwave background. In addition, the galactic contribution to the background was removed – this is radiation found mostly on the galactic plane. One was left with temperature fluctuations frozen to the background at the time of the last scattering. They show up as tiny $\Delta T \sim 10^{-5}$ K fluctuations in the background. By measuring the correlations in these fluctuations over different angular scales over the whole sky down to about 0.3 degree, one can calculate a two-point correlation function. This can then be compared to various cosmological models.

The model that best explains the observed distribution is a Universe that experienced a rapid expansion or inflation starting at 10^{-35} s from the beginning and lasting for 10^{-32} s. Furthermore, the present conditions of the Universe are such that it is expanding at a rate of a Hubble constant, which has a measured value of $H = 71 \pm 2$ km s^{-1} Mpc^{-1} (pc = parsec, a unit of astronomical length, with 1 pc $\approx 3 \times 10^{16}$ m). The average density of the Universe is equal to the critical density within a measurement error of 2%. The critical density is the average density of the Universe at which the geometry of the Universe is flat:

$$\varrho = 3H^2/8\pi G \approx 1.0 \times 10^{-26} \text{g cm}^{-3}, \qquad (2.2),$$

where H = Hubble constant and G = gravitational constant.

The observed statistical properties of the microwave background have been explained by invoking the presence of dark energy. Mathematically, this corresponds to the cosmological constant, which was introduced to the standard Friedman equations by Albert Einstein. He did not like the idea that the Universe could

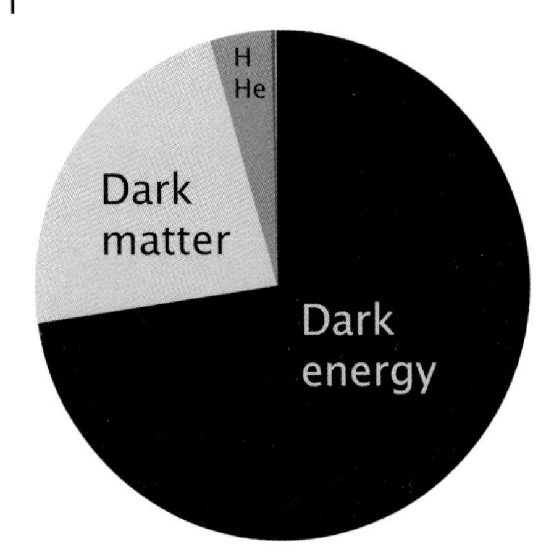

Fig. 2.2 Components of the Universe. Dark energy: 73 %; cold dark matter: 23 %; baryonic matter, which is mostly H and He: 4 %; neutrinos: 0.3 % (dark gray sector); heavier elements such as O, C, and N: 0.04 % (white sector). The last sector is the sector of life.

be non-static and added ad hoc his cosmological constant. For decades, this constant was considered a curiosity, but now, with the latest microwave data, it has become quite useful again.

About 73 % of the energy density in the Universe is in the form of dark energy. At present it is not known what this is physically or theoretically, but it can be mathematically understood as the energy density of vacuum. Its presence is also suggested by the accelerating expansion rate detected from observations of supernovae and some other observations. The remaining 27 % of the energy density is matter. About 85 % of this matter is "dark matter." The physical nature of this matter is not yet known, but its influence can be seen, e.g., in orbits of stars in galaxies or in the velocity dispersion of galaxies in galaxy clusters. The remaining matter (15 % of matter and 4 % of the total energy density) is baryonic matter, mainly H and He. Heavier elements constitute 1 % of the combined H and He mass only. Thus, assuming that the dark energy is real, only about 0.04 % of the density of the Universe is in the form of heavier elements such as C, N, O, or Fe, which are the elements life is made of (Fig. 2.2).

2.3
First Stars: Formation of Small Amounts of C, O, N, S and P and Other Heavy Elements

After the last scattering of the photons, a dark era followed in the Universe (Fig. 2.1). No stars existed for another 100–300 million years. It took about that time for the first gas clouds to collapse and form the first stars. The details of the physics of those stars and the exact timescales of their formation and evolution are still unclear. Using the *Spitzer Space Telescope*, faint infrared (IR) fluctuations in the cosmic IR background were measured by Kashlinsky and his collaborators in late 2005. They argue that the fluctuations are caused by combined light from a large number of the very first stars. No individual stars are seen. The ultraviolet (UV) emission from the stars is now observed as highly red-shifted radiation in the IR. This radiation could well come from the very first stars.

It is likely that these are the only stars that formed along the lines first described by Jeans in 1902. A gas cloud, which was at that time made of H and He, started to collapse under its own weight. The collapse is initiated if the mass of the cloud is larger than the Jeans' mass:

$$M_{cloud} > M_{JM} = 700\ M_{\odot}\ (T\ /\ 200\ \mathrm{K})^{1.5}\ (n\ /\ 10^6)^{-0.5} \qquad (2.3),$$

where n = the number density of particles (H and He) per cm^3, T = the temperature of the cloud in K, and $M_{\odot}$ = the solar mass, a basic mass unit in many astronomical calculations.

Thus, if the cloud has a density of $n = 10^6$ cm^{-3} and a temperature of $T = 200$ K, then a cloud exceeding a mass of 700 $M_{\odot}$ would be able to collapse. If the temperature is higher than $T = 200$ K, then the cloud should be more massive for the total gravitational attraction to overcome the added pressure caused by the increased temperature. Similarly, if the gas density is higher than $n = 10^6$ cm^{-3}, then a smaller cloud could collapse according to this model.

As the collapsing proceeds, the density in the cloud center increases. With the increased density the pressure at the center increases, as does the temperature. Finally, in the central 1% or so, the temperature of the cloud rises to 4×10^6 K, and nuclear reactions begin. There appears to be no clear reason for an upper limit of the mass of a star formed in this way. It is possible that the very first stars had masses of 100–300 $M_{\odot}$, or even as large as 500 $M_{\odot}$. These very massive stars lived short lives of only one million years or so and produced the first heavier elements, such as C, N, O, P, S, Al, and Fe. At the end of their lives, they exploded and shed all these new heavy elements into the surrounding, still quite dark space. These early stars are now referred to as population III stars. The first galaxies formed when the Universe had an age of maybe 600–800 million years. All the stars formed in these new nascent galaxies have a minutely tiny seed of heavier elements in addition to the overwhelmingly abundant H and He.

2.4 Normal Modern Stars, Bulk Formation of C, O, N, S, P and Other Heavy Elements

Stars can be classified by several different means. The most important classification scheme is the Hertzsprung-Russell (HR) diagram (Fig. 2.3). A version of the HR diagram, the color magnitude diagram, shows the brightness of the star, e.g., V magnitude, as a function of its color, e.g., brightness difference in two bands, say the blue and visual (B – V). Several variations of this diagram exist. If the distance to a star is known and its apparent brightness is measured, then its luminosity can be calculated. If a spectrum of the very same star also can be obtained, then the temperature of its photosphere can be determined. By doing this to a multitude of stars, an HR diagram in a form that is more relevant to astrobiology can be drawn.

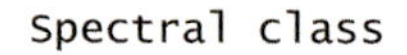

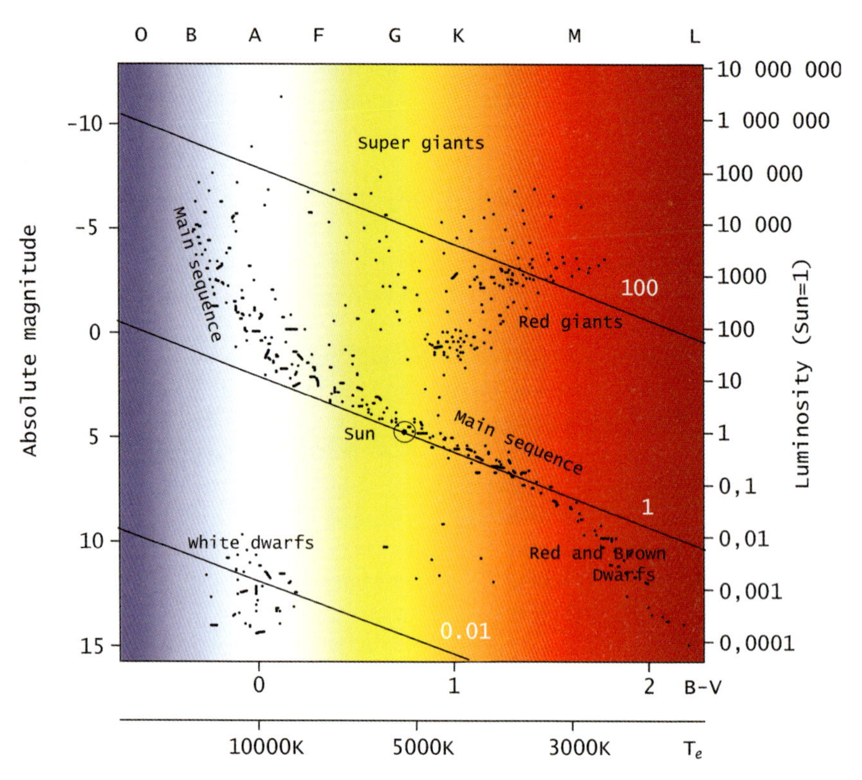

Fig. 2.3 Schematic representation of a Hertzsprung-Russell (HR) diagram; it shows the star temperature or spectral type as abscissa and star luminosity or absolute magnitude as ordinate. Temperature of stars decreases from left to right with the spectral type sequence of stars: O, B, A, F, G, K, M, to L. The diagonal lines at 100, 1, and 0.01 represent the approximate radius of the star in solar radii.

Here, the brightness of the star, given as logarithm of luminosity, is plotted against the logarithm of temperature.

If the star is assumed to behave like a perfect blackbody radiator, which is a reasonable assumption in this context, then straight lines of constant stellar radii can be drawn into this plot. Furthermore, if the luminosity and the temperature of the central star are known, the distance of the habitable zone can be included in this same plot. This is shown in Fig. 2.4 on the right. This assumes the definition of a habitable zone, put forward by Kasting, as the region around the star where the temperature is such that water stays liquid (see Chapter 6). The lower edge of this temperature range is close to 0°C at a range of pressures, but the higher end depends on the atmospheric pressure of the planet itself. At 10^5 Pa (1 atm) it is 100 °C, but at 2×10^5 Pa it is 121 °C.

Because habitability depends on the luminosity of the central star, the distance of the habitable zone is set on approximately horizontal lines on the diagram. Considering the limitations, several areas of the HR diagram can be ruled out as rather unsuitable for life. Even the most optimistic views of habitability can be condensed into a triangle (Fig. 2.4). This may be an area where life could, in principle, form. For life to evolve to a higher evolutionary stage, the conditions on the central star should stay relatively stable for at least one billion years. This would narrow the stars suitable for life as we know it down to a narrow ellipse-like zone (Fig. 2.4).

There is one additional stellar property that is relevant to astrobiology not shown in the HR diagram. It is what astronomers call metallicity (Z), which is the mass fraction of elements heavier than hydrogen and helium. Solar metallicity abundance is about 1–2 %. Hot, bright stars tend to have high metallicity and faint red stars usually have low metallicity. However, this may not represent a proper correlation but rather a manifestation of the age of the average star of these spectral classes. Blue, bright stars are short-lived, but faint red stars are of very different ages, including some that are almost as old as the Universe.

The young stars are called population I stars, and the old ones are called population II stars. Our Sun is a population I star. The different populations of stars have different roles in astrobiology.

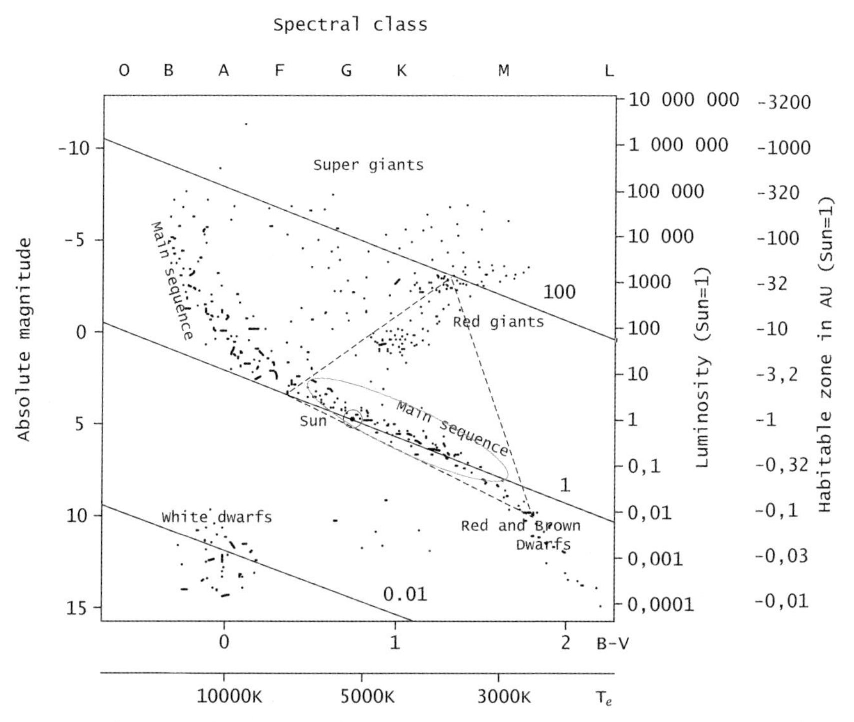

Fig. 2.4 Biological HR diagram; lines of a typical distance of the habitable zone (HZ) are shown on the right of the diagram (scaling that of the Sun at HZ = 1.0). Because the range of the HZ depends on the luminosity of the central star, the distances areset on approximately horizontal lines on the diagram. The triangle (dashed line) describes the range where life in principle could form. The narrow ellipse-like zone describes an area where life might evolve to a higher evolutionary stage.

The first normal stars were formed from very metal-poor gas. There were some metals around from the very first stars, but the amounts were very low. Even with such low concentrations of metals, such as C, the nuclear reactions in stars got an important boost because these metals are used as catalysts, e.g., in the carbon–nitrogen–oxygen (CNO) cycle. These first normal stars were also the primary factories for the heavier elements, which are important for life.

The evolution of a single star is determined mainly by its mass. (Binary stars exist quite commonly, but the theory of binary star evolution will not be considered here.) If the mass of a single object emerging from a collapsing cloud is smaller than about 0.08 $M_\odot$, then its core will not have a high enough temperature to start the fusion of H into He. If the stellar mass is larger than 0.08 $M_\odot$, and less than 0.26 $M_\odot$, the central temperature will rise above 4×10^6 K and He production will start at a very slow pace in the center of the star by the proton–proton chain reaction (p–p chain reaction). Two protons combine to form a deuteron, which then combines with a further proton to form ^3He. Finally, two ^3He combine to form a ^4He and two protons. The final step can also proceed through a loop involving Li, B, and Be, essentially with the same end result. This process will continue throughout the lifetime of the star.

Stellar models suggest that these stars will stay convective and well mixed. The same models indicate that their lifetime is in the range of 10^{11}–10^{13} years. This is much longer than the present age of the Universe. None of them have been witnessed to “die.” It is thought that after a very long time these stars will end up as white dwarfs. The energy within these stars is transferred via convection from the center of the star through buoyant, hot material rising from the central parts of the star and cooler material near the surface sinking down.

In stars of masses larger than 0.26 $M_\odot$, elements heavier than He will also form. Pressure and temperature increase towards the center of a star. It is in the very center of the star where most of the nucleosynthesis takes place. The evolution of stars heavier than about 0.26 $M_\odot$ can be divided into two broad groups with the dividing line at about 1.5 $M_\odot$.

The evolution of a solar type star of mass $0.26\ M_\odot < M < 1.5\ M_\odot$ is approximately as follows. These stars form initially with a central temperature higher than 4×10^6 K. The central parts become radiative, while the outer parts stay convective. The p–p chain reaction is the main source of the stellar luminosity for these stars during the main sequence phase of the star, which lasts about 2–20 billion years. During this time the star is relatively stable. Close to the end of the main sequence phase, the star has developed a helium core that is degenerate, meaning that it is nearly incompressible. Around this core there is a hydrogen fusion envelope that starts to expand. The star becomes a red giant.

The temperature of the He core increases, until suddenly, when the temperature reaches about 10^8 K, the so-called triple-alpha reaction sets in. In this reaction He is transformed into C. Two He nuclei combine into a ^8Be nucleus, which is unstable. If a third He nucleus interacts within 3×10^{-16} s, then a ^{12}C nucleus with an emission of a photon results; otherwise, the ^8Be decays back to two He nuclei. This reaction requires the near-simultaneous collision of the three He nuclei (alpha particles) and is therefore called the triple-alpha reaction.

The onset of the triple-alpha reaction across the core is so sudden (few hundreds of seconds) that it causes a He flash, the inner parts of the star collapse, and the outer parts of the star take the extra energy. Later, around the central parts, triple-alpha reactions continue in a He shell. This He shell is surrounded by a H shell, where He is produced by the p–p reaction. The outer envelope expands further, and eventually the star loses the outer shells as an intense stellar wind. A planetary nebula with a hot star in the center results. (It should be emphasized that physically the planetary nebula has nothing to do with planets or planetary systems. The name "planetary nebula" reflects the appearance of these objects as seen by relatively small telescopes. They tend to look round, not point-like as stars do, and thus are reminiscent of the appearance of planets such as Jupiter or Saturn.) The central star in planetary nebulae is often not visible by small telescopes, but large telescopes reveal that the end of the life of a solar mass it to become a white dwarf. The white dwarf is made mostly of He, but it has a C or O core, and on the surface it may have a thin outer layer made of H.

The gas in the stellar wind during the red giant phase and the planetary nebula phases consists mainly of H and He, but it also contains about 2% C, N, and O, which are released into the interstellar medium (ISM). Elements heavier than O are not produced in these stars. Because these lighter stars are substantially more common than more massive stars, they have been suggested to be a significant source of new C, N, and O in the present galaxy.

Stars more massive than $1.5\,M_\odot$ have such a high central temperature ($> 18 \times 10^6$ K) that a different nuclear fusion mechanism takes over in the conversion of H to He. At these high temperatures the CNO cycle becomes more effective than the p–p cycle. One should note that in a CNO cycle, also known as the carbon cycle, no C is produced; instead, C is needed as a catalyst. The carbon cycle proceeds as follows:

$$^{12}C + {}^{1}H \rightarrow {}^{13}N + \gamma \tag{2.2 a}$$

$$^{13}N \rightarrow {}^{13}C + e^+ + \nu_e \tag{2.2 b}$$

$$^{13}C + {}^{1}H \rightarrow {}^{14}N + \gamma \tag{2.2 c}$$

$$^{14}N + {}^{1}H \rightarrow {}^{15}O + \gamma \tag{2.2 d}$$

$$^{15}O \rightarrow {}^{15}N + e^+ + \nu_e \tag{2.2 e}$$

$$^{15}N + {}^{1}H \rightarrow {}^{12}C + {}^{4}He \tag{2.2 f}$$

Reaction (2.2 d) is the slowest one and it determines the total reaction rate. It is important to note that this bottleneck creates a surplus of N. Although C is a catalyst in this reaction, it effectively transforms some C into N. Even if the carbon cycle is the main source of N in the Universe, the main benefit of this reaction is the efficient conversion of H into He.

In these stars the very center of the star will be convective and the outer parts will remain radiative. As in the less massive stars, a He core will eventually form and the star will become a red giant. Now, however, the He core remains convective, and the onset of He fusion into C by triple-alpha reaction starts at a more leisurely pace; no He flash takes place. A shell-like structure forms, with C in the center surrounded by He and H shells. Stars in the mass range of $1.5\ M_{\odot} < M < 3\ M_{\odot}$ will pass through a red giant phase and end up as white dwarfs.

The evolution of stars in the range of $3\ M_{\odot} < M < 11\ M_{\odot}$ is not well understood theoretically. It appears that in the range of about $3\ M_{\odot} < M < 8\ M_{\odot}$ the C core will not ignite and the star will evolve through a red giant phase and end up as a white dwarf, with the outer parts of the star expelled in a strong stellar wind.

In more massive stars ($M > 8\ M_{\odot}$), the C core will ignite, and a C flash will take place. Here, the main reaction combines two C, creating an O nucleus and two He nuclei. Later, if the degenerate center has a high enough temperature, two O nuclei combine either into a Mg nucleus and two He nuclei or into one Si and one He. In the heaviest stars two Si nuclei combine into one Fe nucleus. All other elements from C to Fe are formed as well. The nuclear reactions producing these elements are more complex. The conditions at the center of the star will eventually determine which are the heaviest nuclei formed. Depending on the mass of the star, an onion-like structure will eventually form with an O, Si, or Fe core. All heavy elements up to Fe and Ni can be formed in these stars as products or side products of the above-mentioned reactions.

If the mass of the star is about $8\ M_{\odot} < M < 11\ M_{\odot}$, the star will end up with an O/Ne/Mg core. This will eventually turn into an O/Ne/Mg white dwarf, with the outer parts of the star blown out in a fierce stellar wind. Other models suggest that the fate of these stars is to end up as a supernova with a neutron star remnant. In either case, the lighter elements in the outer shells will be expelled into the interstellar space.

Stars heavier than about $11\ M_{\odot}$ have been studied theoretically more extensively. They will end up with a shell-like structure with Fe in the center, which implies that all nuclear fuel in the core has been exhausted. The core will collapse on a dynamical timescale, which is on the order of a few hours. The rebounce from this collapse will create a shock traveling outwards, resulting in a supernova explosion. The whole mass of the star and all the elements it contains are thrown into the surrounding space. It is also possible that the center part of the star, the Fe core, will collapse into a black hole or a neutron star, but the outer parts of the star will still be ejected into space. Supernova explosions have been the main means for liberating the heavier elements into the interstellar space.

Elements heavier than Fe are formed in stars in small amounts by the slow neutron-capture (s) process and in supernovae by the rapid neutron-capture (r) process. These two mechanisms have somewhat different isotope distributions. The heaviest elements formed in the s-process are ^{208}Pb and ^{209}Bi. In the r-process the heaviest elements are thought to have an atomic mass of about 270. The formation of some rare isotopes such as ^{190}Pt cannot be explained by the s- or r-processes. These are thought to have formed in a supernova explosion by the proton-capture (p) process. Table 2.1 shows a few properties of some main sequence stars.

Table 2.1 Properties of main sequence stars.

Spectral class	Mass ($M_\odot$)	Temperature (K)	Age (10^9 years)
O5	30	44 000	0.005
A0	3	12 000	0.24
F2	1.5	7 200	2
G2	1	5 700	10
M0	0.5	3 800	30
M7	0.1	2 800	10 000

2.5 The First Molecules (CO and H_2O)

The most common molecule in the interstellar space is molecular hydrogen (H_2). The next most common molecules are carbon monoxide (CO) and water (H_2O). These are very interesting molecules because they already contain the heavier elements important for life. The elements making up these molecules were formed in stars that have already ended their life. Walter and coworkers have observed with the Very Large Array (VLA) situated in New Mexico, USA, the presence of CO in the spectrum of the quasar SDSS J1148+5251, which has a redshift of $z = 6.419$. Based on present models of the Universe, this means that these molecules were already formed 800 millions years after the start of the Universe; thus, at this time the first ingredients for life were available. The presence of CO indicates that substantial amounts of both C and O had already formed. H_2O has not been detected from this quasar, because of observational limitations, but one can infer its presence because it is formed in a way similar to CO, directly from gaseous H and O.

2.6 Interstellar Matter

Aboriginal Australians see a long, dark emu bird in the sky, with the bonfires of their forefathers lighting up the stars of the sky (Fig. 2.5). The "emu" is formed by a multitude of individual dark clouds in the plane of our galaxy. The presence of interstellar matter manifests itself also as bright illuminated patches in the sky viewable even with ordinary binoculars. Some of the best examples are Orion nebula M42 and several clouds in Sagittarius and adjacent constellations.

The space between stars was thought to be void until Trumpler (1930) noted that the calculated linear size of open clusters seemed to increase with the distance. Clearly, because such a correlation should not exist, one could reason that distances to more distant clusters were overestimated. This could only be caused by some

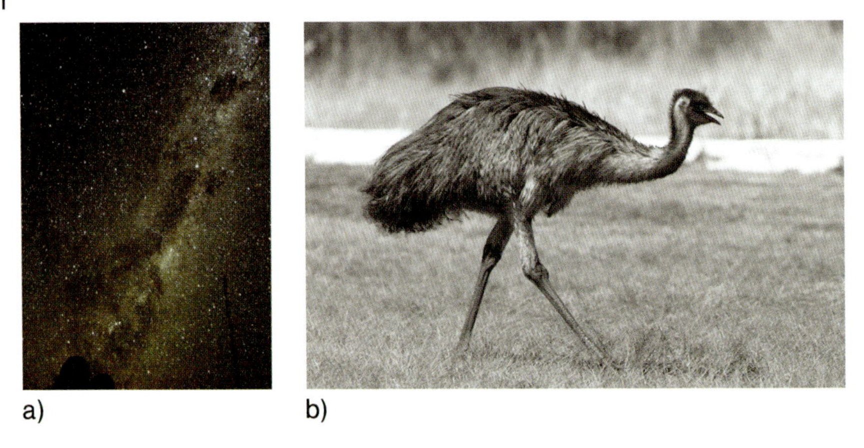

Fig. 2.5 "Emu." (a) Constellation showing the emu bird in the sky (Photo credit: H. Lehto. La Silla, Chile, 5 March 2006). (b) A real emu bird (Photo credit: H. Lehto, Northern Territory, Australia, 8 August 2003).

interstellar dust particles scattering part of the light away on its way from the cluster to our telescopes. At present, it is known that this extinction is on average about 2 mmag pc^{-1} (or 0.06 % per light year).

The presence of interstellar matter is deduced in several ways. In the optical region, *extinction* is one of the most important proofs of interstellar matter. A second indication of matter in interstellar space comes from studying two stars that have similar spectral properties, which means that they have similar intrinsic brightness and temperature. If one then compares the observed spectrum of the more distant star with that of the closer star, then one can note that it is reddened, which means that light in the blue end of the spectrum is scattered away compared to the red end of the spectrum. It is a phenomenon similar to that of the reddening of the Sun as it gets close to the horizon, in which case the scattering is caused by the tiny O_2 and N_2 molecules of the atmosphere (Rayleigh scattering). In interstellar reddening, the scattering is caused by dust particles in the interstellar space, which are on the order of the wavelength (Mie scattering) in size.

The third manifestation of interstellar matter can be observed by studying the *polarization* of the starlight. Interstellar dust particles align themselves preferably along the feeble galactic magnetic field. As light from a star travels towards us, some light is scattered. Because of the shape of the particles, this scattering has a preferred direction: perpendicular to the longer axis of the particle, meaning that light passing through a population of dust particles will have a preferred direction of oscillation, i.e., it will be linearly polarized. By studying this polarization, one may infer the presence of interstellar non-spherical dust particles and even measure the direction of the galactic magnetic field in different parts of the sky. By combining the polarization with the reddening, one obtains the first estimate of the size and shape of interstellar dust particles.

The fourth signature of the presence of intervening interstellar matter are the *interstellar absorption lines* in the spectra of stars. Most of the absorption lines in stellar spectra are formed in the stellar photosphere, but by studying the Doppler shift and the shape of the individual absorption lines, astronomers can identify interstellar lines from the intervening matter and deduce the properties of the interstellar gas. Note that of the four methods described above, extinction, reddening, and polarization measure the properties of the particles, while only one, the spectral absorption lines, indicates the presence of interstellar gas, although the mass of the gas outweighs the dust by about 100 to 1.

The most common gas in the Universe, H, emits radiation at a wavelength of 21 cm. This radio photon is caused by a small change in the H when the single electron orbiting the proton changes its spin direction from being of the opposite sense to being in the same sense as the proton's spin. Technically speaking, the spin axis changes from antiparallel to parallel. It is very unlikely to happen to a single atom, as it happens about once in 10 million years, but because there is so much H along any line of sight, all the added changes shine quite brightly at 21 cm, particularly in the plane of the galaxy. Because practically nothing absorbs this wavelength, it has proven to be one of the most useful tools in mapping the distribution of interstellar H in our galaxy. It were these measurements that showed that gaseous neutral H (HI) is distributed along spiral arms. Because H is the main constituent in the interstellar gas, this study also shows where the bulk of interstellar gas resides. Note that HI refers to neutral H, HII to (singly) ionized H, and H_2 to the molecular H_2. All these are important constituents in different parts of the interstellar medium (ISM).

Several different phases of interstellar gas are known, with the coldest and densest gas being found in molecular gas clouds and the most tenuous and hottest gas, e.g., in supernova remnants. The cold, dense clouds are considered in more detail in the next section.

With the advent of space observatories, extensive IR observations became possible. These have opened a completely new window for the study of interstellar matter in its many forms. These observations are discussed in the context of cool interstellar clouds (Section 2.6.1) and interstellar ice and dust (Section 2.6.3).

2.6.1 Interstellar Clouds

The distribution of interstellar dust is not uniform. It is concentrated into interstellar clouds that are typically located on the inner edges of the galactic spiral arms. Visually, these clouds appear as dark areas in the sky with much fewer stars; in fact, some of them are optically so opaque that they appear as totally starless patches of the sky (Fig. 2.6). Optically, the clouds have such a high extinction that practically the only means for studying them is in radio and infrared wavelengths.

Spectroscopy in radio wavelengths has two functions. One function is the search for different molecules. This is done at a large range of wavelengths. From an astrobiological point of view, the most interesting ones are to be found in the

Fig. 2.6 Pre-collapse black cloud BARNARD 68. Picture taken by the Focal Reducer/Low Dispersion Spectrograph (FORS) team on 10 January 2001 with the ANTU telescope unit of the Very Large Telescope (VLT) located on Cerro Paranal in the Atacama Desert, northern Chile (Image credit: European Southern Observatory [ESO]).

millimeter ranges. The second function, the spectroscopy of CO, OH, SO, SO_2, and other molecules at centimeter and decimeter wavelengths, is critical because it provides the means for measuring the matter distribution, temperature profiles, and internal motions within the clouds. All this has important implications for understanding under which physical conditions complex molecules, dust particles, and ices are formed.

IR spectroscopy is the means for studying the properties of dust grains by looking at signatures of various ices and minerals in the IR emission spectra of interstellar clouds. As the extinction in IR is not as severe as in the optical, one can actually recognize at these wavelengths also newborn stars and their accretion

a)

b)

Fig. 2.7 (a) Orion nebula M42, in the optical range. The bright cloud is illuminated by the stars of the Trapezium cluster, which have an age of about 2 million years. (b) Behind the veils of these clouds, a dense molecular cloud resides, within which hundreds of newly forming stars are seen in IR. (A: Photo credit: H. Lehto, Tuorla Observatory, Finland, 24 January 2004; B: Image credit: ESO).

Fig. 2.8 The dark Horsehead nebula in the Orion cloud, observed by the VLT Kueyen and FORS2 (Credit: ESO).

disks still surrounded by the blown-away dust shrouds in the depths of the gas cloud.

The densest of the clouds, the giant molecular clouds (GMC), are completely opaque at visual wavelengths. The presence of large amounts of H_2 is their characteristic signature, but they contain also atomic H, inert He, and about 1% of heavier elements. The temperature in the densest parts is 10–20 K. The densities, measured in "particles" of mainly H atoms, H_2 molecules, and He, can be as high as 10^9 particles per m^3, and up to 10^{12} particles per m^3, depending on the cloud. A density of 10^{12} particles per m^3 corresponds to about 10^{-11} g m^{-3} or, less conventionally, 10 mg km^{-3}. For comparison with more familiar units, the density of the Earth's air is about 1 300 tons km^{-3}. Still, from an astronomical point of view, these clouds are very dense.

As an example, one may consider the closest GMC, the Orion cloud. It is at a distance of about 500 pc and is about 120 pc in extent. From radio and IR observations, it is known to have an apparent diameter of about 20 degrees in the sky, or 40 times larger than the moon. One can see the outer edges of the Orion cloud also usually, as the young stars in the Trapezium illuminate the edge of the clouds and create a bright patch known as Orion nebula M42 (Fig. 2.7.A). The densest part of the cloud associated with M42 has a diameter of about 25 pc. Examples of the darkness of these clouds can be seen elsewhere in the dense parts of the Orion cloud, e.g., in the dark Horsehead nebula (Fig. 2.8).

2.6.2 Interstellar Grains

2.6.2.1 Formation

Interstellar grains play a critical role in the formation of interstellar molecules. Interstellar grains are formed mainly in the outer shells of stars when they are in their red giant or asymptotic giant branch (AGB) phases (M giants, carbon stars, or radio luminous OH/IR stars). Some dust is formed in supernova remnants, novae, planetary nebulae, and WC stars and in circumstellar shells around supergiants. The presence of the grains is seen in the opacity caused by the dust and in thermal continuum IR emission from the dust. The details of the initial phase of the dust formation, the nucleation, are not well understood. Recent observations suggest that crystallization of SiO_2 takes place directly from the gaseous phase when the temperature in the stellar wind falls below 700 K. The dust particles grow further in the tenuous atmospheres of giants and, because of radiation pressure, are driven away from the star. The terminal velocity depends on the size of the grain and the material – or the mass of the grain. The fastest dispersing grains have a size of about 50 μm. Once formed and ejected into the ISM, the dust particles undergo various kinds of reprocessing, causing both breaking up and growing. The extinction curves observed optically require that a large fraction (>40%) of interstellar grains be >100 nm; therefore, some rebuilding or formation of grains must take place in the ISM, possibly in high-density regions by accretion to preexisting grain cores.

2.6.2.2 **Observed Properties**

IR spectra of ISM show signatures of a population of grains with small radii ($a < 10$ nm) that absorb about half of the energy emitted by the stars. This radiation is later reemitted in the IR. These grains are believed to be carbonaceous in nature. The smallest of them contain less than 100 C atoms and have tiny sizes ($a < 1$ nm). They create the IR emission bands characteristic of aromatic carbonaceous rings at 3.3 μm, 6.2 μm, 7.7 μm, 8.6 μm, 11.3 μm, and 12.7 μm. The details of the bands are well explained by a family of large molecules called polycyclic aromatic hydrocarbons (PAHs). Structurally, the molecules are two-dimensional sets of aromatic carbon rings, consisting of 6 C atoms and each ring connected to each other. The C core is surrounded by an outer shell of H atoms. Structurally, the PAHs appear similar to pieces of chicken wire made of hexagons. However, precise ground-based laboratory spectra of these molecules are still missing for comparison. Furthermore, a spectral feature at 3.4 μm has been attributed to aliphatic hydrocarbons, which are chain-like in structure.

The dust continuum emission in the 25-μm and 60-μm IR bands of the *Infrared Astronomy Satellite* (IRAS) has been attributed to slightly larger carbonaceous very small grains (1 nm $< a <$ 10 nm). The far-IR continuum radiation is further ascribed to big grains ($a \sim$ few 10 to few 100 nm), possibly a mixture of large graphite and silicate grains.

Dusty stellar envelopes show broad featureless bands at 10 μm and 20 μm in both emission and absorption. They have been attributed to amorphous silicates. Recently, additional bands have been observed in the 10–45 μm region. They have been attributed to highly ordered crystalline silicates such as Mg-rich olivine (forsterite, Mg_2SiO_4) or pyroxene (enstatite, $MgSiO_3$). These minerals have not been directly seen in spectra of ISM. As a curiosity, a 21-μm IR emission band in supernova remnants has been attributed to tiny diamonds. Their sizes are in the range 0.005–5 μm.

Depletion of gaseous C, Si, Mg, Fe, and O in ISM suggests that these elements are used in another phase, possibly as silicates and metal oxides in grains. It should be emphasized that graphite has not been unequivocally detected in the ISM, although it is quite often used as a model.

Interstellar matter can be searched for locally in our Solar System. Using isotopic traces, interstellar grains can be identified in pristine meteorites. They differ somewhat from the particles detected spectroscopically in the ISM. The most common pre-solar grains that have been found are diamonds (2 nm in diameter), SiC (0.05–10 μm), graphite (0.4–6 μm), and Al_2O_3.

High-velocity interstellar particles have been observed in the Solar System with the detectors of the International Solar Polar mission *Ulysses* and from radar studies of meteoroids. The sizes of these particles are about 0.8 μm and 15–30 μm, respectively. They are somewhat larger than typical interstellar grains.

2.6.3
Ices

Spectral signatures of ices have been seen in OH/IR stars and other evolved stellar systems, in lines of sight towards field stars located behind molecular clouds, and around embedded protostellar objects. More than 20 ice features have been detected in the IR. Ices create mantles on dust grains. These mantles are important for the formation of molecules.

H_2O ice was detected by Gillett and Forrest in 1973 from the 3-μm waterline. It is the most dominant solid-state species, and abundances of other ices are usually given in reference to H_2O ice. In clouds with substantial extinction, it is comparable in amount or even more abundant than gaseous CO. H_2O ice is usually the second most abundant molecular species after H_2 ($[H_2O]/[H_2] \sim 10^{-4}$–$10^{-5}$). About 15–20% of the cosmic H_2O is locked in ices, and the remainder is in gas phase.

CO_2 ice is difficult to detect with ground-based telescopes because of atmospheric CO_2. Its firm detection was not established until the late 1990 s with the *Infrared Space Observatory* (ISO) mission. At present it appears that CO_2 is the second most abundant ice after H_2O ice, with the CO_2:H_2O ratio being on average about 15–20% .

CO has been detected in icy form. Although it may conform 3–20% of the ice component in the grain mantles, it is much more abundant in the gaseous phase than in the solid phase.

CH_4 has spectral signatures that are difficult to observe from the ground. The abundance of CH_4 ices is typically 1–2% of water ice. The ratio of gaseous to solid state is usually below unity, suggesting that CH_4 is formed directly by hydrogenation of carbon on the surface of the grain.

Formaldehyde (H_2CO), the simplest of the hydroxyl group of ices, was detected from ISO observations towards five massive protostellar envelopes with the typical abundance of 1–2% to water ice.

Methanol (CH_3OH) is one of the key elements in interstellar ices, amounting to up to 10% of water ice content. The formation of methanol ice is still debated. It could be caused by hydrogenation of CO or by energetic processes followed by selective desorption enhancement effects.

Sulfur-containing ices have been searched for, and only OCS has been detected so far, with a very low abundance of about 0.1% to water ice. H_2S has an upper limit of 1% for the abundance.

N_2 and NH_3 ices have been searched for, but they have not been reliably detected. The difficulty with N_2 is that the possible signal is extremely weak; in the case of NH_3, the difficulty is separation from other spectral lines.

O_2 and N_2 are IR-inactive molecules, and the upper limits for O_2 ice are quite high: <6% of water ice. No detection of O_3 or CO_3 has been made, suggesting that O_2 is indeed quite rare. It has not even been found in gaseous form, suggesting that the lack of O_2 is real and is caused by its high reactivity.

Other ices that have been found include OCN^- and NH_4^+ at a maximum level of a few percent. Tentative identifications include $HCOO^-$, CH_3CHO, $HCONH_2$, and $(NH_2)_2CO$.

2.6.4 Molecules in the Gas Phase

2.6.4.1 Observed Properties

The gas phase of interstellar clouds consists of H and He, with H_2 being the most abundant molecule in dense clouds. The second most abundant molecule is CO. O_2 and N_2 have not been found in the ISM. The upper limit for the abundance of gaseous O_2 towards the dense cores of molecular clouds, deduced from observations of the *Submillimeter Wave Astronomical Satellite* (SWAS), is $[O_2]/[CO] < 10^{-7}$.

Radio observations from the sub-millimeter to centimeter wavelengths have recovered signatures of about 130 different molecules. One-quarter of these do not contain C, three-quarters do contain C, and out of these, two-thirds contain both C and H (Table 2.2). Astronomers would call about 100 molecules "organic," whereas organic chemists would give this name to only about 65 molecules detected in interstellar space. For astrobiology, the most relevant molecules found are water (H_2O), formaldehyde (H_2CO), hydrogen cyanide (HCN), sugars such as glycoaldehyde (CH_2OHCHO), and aliphatic hydrocarbon chains up to $H(C{\equiv}C)_5CN$.

Finding longer molecules is in principle possible, but it becomes more difficult with each added atom. One reason for this is that the more complex the molecule is, the rarer it becomes. Another reason is that more complex molecules also have much more complex spectra, and their unequivocal identification becomes more difficult. This has been the case in several claims of the detection of the simplest amino acid, glycine. Some of these have appeared quite convincing. The third problem with complex molecules is that one needs a reference microwave laboratory spectrum for each molecule, and they are not trivial to measure. Many spectral lines in the microwave region still remain unidentified.

Table 2.2 Interstellar molecules.

Number of atoms per molecule	Molecule
2	IF, AlCl, C_2, CH, CH^+, CN, CO, CO^+, CP, CS, CSi, HCl, H_2, KCl, NH, NO, NS, NaCl, OH, PN, SO, SO^+, SiN, SiO, SiS, HF, SH, FeO(?)
3	C_3, C_2H, C_2O, C_2S, CH_2, HCN, HCO, HCO^+, HCS^+, HOC^+, H_2O, H_2S, HNC, HNO, MgCN, MgNC, N_2H^+, N_2O, NaCN, OCS, SO_2, c-SiC_2, CO_2, NH_2, H_3^+, AlNC
4	c-C_3H, l-C_3H, C_3N, C_3O, C_3S, C_2H_2, CH_2D^+(?), HCCN, $HCNH^+$, HNCO, HNCS, $HOCO^+$, H_2CO, H_2CN, H_2CS, H_3O^+, NH_3, SiC_3
5	C_5, C_4H, C_4Si, l-C_3H_2, c-C_3H_2, CH_2CN, CH_4, HC_3N, HC_2NC, HCOOH, H_2CHN, H_2C_2O, H_2NCN, HNC_3, SiH_4, H_2COH^+
6	C_5H, C_5O, C_2H_4, CH_3CN, CH_3NC, CH_3OH, CH_3SH, HC_3NH^+, HC_2CHO, $HCONH_2$, l-H_2C_4, C_5N
7	C_6H, CH_2CHCN, CH_3C_2H, HC_5N, $HCOCH_3$, NH_2CH_3, c-C_2H_4O, CH_2CHOH
8	CH_3C_3N, $HCOOCH_3$, CH_3COOH, C_7H, H_2C_6, CH_2OHCHO, CH_2CHCHO
9	CH_3C_4H, CH_3CH_2CN, $(CH_3)_2O$, CH_3CH_2OH, HC_7N, C_8H
10	CH_3C_5N (?), $(CH_3)_2CO$, NH_2CH_2COOH(?), CH_3CH_2CHO
11	HC_9N
13	$HC_{11}N$

Source: From Al Wootten with permission.

2.6.4.2 Formation of H_2

H_2 is the most abundant molecule in the Universe, yet it is a difficult molecule to observe. All the observations have to be made from space-based observatories. In 1970, the *Copernicus* satellite observed H_2 electronic lines in absorption against UV bright stars. The ISO operating at IR changed the level of knowledge. Rotationally vibrated lines from 2–18 μm could be observed. Combining this with information of other gas-phase atoms, molecular lines, PAHs, silicates, and ices put H_2 into perspective with the other constituents of the ISM.

There is no known efficient gas-phase formation mechanism for H_2 at low temperatures and at the densities of the ISM. It is believed that H_2 is formed on dust particles, but the detailed formation mechanisms are not known because of the limited amount of knowledge about dust particles.

To account for the observed amounts of H_2, very effective means for its production at low temperatures (<30 K) are needed. Models exist whereby H_2 is formed on surfaces of PAHs or on surfaces of very small grains. Several scientists have argued that the surface roughness of particles plays an important role.

The formation of molecules more complex than H_2 has also been studied. The mechanisms include various combinations of molecule building in the gaseous phase or on grain surfaces, sometimes with the aid of UV radiation, which, by charging grain surface, can enhance molecule formation.

2.6.4.3 Formation of CO and H_2O

After H and He, the most common elements in the Universe are O and C. Both CO and H_2O form readily in gaseous form, possibly in the strong stellar winds during the giant phase of stellar evolution. The reservoir of atomic C, which is smaller than that of O, is nearly depleted in forming CO.

The reactions in forming H_2O from H and O are among the most exothermic ones. H or H_2 become ionized and react to form H_3^+. This reacts with atomic O to form HO^+ and H_2 as well as H_2O^+ and H. The ions HO^+ and H_2O^+ react with free electrons to form H_2O and OH radicals. Thermally, the water molecule is stable until a temperature of 3 000 K. This is an important aspect in considering the origin of Earth's oceans. H_2O can also form directly on grain surfaces.

2.7 Generation of Stars: Formation of the Sun and Planets

The formation of a stellar system starts with the collapse of a local dense part of a GMC, the same regions where the most complicated molecules are found. First the density increases and central parts of the condensation heat up. The condensation, which is still starless, acquires a higher angular velocity in the center of the density enhancement. The collapsing system settles into a disk-like geometry. Both the higher angular velocity at the center and the settling into a disk are due to the conservation of angular momentum. Such dusty disks are seen, for example, in the Orion Molecular Cloud (OMC). They are so dense that even IR does not penetrate through them, as they appear as dark ellipses of circles against a brighter IR background.

Some of these "propylads" have a bright star in the center. They have entered an evolutionary phase at which the condensation in the center has reached a high enough temperature and density that it has turned into a new star. As the star begins to shine, it will start to ionize H gas around it and a HII region will form. If the newly born stars are located at the edge of the cloud, as in M42, one can note the typical emission line spectrum, as the tenuous gas emits at a few characteristic wavelengths only. From the spectra one can calculate, among other things, the temperature and the density of the gas in these HII regions.

2.7.1 Accretion Disk of the Sun

When the first generation of normal stars was produced, the building blocks for planets were not readily available. As we will see later, we need ices and rocks to form planets.

It is important to note that the heavier elements C, N, O, Si, and Fe, which are needed for the planetary building blocks, were produced by nuclear fusions in the first generation of normal stars and further processed by the subsequent molecular evolution in the ISM.

Our Sun and its accretion disk formed in a molecular cloud that was enriched by heavier elements from the first generation of normal stars. Traces of the cloud from which our Solar System was formed can be found in meteoritic dust in our Solar System, e.g., in the form of Al_2O_3, diamonds, graphite, and SiC within the "fine-grained" matrix in the most pristine chondrite class meteorites.

The material that would finally form the stars and planets formed a disk similar to the ones seen in the OMC. This disk initially consisted of H and He, but all other elements were also present, as were molecules similar to the ones observed at radio wavelengths in molecular clouds. All the matter began falling towards the central plane of the disk. The dust grains, both ice-coated and iceless, sedimented to a common central plane that was to become the general plane of orbits and equators of the planets and the Sun. The time it took for the embryonic Solar System to achieve that phase was about one million years.

How the planets formed goes along the following line of thought: small dust particles started to stick together and they grew fairly rapidly into meter-size aggregates. Because the aggregates were on similar orbits, their relative velocities were small and thus collisions occurred at low-velocity differences. These aggregates grew to larger sizes, and once they reached about 1 km, gravitation started compacting them. In a few ten thousand years the 1-km particles grew into 1 000-km planetesimals. At this point they were already so large that the gravity started to act on their shape making them round.

The speed of the planet formation processes following next can be estimated by observing dusty disks around solar-type stars of different ages and by correlating the age of the star with the state of the disk. For a solar-type star, a planet-forming disk, the protoplanetary disk, disappears in about 7 million years. After that time, no substantial amount of gas and dust is available for planet formation, although debris disks can last for 400 million years. Some matter remains even for billions of years (e.g., zodiacal light and Kuiper Belt objects). Modeling of disks gives similar timescales.

Models can provide a temperature profile of the disk. The general temperature profiles are dependent on the distance from the central star, usually $r^{-0.75}$ to r^{-1}. In normal air pressure, H_2O would crystallize into ice at 273 K. In space, where H_2O sublimates from the solid phase directly into the gas phase, the sublimation point can be as low as about 150 K. Beyond some distance from the star, at the so-called snow line, H_2O ice stays in a solid phase. Inside the snow line, H_2O ice sublimates into the gas phase. One experimental way to estimate this distance is to study the crystallization temperatures of meteorites, as done by Chris McKay. The separation of ordinary chondrites and carbonaceous chondrites took place at temperatures of 450 K. If the carbonaceous chondrites are identified with the C asteroids and the ordinary chondrites with the S asteroids, then one can place a rough distance for the temperature of 450 K at 2.6 AU. This would now set the snow line at somewhere between about 5 AU and 15 AU. The inner parts of the disk were hot, implying that ices melted and evaporated. The temperature estimate for the accretion disk at 1 AU distance from the above models would be about 500–1 200 K. Clearly, at these temperatures the newly formed Earth was hot, with no water or other volatiles and no atmosphere at the beginning, and was indeed a very inhospitable place for life. Some protoplanetary disk models suggest a flatter temperature profile and a snow line closer to the Sun, but in general, there is reasonable agreement that the snow line was at about 5 AU.

In the inner part of the Solar System, rocky planets were formed. The last impacts in the planetary accretion were often from nearly similar-size objects and were quite severe for the final planet. Examples are the Moon-forming impact on Earth, the impact that possibly caused the retrograde spin of Venus, and the mantle-destroying impact on Mercury.

In the outer parts of the Solar System, beyond the snow line, different processes took place. There was more solid material around in the form of (mainly H_2O) ice. Once the core of Jupiter had grown to about 20 Earth masses, its gravitation became large enough even to hold H and He after capturing them. A runaway accretion took place, whereby Jupiter started to grow rapidly. With its increased sphere of influence, it grew even larger until all the available matter in its vicinity was consumed. A similar process also took place with Saturn, Uranus, and Neptune. Beyond the orbit of Neptune, a large number of icy objects were left. These are now known as Kuiper Belt objects. Several 1000-km objects have been detected in the last 10 years, some of them of the larger than the dwarf planet Pluto.

The large gravitational fields of the outer planets affected the dynamics of the remaining boulders, both the icy ones, the comets, and the iceless ones, the asteroids. Some of these were ejected from the Solar System and others were slingshot into the inner parts of the Solar System. This had two effects. First, comets and asteroids were cleared from the vicinity of the giant planets, and second, the inner planets experienced a shorter, more intensive heavy bombardment of asteroids and comets. Comets and asteroids were important in delivering the ingredients necessary for life (see Chapter 3).

2.7.2
Formation of the Earth

When did the Earth form? Relative abundances of isotopes of various elements in pristine chondrites can be used to determine the age of our Solar System. This gives an age of 4.567 ± 0.002 billion years, which defines the moment when the temperature of the chondrites cooled enough for the minerals to crystallize. The accretion continued for the next 100 million years. The size of large impacts increased until the last one, which stripped the protoatmosphere and possibly melted the whole Earth. This took place about 4.45 billion years ago, as measured by xenon isotope ratios of the atmosphere. The Moon-forming impact was earlier, possibly some 4.52 billion years ago. It was thought to be a grazing impact of a Mars-size body. Because of the impact, both bodies melted and the cores merged to form the core of the Earth. Part of the debris that was ejected into space appears to have formed the Moon. The first landmasses formed quite early on Earth. Zircon crystals collected from the Jack Hills region in western Australia indicate that granitic crust formed as early as 4.42 billion years ago. One can consider that the formation of the Earth ended at this time, although even presently the mass of Earth increases by about 20 000 tons per year due to the fall of micrometerorites.

At the beginning, the Earth was so hot that its surface was covered by molten magma, which means a magma ocean (liquid rock!). The acquired atmosphere was saturated and contained hot H_2O vapor, N_2, and most likely CO_2 and H_2. At present, this is our best guess for the atmospheric composition. It is doubtful that any measurements will be able to determine the exact constituents of the earliest atmosphere. Note that the Earth experienced a strong greenhouse effect of about 2 000 K because of high amounts of both CO_2 and especially H_2O. The total pressure of the atmosphere at the surface was about 1 000 times the present value. H_2 was such a light molecule that Earth was not able to hold it for a long time. The Earth started to cool by IR radiation. Once the atmosphere cooled to the dew point of about 200 °C, a long rain began. Most of the H_2O vapor in the atmosphere rained. The oceans had formed. The water was still hot, the water was fresh, and the mantle below the sea was also hot. The Earth was left with a CO_2 and N_2 atmosphere.

Volcanic activity was high, and volcano islands formed. The hot magma minerals, such as peridotite, interacted with water and formed minerals containing crystal water, e.g., serpentinite. These mineral formations grew larger. They gained weight and collapsed, sinking into the mantle. These hydrated minerals experienced a partial melting at relatively low pressures and temperatures, forming tonalities among others. They were now lighter than the basaltic seafloor and started "floating" on it, eventually becoming the main ingredient for continents. As rocks started to protrude from the ocean, erosion set in. Under high temperature and high CO_2 partial pressure, erosion was enormous, with an estimated rate of about one million times the present value. The CO_2 dissolved into water, creating carbonic acid H_2CO_3. This reacted with wollastonite ($CaSiO_3$), changing it into carbonates ($CaCO_3$) and quartz (SiO_2). The carbonates sedimented onto the ocean floor. This was an effective means for removing CO_2 from the atmosphere.

2.7.3
Early Rain of Meteorites, Comets, Asteroids, and Prebiotic Molecules

The scars of the heavy bombardment of the inner planets by asteroids and comets can be seen on the "face" of the full Moon. The maria are lava-filled impact craters of gigantic proportions. Similar impacts took place on Earth about 23 times more often than on the Moon. Some of those impacts had enough energy to evaporate the oceans.

Asteroids and comets still abound. Asteroids can be described as large chunks (10–1 000 km) of rocks made of darker and lighter materials. The brightness of an asteroid depends on the geometric effects of the asteroid's distance from the Sun and its reflectivity or albedo. The short-term light variations that are seen are usually due to rotation only. Comets are large chunks of a mix of clay, ice, and sand. As they approach the Sun on their elliptical orbits, most of them develop a coma and some also a noticeable dust and/or ion tail. In addition to geometric and rotational effects, brightness variations due to some kind of activity of volatile materials, such as H_2O, CO, are often observed.

Meteorites, which are small pieces of asteroid material surviving the reentry to Earth, have been critical for understanding asteroids, as they have been historically the only means to study these celestial bodies. Radio astronomical spectroscopy opened a new window for cometary molecular studies in the 1970 s, and recent space missions have provided pictures, onsite measurements, and (still unpublished) sample returns from these bodies. The study of these bodies is of the utmost importance for understanding the evolution of the Earth and the delivery of prebiotic molecules to the Earth and other bodies in our Solar System. One has to bear in mind that although comets and asteroids are quite pristine in many respects, most of those bodies have been orbiting our Sun for the last 4.567 billion years and most likely experienced at least some evolution.

While asteroids and comets brought global and local destruction to the early Earth, they brought at the same time key ingredients for building our present Earth and possibly the critical prebiotic molecules for the origin of life.

Asteroids and comets are roughly of the same size. It is thought that the main cause for the difference between these two groups is the site of their formation. Asteroids formed inside the snow line and comets outside of it. Asteroids should have relatively little H_2O, while comets should have a significant portion of solid H_2O. The distinction between these two groups is not sharp because there are many objects that are intermediate in character, but roughly speaking, the classification is valid. Most of the known asteroids are concentrated in orbits between Mars and Jupiter, while most comets have orbits extending far beyond Neptune.

The composition of asteroids has been inferred mainly from studies of meteorites. No onsite measurements exist. Several asteroids have been photographed by flyby space probes, so there is some idea of what they look like and what their colors and albedos are like. This is of great importance in understanding, for example, the modeling of asteroids based on their brightness variations.

Meteorites come in three main types: iron, iron-stony, and stony. Iron meteorites are nearly solid iron-nickel. Stony meteorites usually contain minerals such as olivine and pyroxene and relatively small amounts of iron. The latter are further divided into chondrites and achondrites. The latter are thought to originate from the Moon or Mars (see Chapter 8). About 80% of all recovered meteorites are chondrites. These have small chondrules, which are considered to be the original molten material that crystallized during the formation of the meteorites. They are the oldest rocks in our Solar System. Five percent of chondrites are classified into carbonaceous chondrites, due to their elemental C.

If asteroids are assumed to be made of matter similar to meteorites, then it is quite clear that asteroids delivered iron and nickel and vast quantities of silicates to the Earth. C is an important ingredient in many meteorites (up to 4%), as is H_2O (20% in meteorites). The asteroid Ceres appears to have 25% H_2O by mass, which is more freshwater than found on Earth. The supply of these raw materials was more than sufficient for delivering the quantities found presently in silicates (continents) and carbon (biosphere). Also, a significant portion of H_2O (oceans) may have arrived through asteroids.

The outgassing of volatiles in comets has been studied mainly by radio and IR spectroscopy. Several dozen molecules have been detected, among them H_2O, HCN, H_2CO, CH_4, which all are important ingredients for the origin of life. Recent observations from the *Stardust* return mission to Comet Wild 2 revealed that at least this comet contains particles from the presolar GMC, from the protosolar disk, and also from later reprocessed matter.

The Murchison meteorite fell in Australia in 1969 and was almost immediately collected. It carries a large number – about 70 – amino acids, as well as nucleotides, sugars, and other organic molecules (see Chapter 3). Hence, the ingredients for life can be found in our vicinity in space, and this matter is seen falling even nowadays, as shooting stars. One can easily argue that this would be the easiest means for providing the young Earth with the seeds for life, maybe not in the form of self-destructing shooting stars but rather in the form of lightly floating micrometeorites.

Comets also contain N-bearing ices. Because the newly born Earth was free of volatiles, it seems that comets were the delivery mechanism for N_2, the main component of our present atmosphere. The second most abundant component, O_2, appeared in the atmosphere much later about 2.2 billion years ago, when cyanobacteria started to produce it – maybe – as a waste product of photosynthesis.

2.7.4 D/H Ratio and Oceans

The oceans cover about 70% of the surface of the present Earth. Our planet is often referred to as the "blue planet." If the whole planet is considered, only 0.04% of the mass is in the oceans. Still, the question of oceans is an important one because water has played an important role in the origin and evolution of life (see Chapter 1).

Asteroids and comets delivered H_2O and other volatiles to the young Earth. These volatiles did not have enough time to evaporate significantly, because the asteroids and comets plunged to the inner Solar System from the vicinity of the snow line or from far beyond it. Alternatively, it has been suggested that a significant fraction of the H_2O originated from H_2O within the accretion disk, but it is not clear how this oxygenation of H could proceed in a relatively hot gaseous disk. However, gaseous H_2O formed in the protosolar disk could endure, in principle, temperatures up to 3 500 K in the gas phase. Some geologists have suggested that H_2O was trapped in the minerals that formed the Earth and that H_2O was later outgassed by the hot mantle. One argument that has been raised against the cometary origin of water is the deuterium (D) to H ratio (D/H) of H_2O.

The standard terrestrial reference in geology for the D/H ratio is the Vienna Standard Mean Ocean Water (V-SMOW), which has a value of $D/H = (155.76 \pm 0.05) \times 10^{-6}$. In interstellar molecular hydrogen, the D/H ratio is about 15×10^{-6}: this is the cosmic D/H ratio. Some cold interstellar molecules, however, show a large D/H enrichment, with water having HDO/H_2O showing an enrichment of a factor of 20–40 or a D/H of $300–600 \times 10^{-6}$ and DCN/HCN having an enrichment of up to a factor of 1000 to the cosmic D/H ratio.

The first cometary value was measured by the *Giotto* mission for Comet Halley as $HDO/H_2O = (330 \pm 20) \times 10^{-6}$. This value was in line with what was seen in water from cold interstellar clouds, suggesting that the water in Comet P/Halley was presolar.

This cometary value created a problem in that the oceanic D/H ratio was too low by a factor of about 2 to be explained by a cometary origin if Halley represented a typical value for a comet. The concept of a "D/H problem" was created. The next comets to be measured for the HDO/H_2O ratio were Comet C/1995 O1 Hale-Bopp and Hyakutake. The relative abundances measured, $(330 \pm 80) \times 10^{-6}$ and $(280 \pm 40) \times 10^{-6}$, respectively, were in line with Comet Halley. The controversy between cometary HDO and oceanic HDO appeared to grow stronger.

One can also turn to other planets in our Solar System. The two giant planets have low D/H ratios: a value of $(16 \pm 3) \times 10^{-6}$ for Jupiter and of $(18 \pm 5) \times 10^{-6}$ for Saturn was measured by IRTF telescope, located on top of Hawaii's towering volcano Mauna Kea. They are both close to the cosmic abundance and in agreement with the HDO/H_2O ratios of the comets. Uranus and Neptune appear to have a higher value, at about $D/H = 60 \times 10^{-6}$. The *Huygens* mission (see Chapter 9) measured the D/H ratio for Titan, yielding a fractional abundance of $(250 \pm 50) \times 10^{-6}$. To complete the known HDO/H_2O ratios, Mars has a D/H ratio of $(810 \pm 30) \times 10^{-6}$ or about 5 SMOW, and Venus has a staggering ratio of $(25 \pm 5) \times 10^{-3}$or 160 ± 32 SMOW. Carbonaceous chondrites show typical values of $130–170 \times 10^{-6}$, close to the SMOW value. On Mars and Venus the loss of light hydrogen can be understood by its escape into space. Titan's D/H values are not well understood yet.

As noted, the D/H ratio in the Solar System varies considerably, depending on whether proper D/H is considered or HDO/H_2O is measured. If one considers one of the two types of measurements only, then the variation remains still large in each type.

With a preconceived presumption that the D/H ratios measured in the three comets mentioned above, namely, Halley, Hale-Bopp, and Hyakutake, are representative of the comet population that formed Earth's oceans, and if the D/H ratios measured in the comets and in the seawater represent the D/H at the time of formation of the oceans, then one has a "D/H problem." In this scenario, Earth's water came partially from comets and partially from the water vapor bound to grains in the accretion disk.

The story does not end here. The HDO/H_2O ratio of the comet C/2004 (Machholz) yielded an upper limit of 230×10^{-6} (3σ), obviously below the value from other comets. HDO was also detected in comet C/2001 Q4 (NEAT), but no value has been published so far for the D/H ratio. Recent observations of the comet C/1999 S4 (LINEAR) suggest that this comet has a very different chemistry than comets such as Halley, Hyakutake, or Hale-Bopp, which have chemistry consistent with their formation being beyond the orbit of Neptune. The detailed analysis of the abundances of molecules in comet C/1999 S4 (LINEAR) by Mumma suggests that this comet may have formed in the Jupiter–Saturn region, close to the ice line. In this region the D/H ratio could have been significantly lower than in the colder space beyond Neptune. The delivery of water from these comets could explain the D/H of the terrestrial oceans.

Finally, the recent *Deep Impact* mission to Comet Temple and the *Stardust* sample return mission from Comet Wild 2 have raised questions of whether the comets can undergo substantial evolution during their lifetime and whether the surface of the comet represents the conditions inside the core.

2.8 Further Reading

2.8.1 Books or Articles in Books

Despois, D., Biver, N., Bocklée-Morvan, D., Crovisier, J. Observations of molecules in comets, in: D. C.Lis, G. A. Blake, E. Herbst (Eds.) *Astrochemistry – Recent Successes and Current Challenges*, IAU Symp 231, pp. 19–128, 2005.

Irvine, W. M., Schloerb, F. P., Crovisier, J., Fegley, B. Jr., Mumma, M. J. Comets: A link between interstellar and nebular chemistry in: V. Mannings, A. P. Boss, S. S. Russell (Eds.) *Protostars and Planets IV*, University of Arizona Press, pp. 1159–1200, 2000.

Karttunen, H., Kröger, P., Oja, H., Poutanen, M., Donner, K. J. (Eds.), *Fundamental Astronomy*, 4th ed., Springer, Berlin, Heidelberg, New York, 2003 (a good comprehensive book for astronomy).

Thomas, P. J., Chyba, Hicks, R. D., C. F., McKay, C. P. (Eds.) *Comets and the Origin and Evolution of Life*, 2nd ed., Springer, Berlin, Heidelberg, New York, 2006.

2.8.2 Articles in Journals

Abe, Y. Physical state of the very early Earth, *Lithos* **1993**, *30*, 223–235.

Abergel, A., Verstraete, L., Joblin, C., Laureus, R., Miville-Deschênes, M-A. The cool interstellar medium, *Space Sci. Rev.* **2005**, *119*, 247–271.

Acharyya, K., Chakrabarti, S. K., Chakrabarti, S. Molecular hydrogen formation during dense interstellar cloud collapse, *Mon. Not. R. Astron. Soc.* **2005**, *361, 550–558.*

Barlow, M. J., Silk, J. H_2 recombination on interstellar grains, *Astrophys. J.* **1976**, *207*, 131–140.

de Bergh, C. The D/H ratio and the evolution of water in the terrestrial planets, *Origin Life Evol. Biosphere* **1993**, *23*, 11–21.

Bockelée-Morvan, D., Gautier, D., Lis, D. C., Young, K., Keene, J., Phillips, T., Owen, T., Crovisier, J., Goldsmith, P. F., Bergin, E. A., Despois, D., Wotten, A., Deuterium water in Comet C/1996 B2 (Hyakutake) and its implications for the origin of comets, *Icarus* **1998**, *133*, 147–162.

Ceccarelli, C., Dominik, C., Caux, E., Lefloch, B., Caselli, P. Detection of deuterated water in a young proto-planetary disks, in: D. C.Lis, G. A. Blake, E. Herbst (Eds.) *Astrochemistry –Recent Successes and Current Challenges*, IAU Symp 231, 2005 (abstract)

Cernicharo, J., Crovisier, J. Water in space: The water world of ISO, *Space Sci. Rev.* **2005**, *119*, 29–69.

Charnley, S. B., Rodgers, S. D., Butner, H. M., Ehrenfreund, P. *Astro-ph/0204381* **2002** (14 pages).

Cuppen, H. M., Herbst, E. Monte Carlo simulations of H_2 formation on grains of varying surface roughness, *Mon. Not. R. Astron. Soc.* **2005**, *361*, 565–576.

Dartois, E. The ice survey opportunity of ISO, *Space Sci. Rev.* **2005**, *119, 293–310.*

Dietrich, M., Wilhelm-Erkens, U. Elemental abundances of high redshift quasars, *Astron. Astrophys.* **2000**, *354*, 17–27.

Donahue, T. M. Evolution of water reservoirs on Mars from D/H ratios in the atmosphere and crust, *Nature* **1995**, *374*, 432–434.

Grinspoon, D. H. Implications of the high D/H ratio for the sources of water in Venus' atmosphere, *Nature* **1993**, *363*, 428–431.

Habart, E., Walmsley, M., Verstraete, L., Cazaux, S., Maiolino, R., Cox, P., Boulanger, F., Pineau des Forêts, G. Molecular Hydrogen, *Space Sci. Rev.* **2005**, *119*, 71–91.

Henning, K. Molecular formation in interstellar clouds by gas phase reactions, *Astron. Astrophys. Suppl. S.* **1981**, *44, 405–435.*

Hornekaer, L., Baurichter, A., Petrunin, V. V., Field, D., Luntz, A. C. Importance of surface morphology in interstellar H_2 formation, *Science* **2003**, *302*, 1943–1946.

Jehin, E., Manfroid, J., Hutsemékers, D., Cochran, A. L., Arpigny, C., Jackson, W. M., Rauer, H., Schulz, R., Zucconi, J.-M. Deep Impact: High resolution optical spectroscopy with the ESO VLT and the KECK I telescope, *Astrophys. J.* **2006**, *641*, L145–L148.

Jones, A. P. Interstellar and circumstellar grain formation and survival, *Phil. Trans. R. Soc., Lond. A.* **2001**, *359*, 1961–1972.

Kashlinksy, A., Ardent, R. G., Mather, J., Moseley, S. H. Tracing the first stars with fluctuations of the cosmic infrared background, *Nature* **2005**, *438*, 45–50 (and comments by R. S. Ellis, same issue p. 39).

Kawakita, H., Watanabe, J., Furusho, R., Fuse, T., Boise, D. C. Nuclear spin temperature and deuterium-to-hydrogen ration of methane in Comet C/2001 Q4 (NEAT), *Astrophys. J.* **2005**, *623*, L49–L52.

Kouchi, A., Greenberg, J. M., Yamamoto, T., Mukai, T. Extremely low thermal conductivity of amorphous ice: Relevance to comet evolution, *Astrophys. J.* **1992**, *388*, L73–L76.

Lecar, M., Podolak, M., Sasselov, D., Chiang, E. On the location of the snow line in a protoplanetary disk, *Astrophys. J.* **2006**, *640*, 1115–1118.

Lorenzetti, D. Pre-main sequence stars seen by ISO, *Space Sci. Rev.* **2005**, *119*, 181–199.

Meier, R., Owen, T. C., Jewitt, D. C., Matthews, H. E., Senay, M., Biver, N., Bockelée-Morvan, D., Crovisier, J., Gautier, D. Deuterium in Comet C/1995 O1 (Hale-Bopp): Detection of DCN, *Science* **1998**, *279*, 1707–1710.

Meier, R., Owen, T. Cometary Deuterium, *Space Sci. Rev.* **1999**, *90*, 33–43.

Molster, F., Kemper, C. Crystalline Silicates, *Space Sci. Rev.* **2005**, *119, 3–28.*

de Muizon, M. J. Debris Discs around stars: The 2004 Iso Legacy, *Space Sci. Rev.* **2005**, *119*, 201–214.

Müller, T. G., Ábrahám, P., Crovisier, J. Comets, Asteroids and Zodiacal light as seen by ISO, *Space Sci. Rev.* **2005**, *119*, 141–155.

Mumma, M. J., Dello Russo, N., DiSanti, M. A., Magee-Sauer, K., Novak, R. E., Brittain, S., Retting, T., McLean, I. S., Reuter, D. C., Xu, Li-H. Organic Composition of C/1999 S4 (LINEAR): A Comet formed near Jupiter? *Science* **2001**, *292*, 1334–1339.

Niemann, H. B., Atreya, S. K., Bauer, S. J., Carignan, G. R., Demick, J. E., Frost, R. L., Gautier, D., Haberman, J. A., Harpold, D. N., Hunten, D. M., Israel, G., Lunine, J. I., Kasprzak, W. T., Owen, T. C., Paulkovich, M., Raulin, F., Raaen, E., Way, S. H. The abundances of constituents of Titan's atmosphere from the GCMS instrument on the *Huygens* probe, *Nature* **2005**, *438*, 779–783.

Nisini, B., Kaas, A. A., van Dishoeck, E. F., Ward-Thompson, D. ISO observations of pre-stellar cores and young stellar objects, *Space Sci. Rev.* **2005**, *119*, 159–179.

Podolak, M., Mekler, Y., Prialnik, D. Is the D/H ratio in the comet equal to the D/H ratio in the comet nucleus? *Icarus* **2002**, *160*, 208–211 and comments by Krasnopolsky *Icarus* **2004**, *168*, 200 and rebuttal by original authors in *Icarus* **2004**, *168*, 221–222.

Robert, F. The Origin of water on Earth, *Science* **2001**, *293*, 1056–1058.

Roberts, H., Herbst, E. The abundance of gaseous H_2O and O_2 in cores of dense interstellar clouds, *Astron. Astrophys.* **2002**, *395*, 233–242.

Safarik, D. J., Mullins, C. B. The nucleation rate of crystalline ice in amorphous solid water, *J. Chem. Phys.* **2004**, *121*, 6003–6010.

Sasselov, D. D., Lecar, M. On the snow line in dusty protoplanetary disks, *Astrophys. J.* **2000**, *528*, 995–998.

Schulz, R., Owens, A., Rodriguez-Pascual, P. M., Lumb, D., Erd, C., Stüwe, J. A. Detection of water ice grains after the *Deep Impact* onto Comet 9P/Tempel 1, *Astron. Astrophys.* **2006**, 448, L53–L56.

Speedy, R. J., Debenedetti, P. G., Smith, R. S., Huang, C., Kay, B. D. The evaporation rate, free energy, and entropy of amorphous water at 150K, *J. Chem. Phys.* **1996**, *105*, 240–244.

Talbi, D., Chandler, G. S., Rohl, A. L. The interstellar gas-phase formation of CO2 – Assisted or not by water molecule, *Chem. Phys.* **2006**, *320*, 214–228.

Vanýsek, V., Vanýsek, P. Prediction of Deuterium Abundance in Comets, *Icarus* **1985**, *61*, 57–59.

Varnière, P., Blackman, E. G., Frank, A., Quillen, A. C. Planets rapidly create holes in young circumstellar disks, *Astrophys. J.* **2006**, *640*, 1110–1114.

Zhang, Y. The age and accretion of the Earth, *Earth Sci. Rev.* **2002**, *59*, 235–263.

2.8.3 Web Sites

NASA page on stars and galaxies:
http://www.nasa.gov/vision/Universe/starsgalaxies/index.html

Wilkinson Microwave Anisotropy Probe (WMAP) page, a NASA Explorer mission measuring the temperature of the comic background radiation:
http://map.gsfc.nasa.gov/

European Southern Observatory:
http://www.eso.org/

Al Wootten's page – including a list of known ISM molecules:
http://www.cv.nrao.edu/~awootten/

Cologne database on molecular spectroscopy:
http://www.ph1.uni-koeln.de/vorhersagen/main.html

UC San Diego page about ISM:
http://cassfos02.ucsd.edu/public/tutorial/ISM.html

ISO telescope:
http://sci.esa.int/science-e/www/object/index.cfm?fobjectid=32721

Near Earth Object program:
http://neo.jpl.nasa.gov/

Dave Jewitt's Kuiper belt page:
http://www.ifa.hawaii.edu/faculty/jewitt/kb.html

UC Berkley geology page:
http://www.ucmp.berkeley.edu/exhibit/geology.html

2.9 Questions for Students

Question 2.1

Consider the Si atom in the SiO_2 molecule in your porcelain coffee cup and the C atom in the tiny sugar crystal inside your coffee cup, just before you pour your hot coffee on it. Where were these atoms formed? How did their long way into your coffee cup differ from each other?

Question 2.2

At any given moment, there are about ten 10-meter meteoroids/asteroids in Earth's vicinity, at a distance that is smaller than the Moon's distance from Earth. Assuming a typical velocity of 10 km s^{-1}, make a rough calculation of how long should it take for such an object to hit the Earth. What is the energy liberated in such an impact? Compare it with something sensible. Compare the mass flux from this hypothetical impact with the mass flux from micrometeorites.

Question 2.3

The triple-alpha reaction is a critical step in the formation of C, and thus in the formation of life. Find other cosmic steps that have turned out to be critical for getting together all the important ingredients for life on Earth.

Question 2.4

Interstellar molecules are observed in gaseous and solid phases. What are the differences between these groups? Which of the differences are real and which are due to limitations in observing techniques or due to limiting factors caused by the atmosphere?

3
Basic Prebiotic Chemistry

Hervé Cottin

This chapter is devoted to the study of prebiotic chemistry. It focuses on the abiotic synthesis of key prebiotic compounds – amino acids, purine and pyrimidine bases, and sugars – from simple molecules, such as methane (CH_4), hydrogen cyanide (HCN), and formaldehyde (H_2CO), thereby filling the gap between chemistry and biology. In the first section living systems are dismantled into more elemental fragments: organic molecules. To address the origin of life, the next question is how those fragments can be synthesized in an abiotic manner. After a short historical review, different tracks of prebiotic chemical evolution are explored: an endogenous production of the molecules of life within early Earth's atmosphere or in the depths of primitive oceans, or an exogenous source of prebiotic molecules through space delivery via meteorites and comets. The last section describes chemical pathways of the synthesis of the building blocks of life.

3.1
Key Molecules of Life

Life as we know it is based on a complex chemistry involving highly sophisticated biological molecules such as proteins, ribonucleic acids (RNA), or deoxyribonucleic acids (DNA). Such elaborated structures result from a complex chemical evolution, starting with the simplest organic compounds. At some point, at an as yet undefined stage of complexity and organization of matter, an important step is made and chemistry turns into biology.

Complete Course in Astrobiology. Edited by Gerda Horneck and Petra Rettberg

ISBN: 978-3-527-40660-9

3.1.1
Dismantling the Robots

In Chapter 1 André Brack compares the premises of an emerging life to parts of "chemical robots," which make copies of themselves and are capable of evolution through "mistakes" during reproduction. Following this idea, life could be considered as any type of structure (at a molecular or cellular level or any other form) with some complexity that bears information to make the robot work and that makes copies of itself. From time to time, the copy would not be the exact image of its original. This modified copy could be either more adapted to the environment than its original or less adapted, and therefore either more or less favored to survive. Then it would make more copies of its new model with either more or less efficiency.

Whereas it is quite difficult to find a proper general definition of life (see Chapter 1), it is much easier to answer the question "What is life made of?" The answer turns out to be surprisingly simple. If one considers all living systems on Earth, from unicellular systems such as bacteria to multicellular systems such as plants and animals, they are all built from the same basic components.

Each living system is made of at least one cell. A cell can in turn be broken down into more elemental parts: proteins, RNA, and DNA molecules embedded within the membrane of the cell. These biomolecules are made of mainly the following elements: carbon (C), hydrogen (H), oxygen (O), and nitrogen (N). The membrane is built from long organic amphiphile chains. These chains possess a polar head, which makes them soluble in water (hydrophilic part of the molecule), and an apolar tail, which is insoluble in water (hydrophobic part of the molecule). Such molecules are called amphiphiles (amphi = both), because they consist of both a hydrophobic and a hydrophilic part. These amphiphile structures may self-organize into vesicles (Fig. 3.1).

Inside the cell, which is surrounded by the membrane, a complex molecular engine is working (Fig. 3.2). But, from a simplified chemical point of view, the interior of the cell turns out to be made of mainly proteins and nucleic acids, such as DNA and RNA. Of course, there is a wide range of proteins that play various key

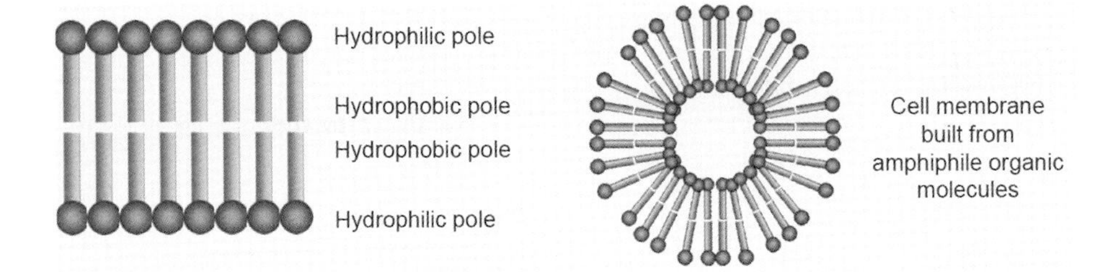

Fig. 3.1 Long organic amphiphile chains with a polar head that makes them soluble in water and a water-insoluble apolar tail (left), which may self-organize into vesicles (right).

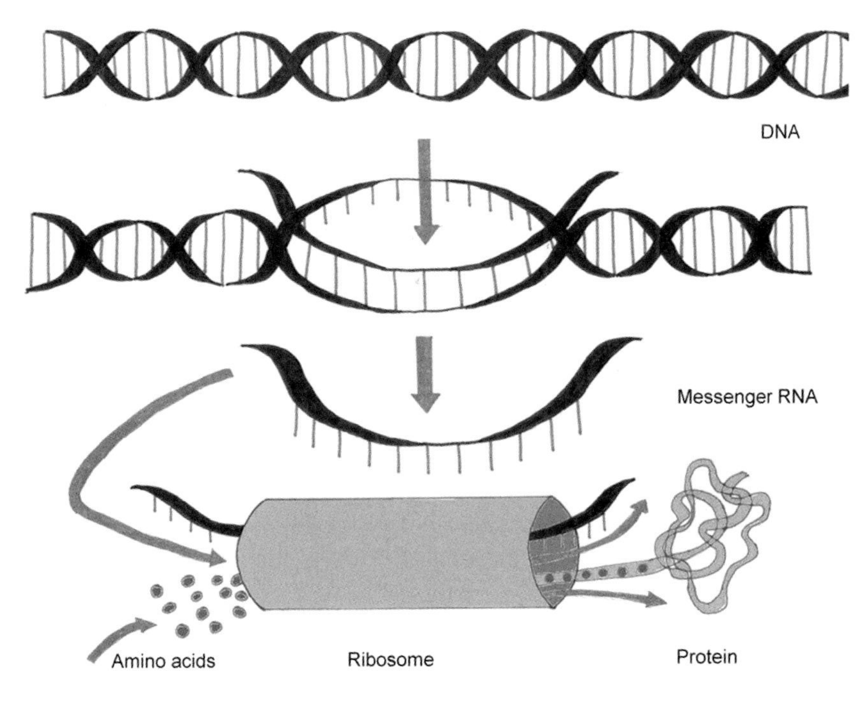

Fig. 3.2 Basic cellular components (DNA, RNA, and proteins) and functions (storage and transfer of information and protein synthesis in the ribosomes).

roles in metabolism and an enormous diversity of DNA and RNA molecules, which is reflected by the large diversity of living forms on Earth. But this diversity results from a combination of 28 molecules only. Every known living organism could be compared with a Lego construction with 28 different kinds of bricks: 20 amino acids to build the proteins, five purine or pyrimidine bases, two sugars, and one phosphate bridge to build DNA and RNA. The following main functions can be attributed to the three families of molecules:

- DNA stores information,
- RNA transmits information, and
- proteins act as catalysts.

Table 3.1 shows the basic composition of a cell exemplified by the bacterium *Escherichia coli*.

Table 3.1 Composition of a bacterial cell as exemplified by *Escherichia coli.*

Component	Percentage of total cellular weight	Molecular mass ($g\ mol^{-1}$)[a]
Water	70	18
Proteins	15	40 000
RNA	6	~ 10^6
DNA	1	2.5×10^9
Carbohydrates and precursors[b]	3	150
Lipids and precursors[b]	2	750
Amino acids and precursors	0.4	120
Nucleotides and precursors	0.4	300
Others[c]	2	–

a Averaged in a mixture
b Parts of the cell membrane
c Other small molecules and inorganic ions

3.1.2
Proteins and Amino Acids

Proteins are organic compounds built from a succession of amino acids linked together by peptide bonds. Proteins can play different roles: some are of importance at a structural level in the cells, while others are enzymes that act as catalysts for chemical reactions. Their properties result from their chemical formula and from their spatial conformation (how the amino acid chain organizes itself in space).

An amino acid is a molecule that contains both an amino group (-NH_2) and a carboxylic acid (-COOH) function (Fig. 3.3). If the carbon that bears the amino

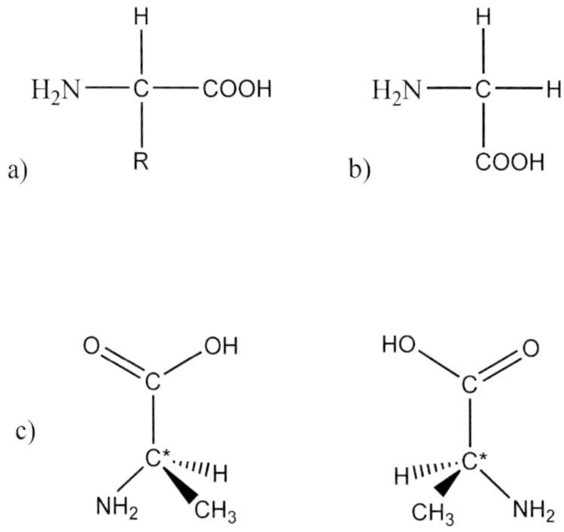

Fig. 3.3 (a) General formula of an α-amino acid; R represents different side chains; (b) glycine; (c) 3-dimensional presentation of an α-amino acid (alanine) with an asymmetric carbon (marked with a star). Note that these molecules, which have an identical molecular formula, cannot be superimposed one over the other.

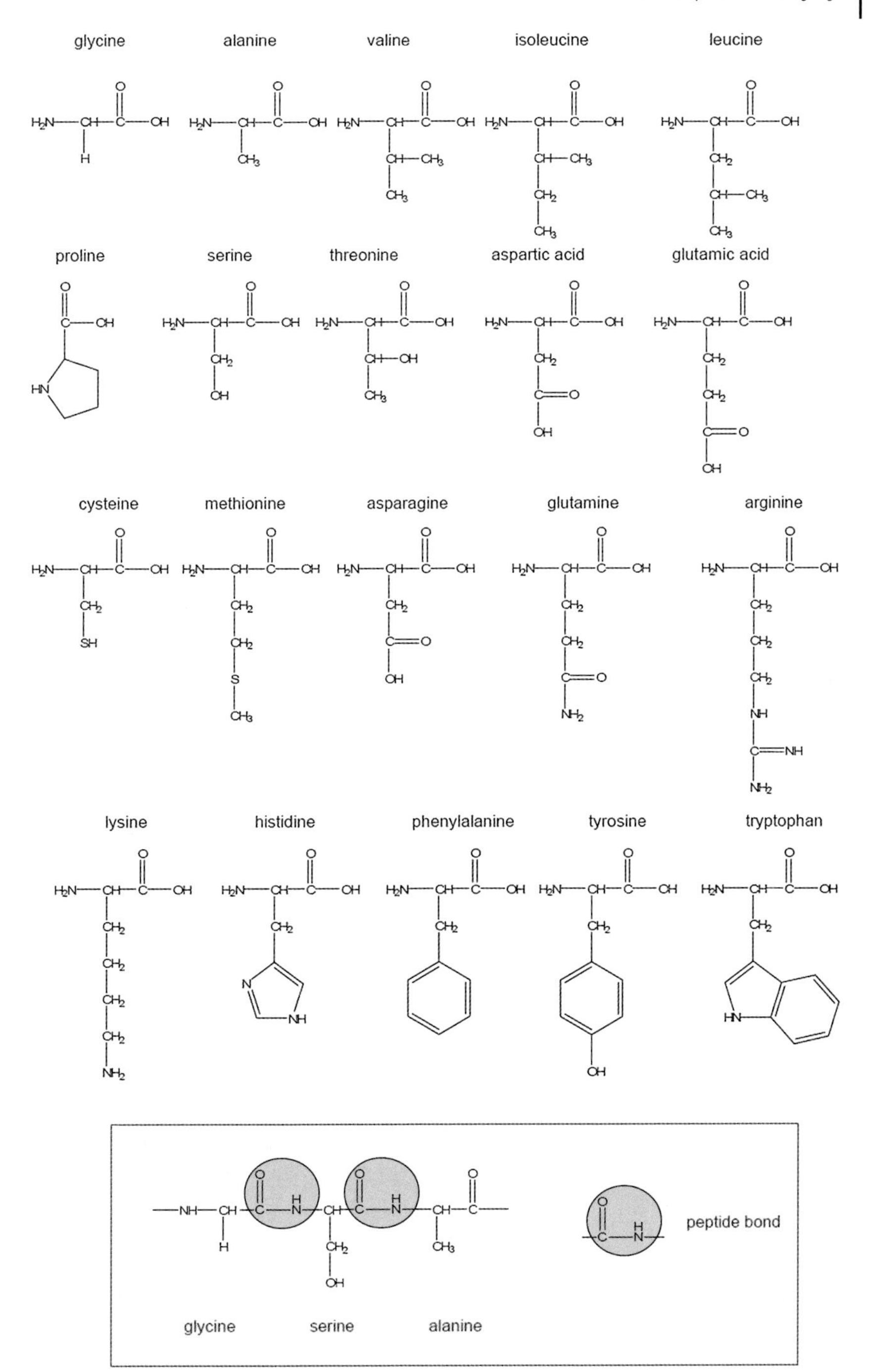

Fig. 3.4 The 20 biological amino acids; association of several amino acids through peptide bonds produces a polypeptide.

group is the same as the one bearing the carboxylic function, then the amino acid is called an α-amino acid. If the amino group is linked to a carbon that is one C away from the carbon bearing the carboxylic function, then the amino acid is called a β-amino acid, and so forth.

The general structure of an α-amino acid is shown in Fig. 3.3 a. R represents a side chain specific to each amino acid. In only the simplest amino acid, glycine, is the R-group hydrogen (Fig. 3.3 b). For all other α-amino acids, R is different from H. In that case the central carbon atom has four different substituents and is then called an asymmetrical carbon. As shown in Fig. 3.3 c, two spatial arrangements (configurations) of these substituents are possible, resulting in two different molecules (called enantiomers) that are mirror images of each other. Molecules with an asymmetric or chiral C atom affect the orientation of polarized light when it passes through a sample of molecules of one configuration. Configuration 1 shifts the orientation of polarized light to the right (dextrorotary), while configuration 2 would shift it to the left (levorotary). These two configurations of amino acids are called L or D (L-compounds rotate the plane of polarized light to the left, D-compounds to the right). A mixture of equal quantities of both configurations is called a racemic mixture. It neutralizes the effect, and polarized light passes unchanged. This property toward polarized light has no direct effect on life, but the configuration of amino acid (D or L) is extremely important for the spatial orientation (conformation) of long chains of amino acids: the proteins. The properties of proteins are directly linked to their conformation, and they have different characteristics if built from L-amino acids, D-amino acids, or a mixture of both (see also Chapter 1).

As mentioned above, in the proteins the amino acids are linked by peptide bonds. Peptide bond formation is a condensation reaction leading to the polymerization of amino acids into peptides and proteins with the concomitant elimination of water (Fig. 3.4). Peptides are small, consisting of a few amino acids. Proteins are polypeptides of greatly divergent length.

To build the proteins, life uses a combination of only 20 amino acids (Fig. 3.4), all of which appear in one enantiomer: the L-form. Why L-enantiomers were selected over D-enantiomers remains unknown, but this is a property shared by all living systems (with a few exceptions in the bacterial domain; see Chapter 5).

3.1.3 DNA, RNA, and Their Building Blocks

DNA (deoxyribonucleic acid) and RNA (ribonucleic acid) are long-chained organic molecules that contain genetic code information (see Chapter 4). From a chemical point of view, in terms of building blocks of life, their composition and structure can be quite simply explained.

The backbone of DNA and RNA is a succession of sugars (deoxyribose for DNA, ribose for RNA) linked by phosphate bridges (PO_4^{2-}) (Fig. 3.5). The "letters" of the genetic code, which are connected to each sugar molecule, are the purine bases adenine (A) and guanine (G) and the pyrimidine bases cytosine (C), thymine (T) (T in DNA only), and uracil (U) (in RNA only) (Fig. 3.6). The association of the sugar

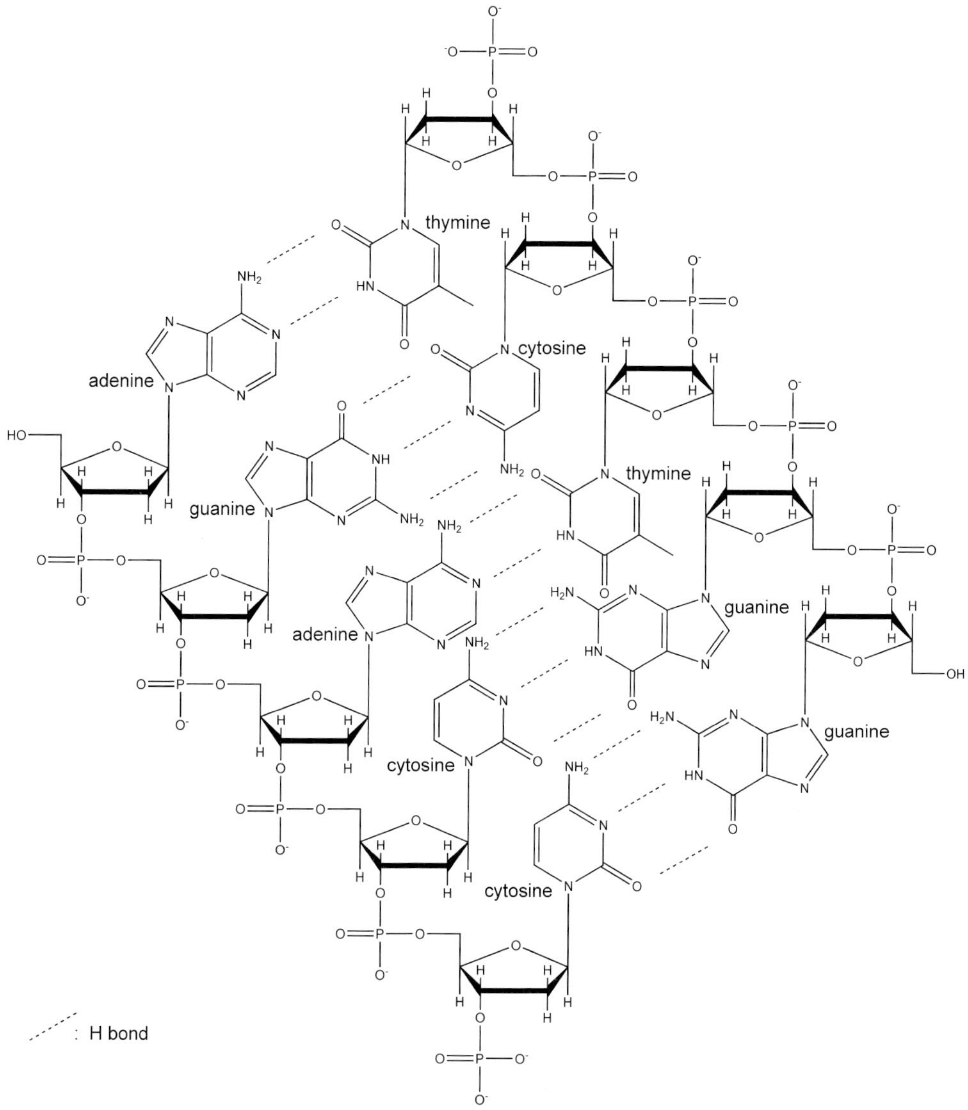

Fig. 3.5 Structure of DNA. DNA and RNA are linked chains of nucleotides, each consisting of a sugar, a phosphate, and one of the five nucleobases (purine bases: adenine and guanine; pyrimidine bases: cytosine and thymine in DNA or cytosine and uracil in RNA)

with one of the bases is called a nucleoside. If a phosphate is linked to a nucleoside, the resulting structure is called a nucleotide. The polynucleotide is named according to the sugar in its structure: DNA if the sugar is a deoxyribose, and RNA if the sugar is a ribose.

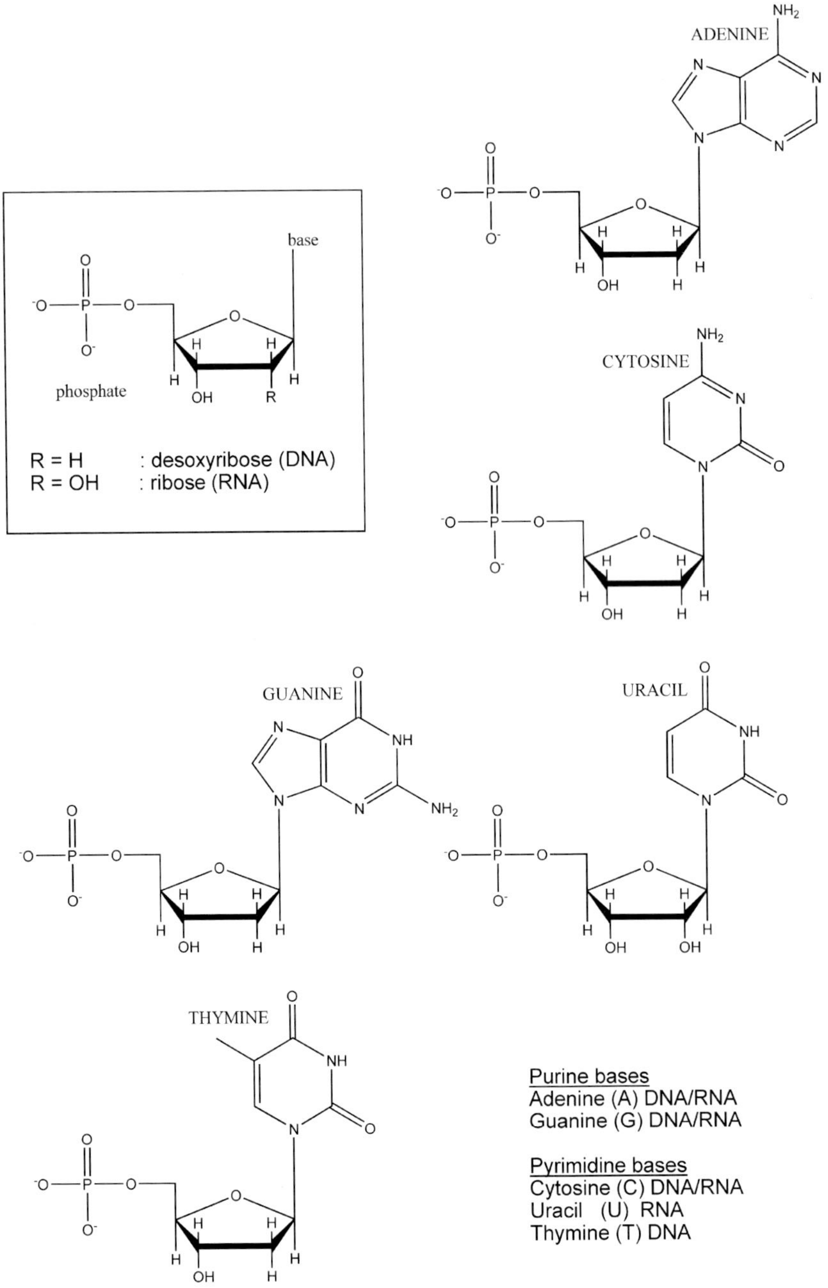
ADENINE
CYTOSINE
GUANINE
URACIL
THYMINE
base
phosphate
R = H : desoxyribose (DNA)
R = OH : ribose (RNA)
Purine bases
Adenine (A) DNA/RNA
Guanine (G) DNA/RNA
Pyrimidine bases
Cytosine (C) DNA/RNA
Uracil (U) RNA
Thymine (T) DNA

Fig. 3.6 Chemical structure of DNA and RNA nucleotides. The difference between the two types of molecules is the substitution of one H (DNA) by OH (RNA) on the sugar and thymine (DNA) by uracil (RNA).

The genetic code is encrypted in the specific succession of purine and pyrimidine bases. It contains information about which succession of amino acids is to be built in which proteins (see Chapter 4). This information is transferred to the ribosomes, the "proteins factory," via RNA (Fig. 3.2). Ribosomes are composed of RNA and proteins.

All living systems function according to this very same mechanism. What makes a mouse different from an elephant or a mushroom is the number and succession of the "letters" in the DNA molecules.

DNA and RNA are usually seen as carriers of genetic information. This picture has changed with the discovery of the self-splicing of certain RNAs. These observations laid the foundation for the concept of catalytic RNA, the ribozymes (see Chapter 4). This catalytic capacity of RNA led to the idea of the *RNA world* that might have preceded life as we know it today: a simplified cellular biochemical machinery in which RNA plays all roles that are essential for life; it carries and transmits information, and it catalyzes the reactions.

3.1.4 First "Prebiotic Robot"

For building the first "prebiotic robot," and then the first living system, the following compounds must be available:

- long organic amphiphile chains for building membranes;
- amino acids for building proteins;
- phosphate for the backbone of nucleic acids;
- purine and pyrimidine bases as "letters" for the genetic code in DNA and/or RNA; and
- sugars for the nucleic acids (ribose in the RNA world scenario).

These molecules need to be synthesized under prebiotic conditions, either on early Earth or in space. Of course, the presence of the building blocks for life is not sufficient to make the first prebiotic robot; however, they are prerequisites in the scenario of the origin of life.

3.2 Historical Milestones

The building blocks of life have been discovered and characterized since the early 19th century (Table 3.2). In these early times, the question of the abiotic synthesis of such molecules in the test tube seemed far beyond the scope of most of the chemists' capabilities. Moreover, it was not even considered as a question of interest. It was the generally accepted opinion that *organic* molecules are the result of the activity of *organized*, i.e., living, systems, and are associated with a mysterious "vital energy." In 1827, the Swedish chemist Jöns Jacob Berzelius (Fig. 3.7) wrote, "Art cannot combine the elements of inorganic matter in the manner of living nature." (Berzelius also invented the words "polymerization," "catalysis," "electronegative," and "electropositive" and discovered the elements selenium [Se], silicon [Si], and titanium [Ti]). At that time, there was no way known to synthesize organic molecules in the laboratory, and no reason to look for such ways.

Table 3.2 Milestones in the discovery and characterization of important biochemical monomers or polymers.

Year	Discoverer	Monomer/polymer
1810	W.H. Wollatson	Cystine[a]
1819	J.L. Proust	Leucine[a]
1823	M. E. Chevreul	Fatty acids
1838	J.J. Berzelius	Proteins
1869	F. Miescher	DNA
1882	A. Kossel	Guanine
1883	A. Kossel and A. Neumann	Thymine
1886	A. Kossel and A. Neumann	Adenine
1894	A. Kossel and A. Neumann	Cytosine
1900	A. Ascoli	Uracil
1909	P.T. Levene and W. A. Javok	Desoxyribose
1906–1936	P.T. Levene et al.	Ribose, ribonucleotides

Source: Adapted from Miller and Lazcano 2002.
a Amino acid

A firm denial to this theory was brought in 1828 by Berzelius' friend and former student Friedrich Wöhler (Fig. 3.7). This German chemist succeeded in the first abiotic synthesis of an organic molecule, urea (NH_2-CO-NH_2), "without the need of an animal kidney," as he said at the time, by simple heating of ammonium cyanate (NH_4OCN). From then on, a breach was made in the "vital energy" theory, and, slowly, organic chemistry became the chemistry of carbon-based molecules.

Fig. 3.7 Famous chemists of the 19th century. From left to right: Jöns Jacob Berzelius (1779–1848), Friedrich Wöhler (1800–1882), Adolph Strecker (1822–1871), and Alexandr Butlerov (1828–1886).

Soon after this pioneering finding, the synthesis of other organic compounds was realized. Regarding the synthesis of molecules of prime interest to the origin of life, two major achievements shall be pointed out. In 1850, Adolf Strecker (Fig. 3.7) succeeded in the first laboratory synthesis of an amino acid: alanine from a mixture of acetaldehyde (CH_3CHO), ammonia (NH_3), and hydrogen cyanide (HCN). A few years later, in 1861, Alexandr Butlerov (Fig. 3.7) performed the first laboratory synthesis of sugar mixtures (also known as the *formose reaction*) from formaldehyde (HCHO) using a strong alkaline catalyst (NaOH) (see Section 3.4.3).

Although these discoveries were very interesting, they were not linked to the origin-of-life problem. Hence, little progress was made in finding a scientific description of the origin of life. One had to wait until 1924 for the publication of the book *Origin of Life* by Aleksandr Ivanovich Oparin (1894–1980). Oparin introduced the concept of chemical evolution, which could be seen as the roots of the Darwinian tree of evolution. In this concept, life was the result of a succession of spontaneous chemical reactions that produce increasingly complex chemical structures. In the first edition of his book (1924) he suggested that such chemical evolution would take place within an oxidizing atmosphere of the primitive Earth (it was the generally accepted view at that time that the early Earth had an oxidizing atmosphere). In the second edition (1938) he changed the early atmosphere to a highly reducing environment. Similar ideas were simultaneously developed by the English biologist John Haldane, who was the first to mention the concept of a "prebiotic soup" where chemical evolution took place, a term that became quite popular in the scientific community studying the origins of life.

Some paragraphs of Oparin's book *Origin of Life* from 1924 show its actuality:

> *At first we found carbon scattered in the form of separate atoms, in the red hot stellar atmospheres. We then found it as a component of hydrocarbons which appeared on the surface of the Earth. ...In the waters of the primitive ocean these substance formed more complex compounds. Proteins and similar substance appeared ....[They]...acquired a more and more complex and improved structure and were finally transformed into primary living beings – the forbears of all life on Earth.*

He concluded quite optimistically, bearing promises of spectacular breakthroughs in the field within a few years:

> *We have every reason to believe that sooner or later we shall be able practically to demonstrate that life is nothing else but a special form of existence of matter. The successes scored recently by Soviet biology hold out the promise that the artificial creation of the simplest living beings is not only possible, but that it will be achieved in the not too distant future.*

Facing the hard reality of facts, he did temper his optimism and wrote as a conclusion of the second edition of his book in 1934:

> *We are faced with a colossal problem of investigating each separate stage of the evolutionary process as it was sketched here. ...The road ahead of us is hard and long but without doubt it leads to the ultimate knowledge of the nature of life. The artificial building or synthesis of living things is a very remote, but not an unattainable goal along this road.*

As elegant as Oparin and Haldane's ideas appeared to be, they were expressed at a theoretical level only. The experimental confirmation of the theory of chemical evolution was provided in 1953 by Stanley Miller, a graduate student in the laboratory of the Nobel Prize laureate Harold Clayton Urey. They conceived and built an experiment to simulate a putative primitive Earth environment (Fig. 3.8).

In this experiment a gaseous mixture of hydrogen (H_2), methane (CH_4), ammonia (NH_3), and water (H_2O) was exposed to an electric discharge that simulated that of storm lightning. The mixture was connected to a bulb filled with liquid water that could be heated. This experiment resulted in the production of a large amount of organic molecules, including several amino acids. Those measurements were experimental proof of the theory of chemical evolution. They showed that the chemistry between simple molecules, which were abundant in the atmosphere of

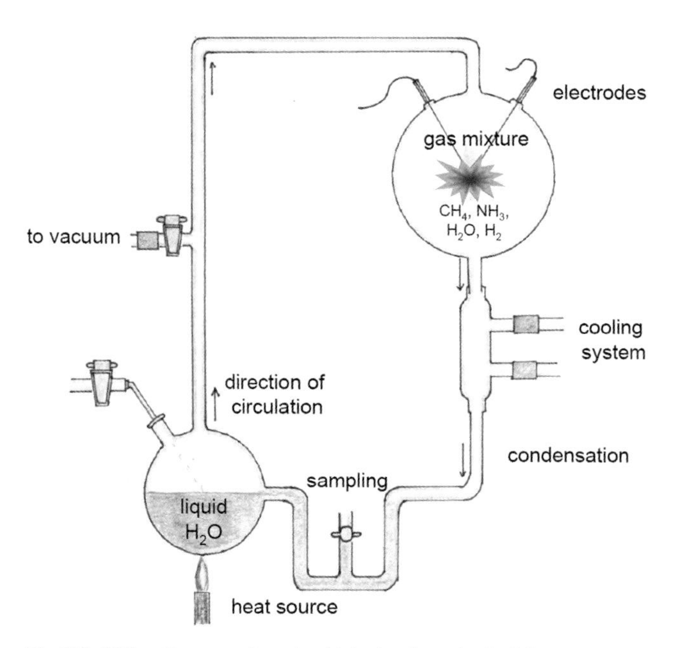

Fig. 3.8 Miller–Urey experiment, which simulates in the laboratory the coupled chemistry between the primitive Earth atmosphere (upper right bulb) and warm oceans (lower left bulb). In the first version, an atmosphere made of CH_4, NH_3, H_2O, and H_2 was considered. A spark discharge simulated storm lightning.

the primitive Earth, led to the synthesis of key compounds that in turn might have led to the forms of life as we know it on Earth.

The choice of a reducing atmosphere with C as CH_4 and N as NH_3 (Table 3.3) was motivated by the following considerations:

- CH_4, NH_3, H_2O, and H_2 had been detected in the atmosphere of giant planets since the 1930 s.
- All the primitive atmospheres of planets were identical, captured from the Solar Nebula.
- Giant planets, cold and distant from the Sun, have kept their original composition.

Therefore, Urey and Miller concluded that the current composition of the atmosphere of giant planets of the Solar System was a good proxy for the composition of the atmosphere of the primitive Earth.

Table 3.3 Classification of planetary atmospheres, in terms of their redox potential, as a function of their composition.

Redox state of the atmosphere	Composition
Reducing	CH_4, NH_3, H_2O, H_2
	CO_2, N_2, H_2O, H_2
	CO_2, H_2, H_2O
Neutral	CO_2/CO, N_2, H_2O
Oxidizing	CO_2/CO, N_2, H_2O, O_2

However, several recent findings disprove this argumentation. First, it is now quite well established that telluric planets do not have sufficient mass to capture the Solar Nebula gas like the giant planets did. To do this, a minimum mass of 10 to 15 Earth masses is required. Only then can a forming planet efficiently trap the volatile elements of the nebula to form its atmosphere. In addition, recent observations have supported the conception that at a distance of one astronomical unit (AU) from the Sun – that is where the Earth accreted – the gaseous component of the Solar Nebula was probably dominated by CO_2 and N_2. Second, Earth, like Venus and Mars, has a secondary atmosphere built from volatile compounds that outgassed from the mantle on one hand or were imported via meteorites and comets on the other hand. Third, the composition of the primitive Earth atmosphere was most probably dominated by CO_2 and N_2, in which organic syntheses are not as efficient as in a reducing atmosphere (see Section 3.3).

Even if the Miller–Urey experiment is not as conclusive as thought at first glance, it was a great achievement because it showed that important prebiotic compounds can be abiotically synthesized in environments simulating "natural" conditions. In Section 3.3 different kinds of such "natural" environments are presented that allow an interesting "prebiotic" chemistry.

Table 3.4 Organic molecules synthesized in Miller-type experiments as a function of the composition of the starting mixture, after electron impact or photolysis as energetic inputs. Compounds in the solid-phase residues are usually detected after acid hydrolysis.

Gaseous mixture	Redox state	Related planetary atmosphere	Families of synthesized organic products	
			Electric discharge	Photolysis
$CH_4 + NH_3 + H_2O$ (+ H_2)	Reducing	Giant planets	RH (saturated and unsaturated) HCN and other nitriles (saturated) RCO_2H H_2CO, other aldehydes Ketones and alcohols Solid: Amino acids and nitrogenated heterocycles after hydrolysis	RH (mostly saturated) HCN RCN (saturated) if N/C <1 RNH_2 if N/C >1 H_2CO, other aldehydes Ketones and alcohols Solid: Amino acids after hydrolysis
$CH_4 + N_2$ (+ H_2O)	Reducing	Titan, Triton	RH (saturated and unsaturated) HCN and other nitriles (saturated and unsaturated) including HC_3N and C_2N_2 H_2CO, other aldehydes Ketones and alcohols Solid: Amino acids and nitrogenated heterocycles after hydrolysis	RH (saturated & unsaturated) H_2CO & other aldehydes with low yields Solid: Carboxylic acids
$CO + NH_3 + H_2O$	Reducing		HCN, oxygenated organic compounds Solid: Amino acids after hydrolysis	HCN, oxygenated organic compounds Solid: Amino acids after hydrolysis
$CO_2 + N_2 + H_2O + CO/H_2$	Neutral	Primitive Earth?	RH (mostly saturated) HCN, other nitriles (saturated) H_2CO, other aldehydes, ketones Solid: Amino acids after hydrolysis	RH (mostly saturated) HCN, other nitriles (saturated) H_2CO, other aldehydes, ketones Solid: Amino acids after hydrolysis
$CO_2 + N_2 + H_2O$	Neutral	Primitive Earth? Venus, Mars	No synthesis	No synthesis

Source: Adapted from Raulin 2001.
RH = hydrocarbons

3.3 Sources of Prebiotic Organic Molecules

3.3.1 Endogenous Sources of Organic Molecules

3.3.1.1 Atmospheric Syntheses

Miller and Urey thought that the first building blocks of life were synthesized in Earth's primitive atmosphere. For the reasons discussed above, the gaseous mixture they chose is no longer considered to be representative of Earth's early atmosphere. Table 3.4 shows some results obtained in different "Miller-type" experiments, using different gaseous mixtures and energy sources.

Energy inputs through electric discharges represent lightning or electron inputs in the atmosphere. For example, in the case of Saturn's moon Titan, electrons trapped in the magnetosphere of Saturn cascade down into Titan's atmosphere (see Chapter 9). Photolysis experiments simulate the energetic inputs of the Sun via energetic UV photons that initiate chemical reactions. Using different energy sources but the same atmosphere led to minor differences in the resulting products.

In contrast, a high sensitivity of the resulting syntheses to the starting gaseous mixture was observed. Reducing gaseous mixtures, such as the one chosen by Miller and Urey, are the most efficient ones in the formation of complex organic structures. However, current models of the primitive atmosphere of Earth suggest a rather neutral atmosphere, and such environments are not as efficient in initiating chemical evolution as reducing ones. Figure 3.9 shows the yield in amino acids (percentage of initial C) as a function of the composition of the simulated atmosphere when performing Miller-type experiments. In a reducing atmosphere, if carbon is in a reduced state (in the form of CH_4), a large diversity of amino acids is synthesized in a rather efficient way. On the other hand, if carbon is in an oxidized

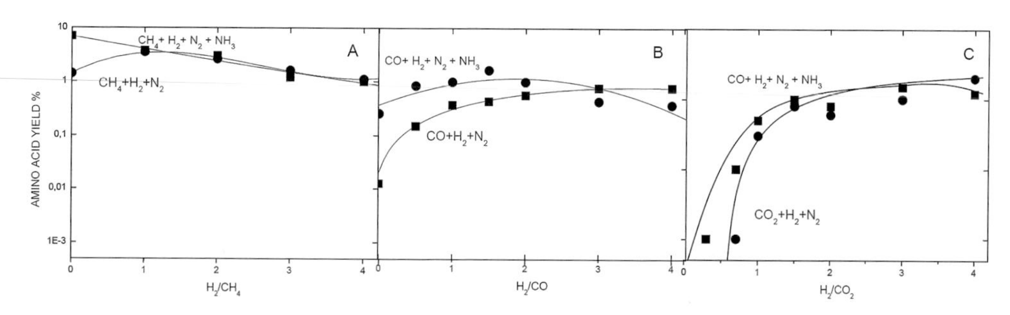

Fig. 3.9 Amino acid yield (percentage of initial C) as a function of the simulated atmosphere composition in Miller-type experiments, performed at room temperature and in the presence of liquid water. Results were obtained after 2 days of electric discharge experiments. Partial pressure of N_2, CO, or CO_2 was 100 Torr (131 hPa). (A) A great diversity of amino acids was produced. (B, C) Glycine predominated; little else was produced except a small amount of alanine (adapted from Miller and Lazcano 2002).

state (CO or CO_2), the amino acid glycine predominates, and the reaction yields are much lower than those using CH_4. Moreover, in the absence of CH_4, the yield of amino acids drops significantly if the atmosphere is depleted in H_2 and becomes more and more neutral.

Therefore, if the Earth mantle was less oxidized 3.8 billion years ago than today, an important amount of CH_4 could have been emitted through volcanism. In this case, endogenous Miller-type syntheses are possible. It has been estimated that depending on the redox state of the atmosphere, the endogenous production could vary by several orders of magnitude, leading to a steady-state concentration of organics in primitive oceans ranging from $0.4\ 10^{-3}$ g L^{-1} to 0.4 g L^{-1}. However, it must be stressed that to date the actual amount of reduced gas in Earth's primitive atmosphere is not known.

3.3.1.2 Hydrothermal Vents

The synthesis of organic compounds at the bottom of the ocean in hydrothermal vents (also known as black smokers; Fig. 3.10) has to be considered as another endogenous source of prebiotic molecules. Where oceanic plates drift apart, seawater circulating through the ocean crust is heated up, and it dissolves and exchanges chemicals with the rock. At some places it springs back to the ocean from black smokers at a high temperature, enriched in gas, ions, and minerals. Catalytic clays and minerals interact with the aqueous reducing environment rich in H_2, H_2S, CO, CO_2, and CH_4 (and possibly HCN and NH_3). When exhausted from the vent, the dramatic drop in temperature, from 350 °C to about 2 °C, favors chemical reactions and polymerizations.

It has been shown experimentally that amino acids can be synthesized in such conditions (high temperature and pressure, reduced environment, and rich in minerals, which can act as catalyst). But one has to take into account that molecules synthesized in the vicinity of black smokers can also be destroyed because of the high temperature. Nowadays, a volume equivalent to the world's entire ocean

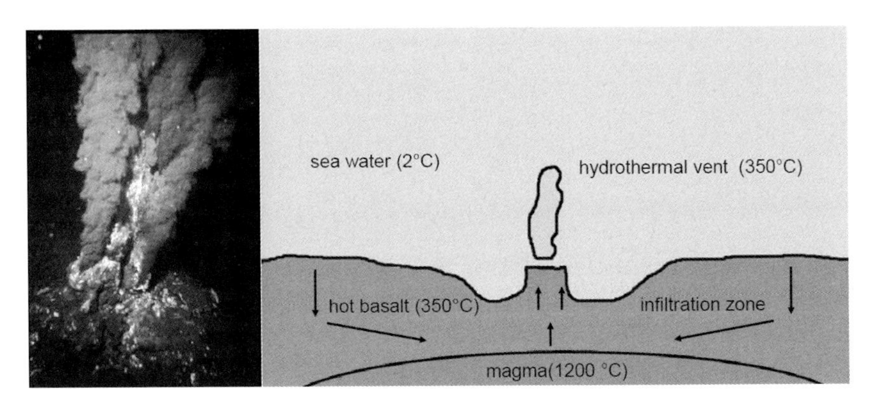

Fig. 3.10 Hydrothermal vents (photograph on the left courtesy of NOAA).

system circulates through hydrothermal vents every 10 million years. Their presence would have fixed an upper limit on the concentration of organics in the primitive ocean. More experimental data and field measurements are required to assess the feasibility of this mechanism.

3.3.2 Exogenous Delivery of Organic Molecules

Remote sensing observations have given evidence that organic chemistry is very active in interstellar molecular clouds, as much in the gaseous phase as in the solid phase (interstellar ices) (see Chapter 2). HCN, HC_3N, and HCHO have been detected. Such molecules are of great astrobiological interest, as will be discussed in Section 3.4. Our Solar System is thought to be the result of the collapse of such a cloud.

Laboratory experiments simulating the conditions in the molecular clouds predict the existence of molecules that are much more complex than the ones already detected by remote sensing. These experiments aim to reproduce the chemistry occurring in interstellar ices. For this, gaseous mixtures consisting of simple volatile compounds (H_2O, CO, CO_2, CH_4, NH_3, CH_3OH, H_2CO, and others), which have been detected in interstellar environments, are introduced into a cryostat. These molecules then condense onto a cold finger where they form an icy mixture. If exposed to irradiations (either photons or charged particles) or to thermal cycles, chemical reactions between these simple compounds, which the ice is made of, lead to the formation of much more elaborated organic structures, which remain solid at room temperature. This refractory residue was called "yellow stuff" by Mayo Greenberg, the astrophysicist who conceived these laboratory experiments in the late 1970 s. From those simulations, one can infer that in molecular clouds a large amount of organic matter should be frozen on condensation nuclei made of silicates (Fig. 3.11).

3.3.2.1 Comets

It is generally accepted that during the formation of our Solar System the original composition of interstellar grains was lost, because they were either incorporated in the Sun or planets or pyrolyzed in locations close to the Sun. Turbulent radial

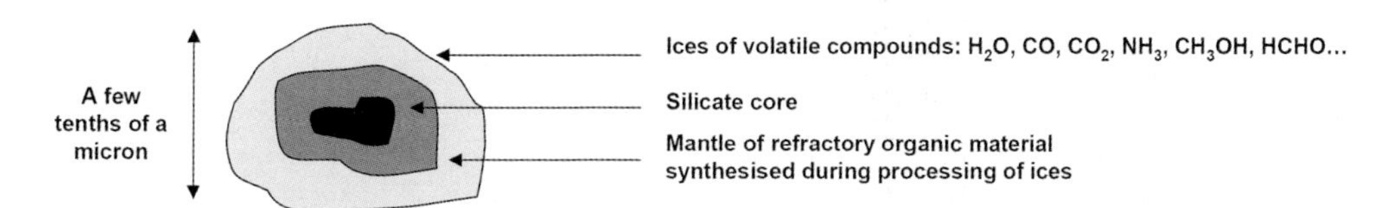

Fig. 3.11 Interstellar grain model: silicate core embedded in a mixture of complex refractory molecules, coated with ice of volatile compounds.

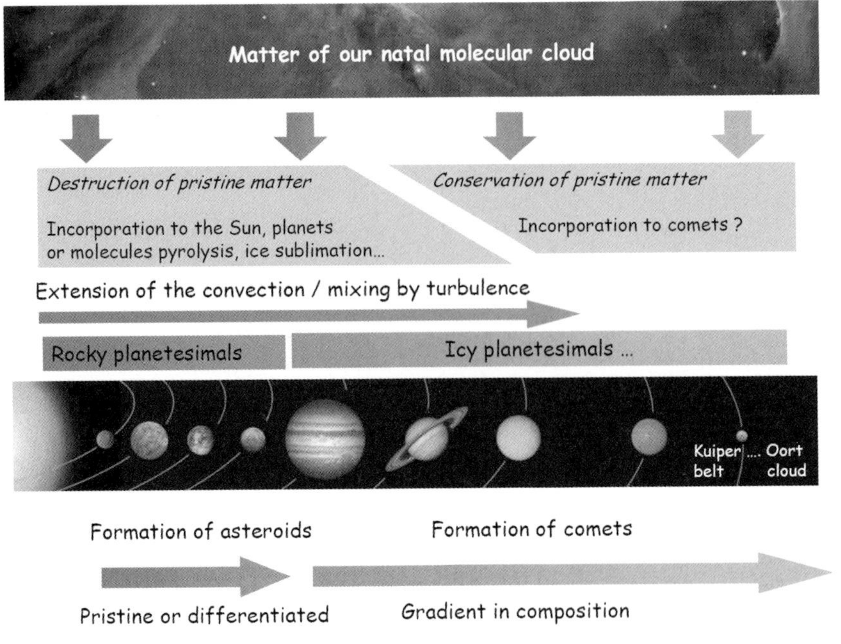

Fig. 3.12 Scheme of the evolution of matter in our Solar System: from a natal molecular cloud, the Solar Nebula, until its incorporation into Solar System bodies (Molecular cloud background courtesy of NASA, Solar System background courtesy of ESA).

mixing in the Solar Nebula can bring interstellar ices to the warmest parts of the nebula where they would sublimate (Fig. 3.12). Inside the turbulent region, radial mixing brings interstellar grains close to the Sun, resulting in a loss of their initial composition. According to models, this turbulent region of the Solar Nebula might have extended up to the orbit of Neptune (about 30 AU). Beyond this region, grains might have remained in a cold environment and thus might have kept an unaltered interstellar composition. Hence, the extension of the turbulent region determines the chances of keeping pristine interstellar matter in small, undifferentiated icy bodies: the comets.

Cometary models consider that interstellar organic matter undergoes different levels of transformation before it will be stored in comets. In some models, comets are considered to be aggregates composed of unaltered interstellar grains. From a rather different point of view, comets are considered to be made of matter completely reprocessed in the Solar Nebula. Other models consider an intermediate scenario, which is probably the more realistic one.

Putting aside the discussion about the origin of cometary matter, observations show an undeniable abundance of a large variety of organic compounds in comets. Molecules such as HCN, HCHO, or HC_3N, which have been detected in several comets, might have an origin in the interstellar medium or might have been produced inside the Solar Nebula. In either case, they are of the same astrobiological interest.

In 1961, John Oro calculated the amount of organic molecules that could have been delivered to the Earth via comets as follows: if one considers a typical cometary nucleus with a density of 1 g cm^{-3}, a comet 1 km in diameter would contain about 2×10^{11} moles of HCN. This amount is equivalent to about 40 nmoles cm^{-2} of HCN distributed over Earth's surface. It corresponds to an amount comparable to the yearly production of HCN by electric discharge in a CH_4-rich reducing atmosphere.

As predicted in laboratory simulation experiments, *in situ* measurements in close vicinity to the comets 1P/Halley (in 1986) and 81P/Wild 2 (in 2004) bear witness to the existence of more-complex structures that – because of their high molecular mass – remain in the solid phase on dust grains when ejected from the cometary nucleus. It must be stressed that the lack of liquid water in comet nuclei over long periods of time excludes the possibility of the existence of life in or on comets – even if short periods of ice melting occur after nucleus formation caused by the decay of radioactive elements.

Summing up, comets are considered to be exogenous sources of prebiotic molecules of great astrobiological potential. This has been inferred from measurements in space as well as from laboratory simulation experiments. Therefore, comets are the target of several past, current, and future space missions, such as NASA's *Stardust* (http://stardust.jpl.nasa.gov/) and *Deep Impact* missions (http://deepimpact.jpl.nasa.gov/) and ESA's *Rosetta* mission (http://rosetta.esa.int/).

3.3.2.2 **Meteorites**

By definition, meteorites are celestial bodies that reach the surface of the Earth. Meteorites – more specifically, those belonging to the carbonaceous chondrite family (the most famous among them are the Murchison, Murray, and Orgueil) – are another exogenous source of organic molecules. Unlike comets, for which no direct analysis of the nucleus composition has ever been made, a large number of meteorites have been studied with the most sensitive laboratory instruments.

The current flux of meteorites is estimated to be about 10 tons per year and was probably higher on the primitive Earth. A great number of organic molecules have been detected in meteorites, such as hydrocarbons; alcohols; carboxylic acids; amines; amides; heterocycles including uracil, adenine, and guanine; and more than 70 amino acids in the Murchison meteorite. Recently, diamino acids have been found in the same object. This shows that molecules once synthesized in space are able to survive a meteorite impact.

Enantiomeric excess at the level of a few percent has been measured for some amino acids in the Murchison and Murray meteorites. This could give us a key to understanding the origin of homochirality in living organisms on Earth. Unlike comets, the parent bodies of meteorites might have gone through periods with liquid water, which might have led to more advanced stages of chemical evolution. It is interesting to note that some organic compounds extracted from carbonaceous chondrites would spontaneously organize to form vesicles. They can be considered as a type of primitive membrane that might have been used by the very first organisms.

No space mission is currently scheduled to analyze the composition of carbonaceous asteroids, which are probably the parent bodies of chondrites. However, projects to explore a carbonaceous asteroid could certainly be planned in the years to come, and this will provide important information about the evolutionary stage of those bodies and the origin of their organic components.

3.3.2.3 Micrometeorites

Micrometeorites are another vector for exogenous delivery of organic molecules. With an asteroidal or cometary origin, their current flux is estimated to be about 20 000 tons per year. They slowly sediment in the terrestrial atmosphere and thus undergo little warming that could destroy their organic content. Amino acids have been detected in micrometeorites collected in Antarctica. Therefore, they could also have played an important role in the origin-of-life process.

3.3.3 Relative Contribution of the Different Sources

The three different types of exogenous delivery of organic matter to the early Earth, i.e., by comets, meteorites, or micrometeorites, occurred not only on the Earth but also throughout the Solar System. However, in order to reach an increased level of complexity that could have led to life, those ingredients require liquid water.

Unlike endogenous sources of organics, as inferred by Miller-type experiments and the hypothesis of black smoker syntheses, which are still at an experimental and conceptual level, exogenous deliveries are actually observed even nowadays. The contribution of the atmosphere as an endogenous source of organics is still model dependent, and the contribution of hydrothermal vents to produce and destroy organics lacks experimental data, whereas the exogenous delivery of organics is an ongoing process.

A clear distinction of the relative contribution of each source cannot be established and seems to be out of reach as long as no other life forms are detected on another planet that would be deprived of one or two of these factors. Only then could it be said that one source is not essential. On the other hand, non-detection of life on a planet with liquid water and at least one of the three factors could lead to the conclusion that one specific source of organics is not sufficient for the origin of life. But from an Earthling point of view, we can say that too many cooks can't spoil the prebiotic soup and that the processes leading from the rather simple building blocks of life discussed in this section to more elaborated structures are more interesting.

3.4
From Simple to Slightly More Complex Compounds

In this section, the chemical pathways to synthesize the building blocks of life in a test tube are described. Some of the mechanisms presented here are already familiar to students of chemistry or biology, although they generally are not presented from an astrobiological perspective. Students in other fields should not be afraid of the reactions presented here and should retain that simple mechanisms explain the production of most of the molecules considered essential to life.

3.4.1
Synthesis of Amino Acids

The synthesis of amino acids has been known since 1850 and was discovered by Adolf Strecker. The mechanism is now well established (Fig. 3.13) and proceeds in liquid water as follows:

- In the first step, the addition of ammonia to aldehyde produces an imine.
- In the second step, the addition of HCN onto the imine produces an α-aminonitrile.
- In the final step, the hydrolysis of the -CN function (nitrile) into $-CO_2H$ (carboxylic acid) leads to the production of an amino acid.

In works published after the release of his first results in 1953, Stanley Miller showed that the production of amino acids during his experiments followed the

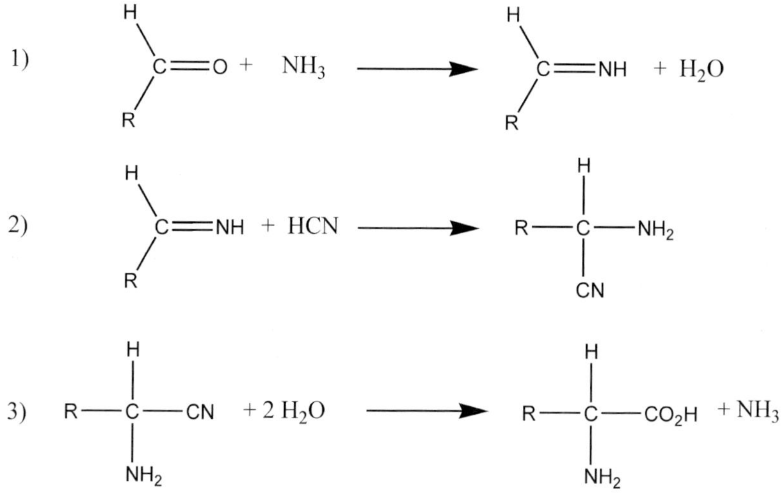

$$1)\quad \mathrm{R(H)C{=}O + NH_3 \longrightarrow R(H)C{=}NH + H_2O}$$

$$2)\quad \mathrm{R(H)C{=}NH + HCN \longrightarrow R{-}CH(CN){-}NH_2}$$

$$3)\quad \mathrm{R{-}CH(NH_2){-}CN + 2\,H_2O \longrightarrow R{-}CH(NH_2){-}CO_2H + NH_3}$$

Fig. 3.13 Synthesis of an amino acid through Strecker synthesis, with H_2CO, NH_3, and HCN.

same synthesis processes (see Fig. 3.13). Other pathways are also possible, because it has been shown that an acid hydrolysis of HCN polymers results in the production of various amino acids.

3.4.2 Synthesis of Purine and Pyrimidine Bases

Synthesis of the purine base adenine follows a spectacular prebiotic synthesis pathway that proceeds by oligomerization of HCN in aqueous solution under the influence of photons of 350 nm. This process was discovered by James P. Ferris and Leslie Orgel in 1966 (Fig. 3.14). It proceeds as follows:

- In the first step, the addition of four HCN compounds results in the formation of diaminomaleonitrile (DAMN).
- In the second step, a first-photon-step rearrangement of DAMN results in the formation of diaminofumaronitrile and then a second one in the formation of aminoimidazole carbonitrile.
- In the last step, the addition of a fifth HCN molecule leads to the production of adenine.

It is quite stunning to imagine that one of the most poisonous organic compounds, HCN, reacting with itself may result in the formation of one of the "letters" of the genetic code, i.e., adenine, which is essential for life. Such are the ways of chemistry.

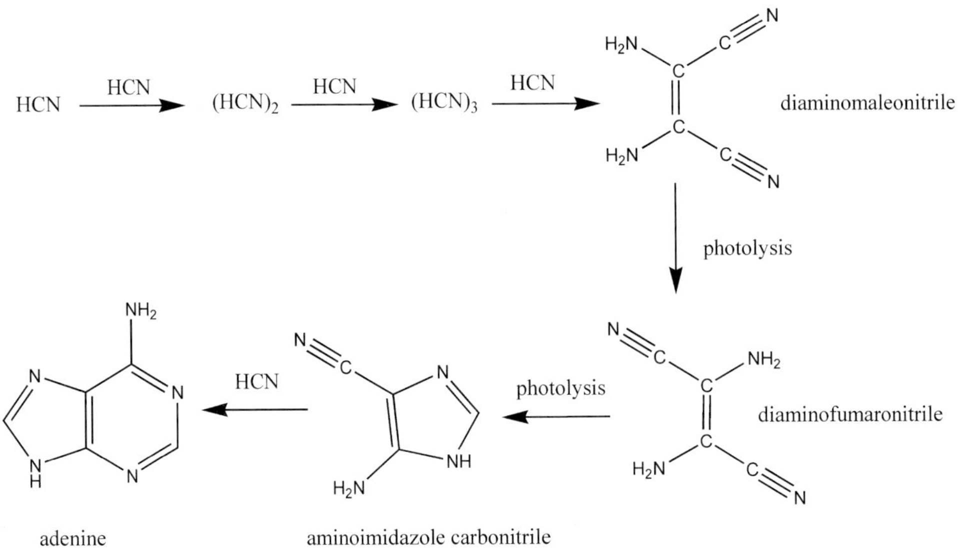

HCN —HCN→ $(HCN)_2$ —HCN→ $(HCN)_3$ —HCN→ diaminomaleonitrile —photolysis→ diaminofumaronitrile —photolysis→ aminoimidazole carbonitrile —HCN→ adenine

Fig. 3.14 Synthesis of adenine from HCN.

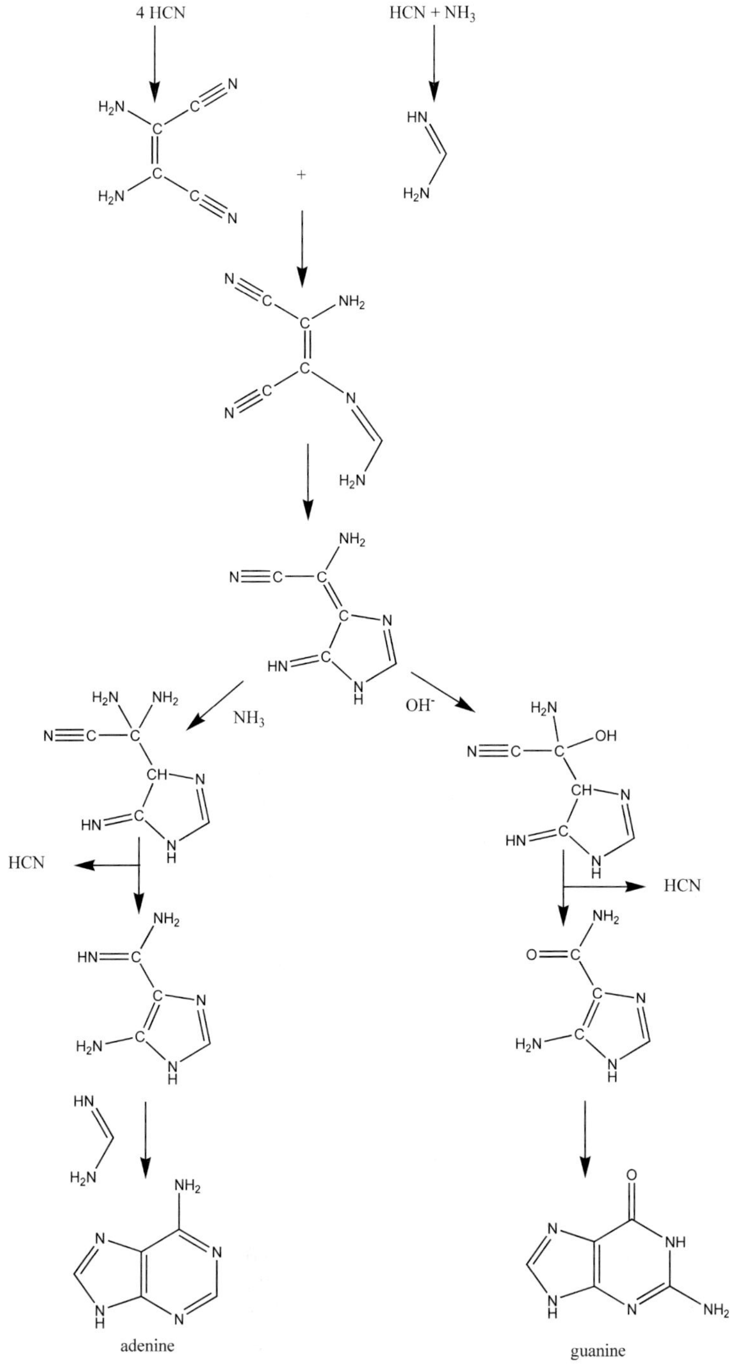

Fig. 3.15 Synthesis of adenine and guanine from HCN and NH_3.

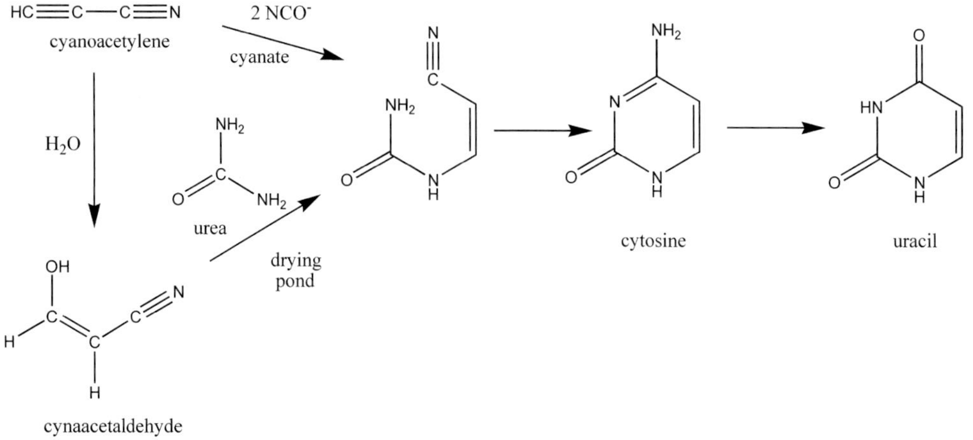

Fig. 3.16 Synthesis of cytosine and uracil from cyanoacetylene (HC_2CN) and cyanate (NCO^-).

Other pathways leading to the synthesis of purine bases are now known, such as the ones presented in Fig. 3.15. In this example, a mixture of HCN and NH_3 leads to the production of adenine and guanine.

Prebiotic pathways leading to the formation of pyrimidine bases have also been investigated, such as the ones presented in Fig. 3.16. It is interesting to note that cyanoacetylene (HC_3N), the main reactant of the synthesis, is a compound detected in the interstellar medium and comets and is also efficiently produced by the action of spark discharges in CH_4/N_2 mixtures.

The direct reaction of HC_3N with cyanate (NCO^-) gives quite low yields of pyrimidine products, or it requires a quite large concentration of cyanate (>0.1 M) for higher yields. In 1995, Robertson and Miller showed that after hydrolysis and formation of cyanoacetaldehyde, the reaction with urea in an evaporating solution (simulating primitive drying lagoons on Earth) forms a noticeable amount of cytosine.

These are only a few illustrations of reactions that lead to the production of purine and pyrimidine bases in prebiotic conditions. It is quite reasonable to believe that not all the pathways have been discovered yet, but the synthesis of these important compounds is not yet considered a problem for the chemists studying the origin of life.

3.4.3 Synthesis of Sugars

Synthesis of the sugars required for the backbone of RNA and DNA molecules is more difficult than that of the previously discussed compounds. Actually, it is not complicated to abiotically produce sugars. Butlerov found a simple way to obtain sugars as early as in 1861. This reaction is now known as the “formose reaction”

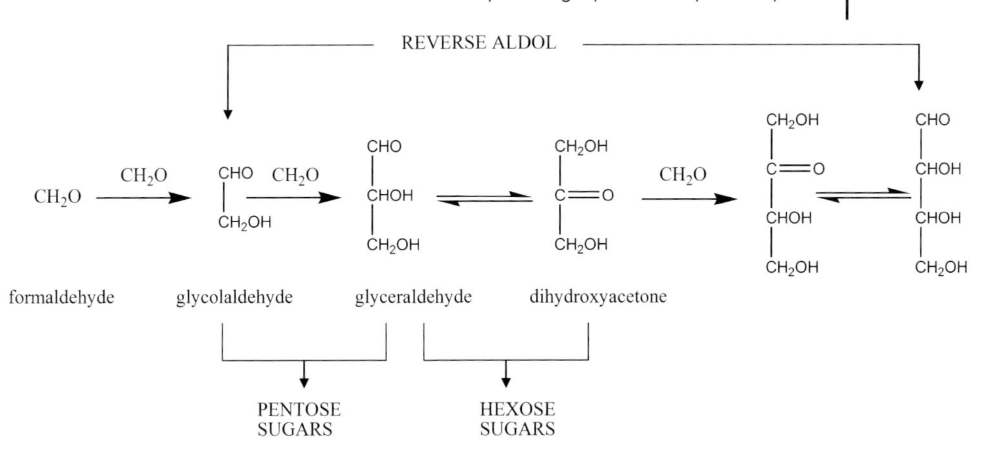

Fig. 3.17 Synthesis of sugars through the formose reaction from H_2CO.

(Fig. 3.17). It requires a concentrated formaldehyde solution and a suitable catalyst, such as calcium hydroxide $Ca(OH)_2$, calcium carbonate $CaCO_3$, or clay minerals.

The formose reaction results in a mixture of many kinds of sugars (more than 40 species), in which ribose is present at only a very low level. Hence, the formose reaction, although highly efficient, leads to an unspecific production of sugars.

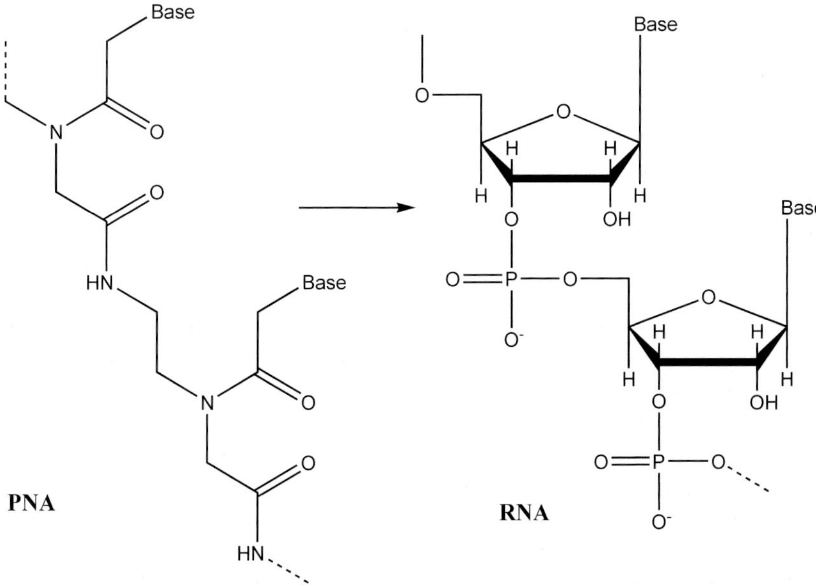

Fig. 3.18 Structure of PNA, a hypothetical information-bearing molecule based on peptides, and of RNA.

This makes it less interesting from an astrobiological point of view. Moreover, the formose reaction needs a concentrated solution of formaldehyde, which is not very relevant for the conditions of the primitive Earth. However, with the use of a proper catalyst, e.g., the clay mineral kaolin, sugars including ribose can be obtained from lower concentrations (0.01 M) of formaldehyde.

Therefore, in the frame of the RNA world hypothesis (see Chapter 4), the question of why ribose among other sugars has been selected remains open. Because it is so complicated to find a suitable pathway for the prebiotic selective synthesis of ribose and to understand how ribose was selected among other sugars, alternative solutions have been suggested, such as a pre-RNA world. This could be based, for example, on peptide analogues of nucleic acids, which are called "peptide nucleic acids" (PNAs) (Fig. 3.18), in which the sugar would be replaced by amino acids. In this scenario, ribose would be incorporated into the structure of life's informational molecule at a later stage.

3.4.4
Synthesis of Polymers

It has been shown in the previous sections that chemistry is not without solutions to explain how the building constituents of life could have been synthesized abiotically. The next step would be to associate the prebiotic elements into complex biochemical material such as polypeptides (proteins) and polynucleotides (RNA and DNA). Here, prebiotic chemistry faces a further challenge.

The linkage of amino acids to peptides is not favored from an energetic point of view. The free enthalpy for the association of alanine (Ala) with glycine (Gly) is $\Delta G° = 4.13\ \text{kcal mol}^{-1}$. This means that the equilibrium is displaced to left and the dissociation of the di-peptide:

$$\text{H-Ala-OH} + \text{H-Gly-OH} \rightleftharpoons \text{H-Ala-Gly-OH} + H_2O \qquad (3.1)$$

$$\Delta G° = 4.13\ \text{kcal mol}^{-1}$$

$$37\,°\text{C},\ \text{pH} = 7$$

Reaction 3.1 shows that peptide bond formation is energetically not favored. The reaction occurs in water and the equilibrium tends to be displaced to the left, because a water molecule is released during peptide bond formation.

A thought experiment might illustrate this point. If one considers starting peptide bond formation from a molar solution ($1\ \text{mol L}^{-1}$) of 20 natural amino acids, thermodynamics says that a resulting protein of about 100 monomer units would have a concentration equal to $10^{-99}\ \text{mol L}^{-1}$. This means that 50 times the volume of the Earth is required to get at least one protein!

Other promising ways of polypeptide formation include catalytic synthesis on mineral surfaces or evaporation–hydration cycle processes with activated amino acids, e.g., with compounds, in which the acid function of the amino acid is replaced by an ester function. However, these pathways are still under investigation.

The abiotic production of polynucleotides is even more complicated than that of polypeptides. The reaction of a purine base with ribose in the solid state at 100 °C results in their association into a nucleoside with a yield of about 3%. But in this case, the base is not linked to the sugar at its "natural" place (Fig. 3.6). Moreover, nucleoside synthesis with a pyrimidine base has not yet been accomplished.

As the next step, the reaction of the nucleoside with a phosphate needs to be considered. Phosphorus is present in igneous rocks as fluorapatite ($Ca_5(PO_4)_3F$) and in meteorites as chlorapatite ($Ca_5(PO_4)_3Cl$). Heating at temperatures above 100 °C leads to the association of the compounds into a nucleotide, but, again, the linkage of the phosphate does not occur specifically at the "natural" place. To date, the polymerization of nucleotides from their building blocks has not been achieved under prebiotic conditions.

3.5
Conclusions

In this chapter, the basic building blocks of life on Earth and simple ways to synthesize them under prebiotic conditions have been discussed. But, having the blocks of life does not give a living system. Jacques Reisse, a Belgian chemist working on the origin of life, says that at the present stage trying to understand the chemistry of the origin of life is like considering building a cathedral (life) but having only a pile of bricks (amino acids, sugars, purine and pyrimidine bases). Getting from the bricks to the cathedral, without a plan, an architect, or any workers (for the stones have to spontaneously combine themselves), is the challenge chemists are facing. Even if the road is tricky, many chemists have decided to devote their studies to that complicated task. Year after year new discoveries are made, but it is difficult to know how long it will take before, somewhere in a laboratory, an organic molecule synthesized under prebiotic conditions will behave like a "robot," spontaneously making copies of itself and capable of evolution. To date, it is not even possible to say if such an achievement will ever be made. What does it take to turn chemistry into biology? Many tracks have to be investigated, and crucial breakthroughs may not come from the laboratory studies but rather from the observation of other bodies in our Solar System. For example, how far does the chemistry go on Mars, Saturn's moon Titan, or the Jovian moon Europa? What does it tell us about the origin of life on Earth? These are a few of the many fascinating questions that need to be addressed in future studies.

3.6
Further Reading

3.6.1
Books or Articles in Books

Cronin, J., Reisse, J. Chirality and the origin of homochirality, in: M. Gargaud, B. Barbier, H. Martin, J. Reisse (Eds.) *Lectures in Astrobiology*, Vol. 1, pp. 473–515, Springer, Berlin, Heidelberg, New York, 2005.

Despois, D., Cottin, H. Comets: Potential sources of prebiotic molecules for the early Earth. In: M. Gargaud, B. Barbier, H. Martin, J. Reisse (Eds.) *Lectures in Astrobiology*, Vol. 1, pp. 289–352, Springer, Berlin, Heidelberg, New York, 2005.

Greenberg, J. M. What are comets made of? A model based on interstellar dust, in: L. L. Wilkening (Ed.) *Comets*, pp. 131–163, University of Arizona Press, 1982.

Maturana, H., Varela, F. J. *Autopoiesis and Cognition: The Realization of the Living*, Reidel, Boston, 1980.

Miller, S., Lazcano, A. Formation of the building blocks of life, in J. W. Schopf (Ed.) *Life's Origin*, pp. 78–112, University of California Press, 2002.

Oparin, A. I. *The Origin of Life*, Foreign languages publishing house, 1924.

Oparin, A. I. *Origin of Life*, Dover publication, 1938.

Oro, J., Cosmovici, C. B. Comets and life on the primitive Earth, in: C. B. Cosmovici, S. Bowyer, D. Werthimer (Eds.) *Astronomical and Biochemical Origins and the Search for Life in the Universe, Proceedings of the 5th International Conference on Bioastronomy* pp. 97–120, Editrice Compositori, Bologna, 1997.

Raulin, F. Chimie prébiotique: expériences de simulation en laboratoire et "vérité terrain," in: M. Gargaud, D. Despois, J.-P. Parisot (Eds.) *L'environnement de la Terre Primitive*, pp. 343–360, Presses Universitaires de Bordeaux, 2001.

Raulin, F., Coll, P., Navarro-Gonzalez, R. Prebiotic Chemistry: Laboratory Experiments and Planetary Observations, in: M. Gargaud, B. Barbier, H. Martin, J. Reisse (Eds.) *Lectures in Astrobiology*, Vol. 1, pp. 449–471, Springer, Berlin, Heidelberg, New York, 2005.

Schopf, J. W. *Life's Origin, the Beginnings of Biological Evolution*, University of California Press, 2002.

3.6.2
Articles in Journals

Botta, O., Bada, J. L. Extraterrestrial organic compounds in meteorites. *Surveys in Geophysics* **2002**, *23*, 411–467.

Cottin, H., Gazeau, M. C., Raulin, F. Cometary organic chemistry: a review from observations, numerical and experimental simulations. *Planet. Space Sci.* **1999**, *47*, 1141–1162.

Ferris, J. P., Hagan j., W. J. HCN and chemical evolution: The possible role of cyano compounds in prebiotic synthesis. Tetrahedron **1984**, *40*, 1093–1120.

Hennet, R. J.-C., Holm, N. G., Engel, M. H. Abiotic synthesis of amino acids under hydrothermal conditions and the origin of life: a perpetual phenomenon. Naturwissenschaften **1992**, *79*, 361–365.

Miller, S. L. The production of amino acids under possible primitive Earth conditions, *Science* **1953**, *117, 528–529.*

Pascal, R., Boiteau, L., Commeyras, A. From the prebiotic synthesis of á-amino acids towards a primitive translation apparatus for the synthesis of peptides, *Top. Curr. Chem.* **2005**, *259, 69–122.*

Shapiro, R. Prebiotic ribose synthesis: a critical analysis. *Orig. Life Evol. Biosphere* **1988**, *18*, 71–85.

3.7 Questions for Students

3.7.1 Basic-level Questions

Question 3.1

Describe the Miller–Urey experiment. What did we learn from its results? What are its limitations?

Question 3.2

What are the main sources of organic compounds that are currently considered for understanding the first steps of chemical evolution towards the origin of life? Could they be the same on Jupiter's moon Europa? Why or why not?

3.7.2 Advanced-level Questions

Question 3.3

Draw a scheme describing an experiment attempting to simulate the chemistry of interstellar ices.

Question 3.4

Describe the history of a C atom, from its synthesis in a star to its incorporation into a living system on an imaginary planet harboring life.

Question 3.5

Describe the extent to which RNA molecules could be considered living systems.

4
From Molecular Evolution to Cellular Life

Kirsi Lehto

In this chapter, contemporary cellular life is analyzed in order to see which components are the most ubiquitous and conserved in all life forms and therefore assumed to be the most original functional units of life. Molecular analysis of these structures gives suggestions of how they were involved in early molecular evolution and what functions they may have provided to the early (pre-cellular) replicators. By going backwards, step by step, one sees which inventions had to precede each stage of early evolution: genetically encoded proteins had to exist prior to the establishment of the Last Universal Common Ancestor (LUCA), and RNA replicators, or replicating RNA analogues had to predate genetically encoded proteins. Special attention is given to the most conserved components of the present life forms, considered to be relicts of the early RNA world.

4.1
History of Life at Its Beginnings

Our home planet was formed about 4.6 billion years ago. The initial conditions on the young planet Earth were very harsh. The accretion process had heated the protoplanet, and the temperature of the young planet was further increased by the greenhouse phenomena caused by the dense atmosphere formed by H_2O and CO_2 and by continuing impacts from the colliding comets and meteorites. The initial planet temperature was hot enough to keep the Earth's crust in liquid form, forming a magma ocean on the surface. As the planet radiated heat, the temperature was reduced; when it came down to about 200 °C, with the prevailing high atmospheric pressure, the water vapor precipitated and rained down to form oceans. The dating obtained from the oldest zircon crystals indicates that liquid water was present on the planet 4.4 billion years ago. The first Earth crust formed

Complete Course in Astrobiology. Edited by Gerda Horneck and Petra Rettberg

ISBN: 978-3-527-40660-9

underneath the water. Because of heavy cometary and asteroid impacts, the conditions on Earth remained hot and violent for perhaps 500 million years; this time is called the Hadean period.

Conditions during the Hadean period are difficult to investigate because all the rocks formed during that era have been thoroughly metamorphosed by the tectonic processes of the later times. The oldest still existing sedimentary rocks are located at Isua, Greenland, and date back to about 3.8 billion years. These rocks contain small granules of sedimented carbon, which is specifically enriched in the ^{12}C rather than the heavier ^{13}C isotope. The enrichment of the lighter carbon isotope in these sediments suggests that the carbon was once bound in biologically produced compounds (see Chapters 1 and 2). The real biological origin of these carbon deposits has been questioned, but other nearly as old samples containing ^{12}C-enriched carbon and fossilized microbial structures have been detected at two other locations: in the 3.5-billion-year-old sediments at Pilbara, western Australia, and at Barberton, South Africa. These findings suggest that life was well established, and apparently widely distributed on the Earth, at this very early stage – or soon after the heavy bombardment ended and the planet cooled to temperatures conducive for life. It is not known at what temperature regimes life started. The most thermotolerant contemporary life forms are known to propagate in temperatures up to 113 °C (see Chapter 5), but the prebiotic replicating macromolecules, and also the earliest cellular structures, may have been rather fragile and sensitive for hydrolytic conditions and most likely were not durable in very high temperatures.

Since its early start, life has spread and diversified and now occupies all conceivable ecological niches on the Earth (see Chapter 5). During its spread and accumulation, life has also modified the environment and has profoundly changed the conditions on planet Earth, including the chemical composition of the atmosphere, the oceans, and the Earth's crust. The long-term geological processes of tectonics, volcanism, weathering and erosion, as well as degrading processes caused by life itself, have completely reshaped and mixed multiple times the conditions that prevailed on the early Earth and have abolished most of the chemical and geochemical traces that were formed at this early time. Therefore, there are not any direct data of the conditions that once allowed or induced the origin of life. It is even not know in what type of locations on Earth – or, potentially, on nearby planetary bodies – life did start. Thus, trying to trace back the very earliest stages of this process becomes difficult and speculative.

However, different scientific approaches may be used to try to resolve what kind of conditions and what chemical compounds might have driven the life-producing reactions. Clearly, the origin of life has been a very complex process and needs to be studied in many different fields of science. Astronomy is needed to resolve how the elements, the Solar System, and the planet Earth were formed in the first place. Geosciences are needed to resolve the conditions on the early Earth and the composition of the early atmosphere, the ocean, and the crust. Inorganic and organic chemistry are needed to resolve how the elements, inorganic minerals, and volatiles may have reacted to form small organic compounds and more complex organics such as nucleotides and amino acids, which are the building blocks of life.

Further on, the quest is to resolve how these organic molecules may polymerize and, in particular, how these reactions might conceivably have happened in prebiotic conditions in the absence of genetically encoded enzymes and biochemical reactions.

Present-day life is solely driven by multiple protein-catalyzed (-mediated) biochemical reactions and complex molecular interactions. The fields of biochemistry and bioinformatics study how these enzymes and other proteins now function and how they have changed over the evolution of life. These studies can identify the genetic sequences that are the most conserved (i.e., the oldest and most original ones) in different life forms. Of particular interest here are the reactions that still are catalyzed not by proteins but by different RNA molecules. Such RNA enzymes (or ribozymes) may be derived from the very oldest life forms or from the era predating the DNA-protein-based life forms.

As a central part of the study of life, different fields of biology study the essential structures and functions of life; how life reacts and adapts in its environment; and how all functions are mediated, regulated, and coordinated on the cellular and molecular level. All these integrated efforts are needed to construct an understanding of what life is, how it has evolved through times, and, maybe, how it started in the first place.

It may be assumed that the chemical and molecular evolution towards life proceeded through a long cascade of different stages, initiated by spontaneous reactions between abundant chemical elements, or the chemical constituents of life: carbon, hydrogen, oxygen, nitrogen, sulfur, and phosphor (C, H, O, N, S, P). In some high-energy conditions these elements formed reduced carbon and nitrogen compounds, such as hydrogen cyanide (HCN), ammonia (NH_3), methane (CH_4), and formaldehyde (HCHO). In suitable conditions these small organic molecules, together with water (H_2O), reacted to form the building blocks of life: the nucleic acid bases (purines and pyrimidines), sugars and other hydrocarbons, and amino acids (see Chapter 3 and Section 4.2). These, again via multiple chemical reactions and interactions (which are not well understood as yet), formed polymers and more-complex molecules. Some of the polymers gained the chemical potential to copy themselves or to replicate, which was the first major transition towards the origin of life. Replication initiated molecular evolution, which led to the invention of genetically encoded proteins and, eventually, to the appearance of the first cellular life forms. From then on, the diversification of life into all the species of the biosphere has been covered by Darwinian evolution.

This chapter aims to trace back this process in order to visualize the different stages of the progress from early organic molecules to cellular life forms. Instead of starting from the very beginning and progressing upwards, the chapter starts from the central features of cellular life as they are known now, because it is known for certain that this is what eventually came out of early evolution (at the time it happened, evolution was random and not directed – the outcome could be known only after it had happened). Then proceeding backwards, step by step, it discusses what types of simpler forms could have given rise to each of the new and more complex innovations.

4.2
Life as It Is Known

4.2.1
The Phylogenetic Tree of Life

Life is characterized by endless complexity. On the visible level one can observe the multitude of genera, species, forms, and individuals, each thriving in their own typical ecological niches. Cells of higher organisms are differentiated into organs and tissues, which each present their specific form and function even that they all are regulated by the same genome of the same individual. Variation of species and subspecies in the microbial world is vastly larger than that of the macrospecies. The microbial species can be separated into three different domains of life. Single-celled organisms such as yeasts, ciliates, and flagellates, which do posses a nucleus in their cells, together with all the higher organisms belong in the domain of Eukarya. The cells that do not posses a nucleus belong to the prokaryotes, which again belong to two distinct domains, the Bacteria (or by older nomenclature, the Eubacteria) and the Archaea (older name, the Archaebacteria). These three domains of life (Eukarya, Bacteria, and Archaea) each constitute their own and separate lineages of life and can be distinguished from each other by multiple genetic and biochemical markers.

The three domains of life share a large number of common features that are ubiquitous to life, or are the same or very similar in all living organisms. It is believed that these features were inherited from a common ancestor of all life forms, the Last Universal Common Ancestor (LUCA). However, the properties of LUCA at the time of diversification and the exact sequence of the separation of the three lines are still unclear. Clearly, the common ancestor has developed through a cascade of gradually increasing complexity, and it is not quite clear, which stage was The Last Common One. As reviewed by Delaye and coworkers (2005), it might have been a community of simple precursors of life, the so-called progenotes (see Section 4.3.2), or even a community of free-living replicating molecules, predating the establishment of cellular life. However, recent phylogenetic studies (i.e., studies of the lineages and relatedness of known organisms through statistical comparison of their genetic sequences) by, e.g., the groups of Lazcano and Doolittle, indicate that LUCA was a well-established cell line. The properties of LUCA will be discussed below.

4.2.2
Life is Cellular, Happens in Liquid Water, and Is Based on Genetic Information

From unicellular microorganisms to all higher life forms, all life on Earth is cellular (Fig. 4.1). Prokaryotes are single-celled organisms. The cells are simple and usually devoid of any internal membrane structures (except for the photosynthetic cyanobacteria, which possess intensive photosynthetic membranes). The cells are spherical, ellipsoid, elongated, or rod-shaped (see Chapter 5) and of sizes varying from 0.1 μm to 600 μm, depending on the species.

A

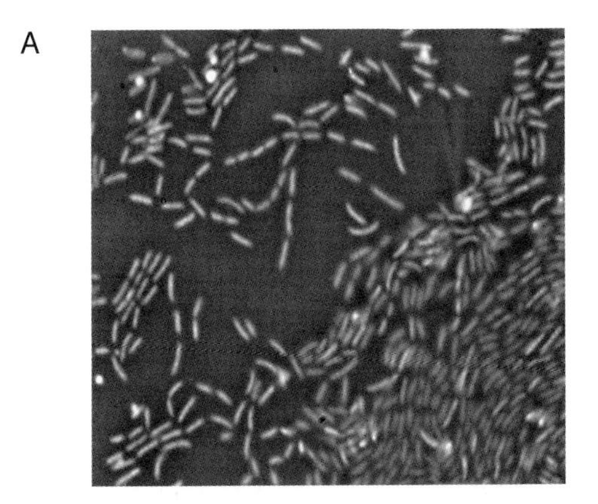

B

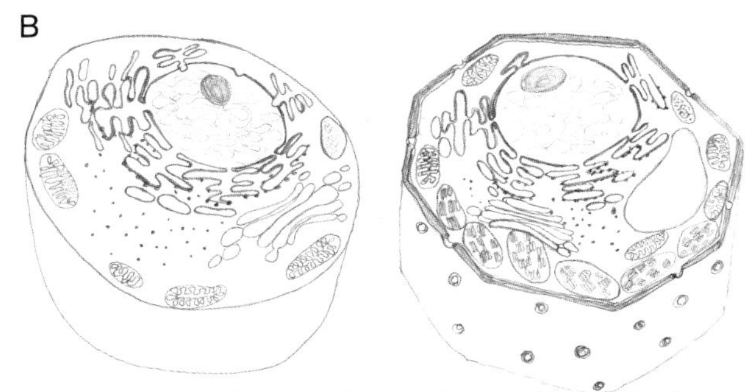

Fig. 4.1 All life is cellular. (A) Microscopic view of (unidentified) bacterial cells. (B) Scheme of a eukaryotic cell (animal cell on the left, and plant cell, with a rigid cell wall, on the right) showing the intensive internal structure.

All eukaryote cells have intensive internal structure. Cell organelles include

- the nucleus (contains the chromosomes or the genomic DNA);
- mitochondria (energy-producing organelles);
- the endoplasmic reticulum (internal membranes),
- the Golgi apparatus (membranous vesicles for protein processing and transport),
- lysosomes and peroxisomes (for containment of degradative enzymes and wastes); and
- chloroplasts (photosynthetic organelles; in plant cells only).

The cells of all eukaryotes are very similar to each other, except that the plant cells have thick and rigid cell walls and contain numerous chloroplasts (or other types of plastids). Typical sizes vary from 10 μm to several hundred micrometers, but some large cell types occur (of a size up to more than 1 m for some nerve cells).

Multicellular organisms may be conceived as tightly bound colonies of individual cells. The cells are entities contained within the cell membrane, which separates the cell from its environment but also allows sophisticated communication with the environment. The communication includes the intake of water and necessary nutrients, excretion of waste products, receiving of chemical signals via different receptor molecules, and sending off signals via secretion of different molecules. Thus, the cells dynamically communicate with their environment. All communication through the membrane is mediated via specific proteinaceous channels and structures embedded in the membrane, because the membrane itself is a hydrophobic barrier and therefore impermeable to water-soluble molecules.

Inside the cells, life is a chemical process where multiple biochemical reactions take place in aqueous solution. Water functions as the solvent and carrier for ionic and polar molecules, and also is a reacting component in many reactions. Water also serves as the exclusion medium to allow the lipid molecules to assemble and form membranes, and as the medium that interacts with proteins, helping them to fold into their three-dimensional forms. Therefore, water provides a suitable environment for the different cellular reactions and for the assembly of important cellular structures. It provides turgor (internal pressure) and strength to the cells. Water is such an essential environment and requirement for all the life functions that it seems likely that the origin of life had to happen in an aqueous environment (see Chapter 1).

Proteins, or gene products, have a multitude of essential roles in the cells. They form several different structures, such as the cytoskeletons in eukaryotes, components of cell walls, and the porous structures in cell membranes. They regulate reactions such as gene expression and genome replication, and they function in signal transduction by interacting with different target molecules. Still another area where proteins are essential for life is in the (regulated) catalysis of all the reactions going on inside the cells. Most cellular molecules are both synthesized and degraded via multi-step reaction pathways – and none of these reactions would happen, at least at any desired rate, without the assistance of proteinaceous enzymes. Even the energy conversion and -releasing reactions have to be mediated via enzymatic catalysis so that they happen in a controlled and regulated

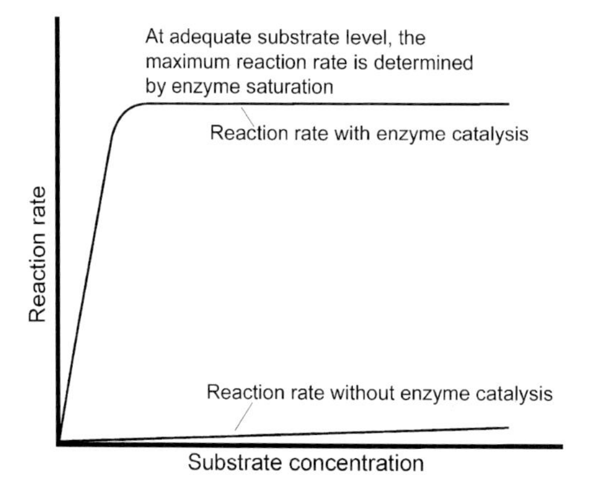

Fig. 4.2 Chemical reaction rates in dependence of substrate concentration, with and without enzyme catalysis. Enzymes speed up reaction rates very strongly as compared to non-catalyzed reactions and make them happen efficiently even at very low substrate concentrations. The maximum rate of reaction is reached when the enzyme becomes saturated with the substrate molecules.

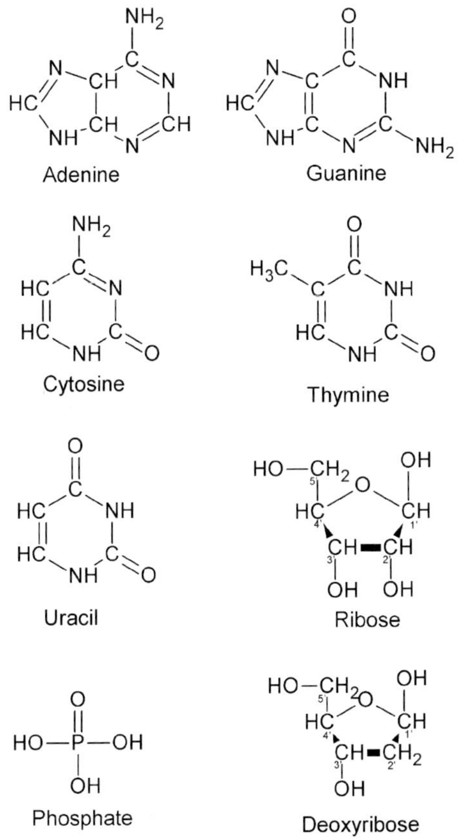

Adenine

Guanine

Cytosine

Thymine

Uracil

Ribose

Phosphate

Deoxyribose

Fig. 4.3 Structure of the components of genetic material: the nucleic acid bases adenine, guanine, cytosine, thymine, and uracil; the deoxyribose and ribose sugar molecules; and the phosphate moiety.

fashion. The enzymes themselves do not participate in the reactions but associate with the reacting molecules and hold them in such a conformation that the reaction can proceed. Enzymatic catalysis is capable of raising reaction rates by many orders of magnitude, from nearly zero to very high, and in practice most biological reactions would not happen in physiological conditions without enzyme catalysis (Fig. 4.2).

Proteins form the tools and the structures in the cells, and genetic information, encoded in the genomic deoxyribose nucleic acid (DNA), specifies how these tools and structures are made. Life depends on the function of the proteins and thus on the genetic information coding for them. Proteins are composed of long chains of different amino acids, and the sequence of the amino acids in these chains is determined by the genes coding for them. From the DNA, the protein-coding sequences are first copied into ribose nucleic acid (RNA) and from this are converted into amino acid sequences in peptide chains. RNA is a polymer molecule very similar to DNA; it is composed of similar nucleotide components, with only the small difference of an OH group, rather than H, being connected to the carbon in position 2 (2') in the pentose sugar ring (thus, the sugar is ribose instead of deoxyribose) (Fig. 4.3). However, this OH group is very reactive and, differing from the DNA, makes the RNA polymers very fragile, unstable, and short-lived.

4.2.2.1 Genetic Information

The complexity of life extends strongly to the molecular level, i.e., to the nucleic acid and protein level inside the cells. The genetic information is encoded in the genomic DNA in the sequence of four different deoxyribonucleotides (or, in short, nucleotides), which are the phosphorylated form of adenosine (A), guanosine (G), cytosine (C), and thymidine (T) nucleosides. In RNA, the uridine nucleoside is used instead of the thymidine nucleoside. The nucleosides themselves are formed of the corresponding bases (adenine, guanine, cytosine, thymine, and uracil); of these, A and G bases are purines and C, T, and U bases are pyrimidines (Fig. 4.3). The bases are covalently bound to a deoxyribose (in DNA) or ribose (in RNA) sugar to form nucleosides. These are phosphorylated at the carbon at their 5' position to form nucleotides, and these again are connected into long chains via phosphodiester bonds between the 5' and 3' carbons (these are the carbons in positions 3 and 5 in the ribose ring) of the adjacent sugars (Fig. 4.4).

The number of nucleotides in the genomes varies largely between different species. The genomes of the smallest prokaryotes are composed of about 0.5×10^6 nucleotides, while the genome of the bacterium *Escherichia coli* is composed of 4.8×10^6 nucleotides. The smallest eukaryotic genomes are about an order of magnitude larger, but the more complex ones are many orders of magnitude larger than the prokaryote genomes. The genome of wheat, for instance, is composed of

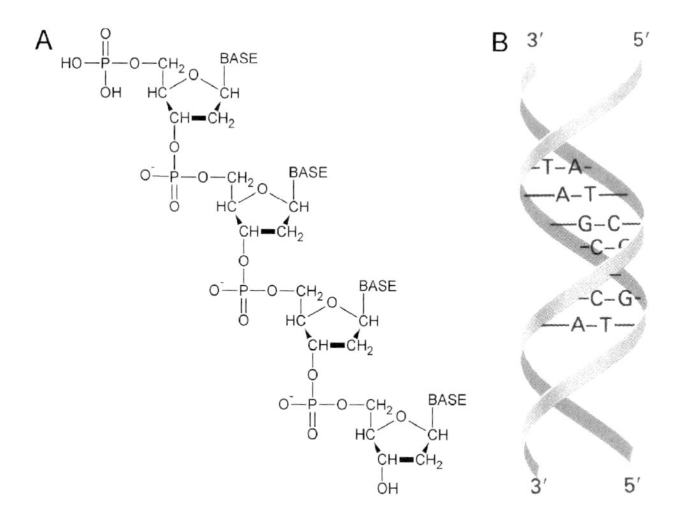

Fig. 4.4 Schemes of DNA. (A) The deoxyribonucleotide polymer formed by diester bonds between the 3' and 5' carbons of adjacent nucleotides; (B) the double helix of two DNA strands showing the base pairing of the nucleic acid bases.

about 17 000 × 10^6 nucleotides, the human genome of 3 400 × 10^6 nucleotides, and that of the fruit fly of 97 × 10^6 nucleotides (Table 4.1). Further, the number of genes varies a lot, being about 400 in the smallest endosymbiotic bacteria (*Buchnera aphidicola*) and some hundreds of thousands in the largest genomes.

Table 4.1 Genome size and coding content of different types of organisms.

Species	Domain	Genome size (million bp)[a]	No of genes	Coding (%)
Mycoplasma genitalium	Bacteria	0.58	470	100
Escherichia coli	Bacteria	4.8	4 288	100
Methanococcus jannaschii	Archaea	1.7	1 738	100
Saccharomyces cerevisiae	Eukarya	12	6 144	70
Caenorhabditis elegans	Eukarya	100	18 266	25
Arabidopsis thaliana	Eukarya	100	25 498	31
Homo sapiens	Eukarya	3 400	32 000	28[b] 1.4[c]
Triticum aestivum	Eukarya	17 000	60 000	Not determined
Amoeba dubia	Eukarya	670 000		Not determined

a Million base pairs = number of nucleotides.
b Total.
c For proteins.

The very large genomes, e.g., those of wheat or amoebas, have not been sequenced. Therefore, their accurate sizes and gene contents are not known as yet but can be estimated from their physical properties. It is clear that genome sizes are correlated with neither the size nor the complexity of the different species. This seeming paradox is explained by the fact that the large eukaryotic genomes contain large amounts of non-coding DNA. In humans, for instance, only about 1.5% of the genomic DNA codes for protein products, and only about 28% of it is transcribed into RNA. A large part of the remaining DNA is composed of highly repetitive sequences, which are often referred to as "junk-DNA." However, the non-coding DNA is not all insignificant, because some of those sequences have important functions as structural elements of the DNA or have contributed (or may contribute in the future) to the evolution of the genome.

Although the cellular genomes are very large, they function in very controlled manner. Among all the multiple genes, and in the middle of the very large non-coding sequences, the functional genes are expressed in a highly coordinated fashion: the amount, timing, and location of the expression of each gene are accurately regulated. The gene regulation is mediated mostly by different proteins – but in many cases also by small, non-coding RNAs.

Different proteins, RNAs, membranes, and small soluble molecules such as sugars and salts fill up the cellular contents, up to a level of molecular crowding in

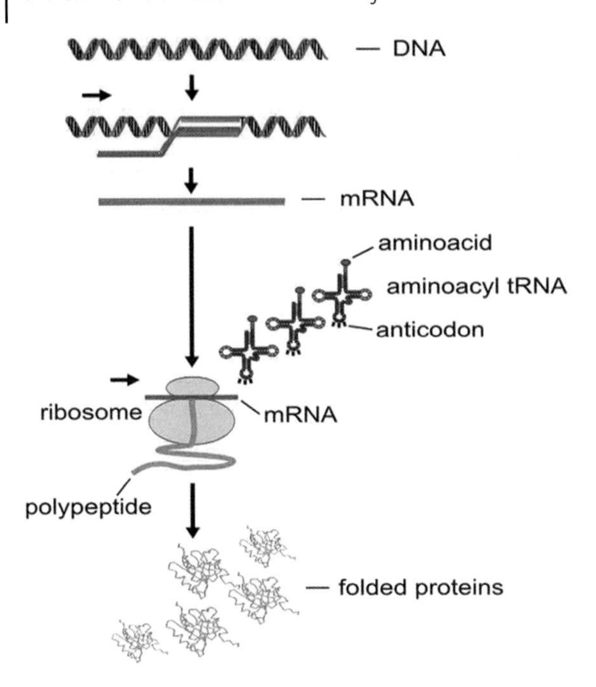

Fig. 4.5 Flow chart of the expression of genetic information.

the interior space. Although the cellular contents are very complex, their central functions are uniform between all different cells and between all species. Particularly, the genetic code itself and its expression mechanism are the same, in principle, in all species (Fig. 4.5).

4.2.2.2 The Genetic Code and Its Expression

The genetic information is encoded in the nucleotide sequences in groups of three adjacent nucleotides, the codons, each corresponding to specific amino acids (or a stop sign). Although the four nucleotides (A, G, C, and T in DNA; in RNA, T is replaced by U) can be organized into a total of $4^3 = 64$ different triplet combinations, only 20 different amino acids are utilized and coded for in the current life forms – although some prokaryotes use 22 different amino acids. Thus, the coding system includes some redundancy: several codons code for the same amino acid (Table 4.2). Each gene is composed of a set of triplet codons. The code of each gene (or the open reading frame, ORF) starts at a specific ATG triplet in the DNA and continues until it reaches one of the following triplets: TAG, TAA, or TGA, which do not code for any amino acid. These stop codons mark the end of the gene (Table 4.2). In eukaryotes the genes are often interrupted by intervening sequences (introns) that must be removed from the sequence during its expression process.

Table 4.2 The genetic code, composed of triplets of the four nucleic acid bases uracil (U), cytosine (C), adenine (A), and guanine (G), and their information (amino acid or start or stop codon).

First base	Codon Second base U	Information	Codon Second base C	Information	Codon Second base A	Information	Codon Second base G	Information	Third base
U	UUU	Phenylalanine	UCU	Serine	UAU	Tyrosine	UGU	Cysteine	U
U	UUC	Phenylalanine	UCC	Serine	UAC	Tyrosine	UGC	Cysteine	C
U	UUA	Leucine	UCA	Serine	UAA	Stop codon	UGA	Stop codon	A
U	UUG	Leucine	UCG	Serine	UAG	Stop codon	UGG	Tryptophan	G
C	CUU	Leucine	CCU	Proline	CAU	Histidine	CGU	Arginine	U
C	CUC	Leucine	CCC	Proline	CAC	Histidine	CGC	Arginine	C
C	CUA	Leucine	CCA	Proline	CAA	Glutamine	CGA	Arginine	A
C	CUG	Leucine	CCG	Proline	CAG	Glutamine	CGG	Arginine	G
A	AUU	Isoleucine	ACU	Threonine	AAU	Asparagine	AGU	Serine	U
A	AUC	Isoleucine	ACC	Threonine	AAC	Asparagine	AGC	Serine	C
A	AUA	Isoleucine	ACA	Threonine	AAA	Lysine	AGA	Serine	A
A	AUG	Methionine, start codon	ACG	Threonine	AAG	Lysine	AGG	Serine	G
G	GUU	Valine	GCU	Alanine	GAU	Aspartic acid	GGU	Glycine	U
G	GUC	Valine	GCC	Alanine	GAC	Aspartic acid	GGC	Glycine	C
G	GUA	Valine	GCA	Alanine	GAA	Glutamic acid	GGA	Glycine	A
G	GUG	Valine	GCG	Alanine	GAG	Glutamic acid	GGG	Glycine	G

The genetic code is interpreted (or expressed) in a multi-step process (Fig. 4.5) that converts the sequence of the nucleotide triplets into a linear sequence of the corresponding amino acids (Table 4.2). First, the sequence of the target gene is copied (transcribed) into a messenger RNA (mRNA). The mRNA contains the same sequence as the corresponding DNA, but in contrast to DNA, the RNA polymer is short-lived and is used only for short-term expression of the code. The mRNA associates with the two subunits of the ribosome, which then moves along the mRNA and recognizes the triplet code, starting from the first AUG codon on the mRNA (corresponds to the ATG on the DNA). The nucleotide triplets are recognized by the anticodon triplets of the aminoacylated transfer RNAs (tRNAs), which enter into the ribosome complex and bring in the specific amino acids corresponding to each triplet codon. The peptide bonds between the adjacent amino acids are formed by the catalytic RNA component of the ribosome (ribosomal RNA, rRNA). As these amino acids are brought into the reaction complex, the ribosome joins them into a long chain or "translates" the nucleotide sequence into an amino acid sequence (Fig. 4.5). The produced amino acid chain is initially called a polypeptide, but after it folds into the accurate three-dimensional structure, it becomes a functional protein.

Thus, the components needed for the expression of the genetic code are the mRNA, the tRNAs and the amino acids, and the ribosomes. The ribosome itself is composed of multiple proteins and RNAs (varying from 52 to 79 proteins and from three to four RNAs in different species). Each of the steps of the gene expression (transcription, tRNA amino acylation, and the translation itself) requires multiple accessory proteins for correct function and regulation. This strong protein dependency of the whole gene expression process creates a dilemma similar to the one of the chicken and egg. Proteins are required for expression of the genetic code as well as for the maintenance (or replication) of the code, and the code is needed for the expression of the proteins. Thus, how could the genetic DNA first be generated and replicated, and how was its expression machinery first established – how did this cyclic process get started?

It is assumed that the storage and expression of the genetic information were the first hallmarks for life as we know it and that these were the key elements in the molecular evolution that led to the birth of the LUCA. They existed already at the time of the LUCA, and they required functions such as genome replication, transcription and translation, and pathways for amino acid and nucleotide synthesis. These functions are so essential for present life that their origin is the central part of the question of the origin of life as a whole.

4.3
Last Universal Common Ancestor (LUCA)

The features that occur in all life forms are essential to life. Therefore, they should have been present already in the earliest established cellular life, i.e., in the LUCA. Such conserved features include

- the enclosure or containment in cell membranes,
- the genetic code, and
- the translation machinery for the expression of the code (the ribosomes, tRNAs, and amino acids).

Some enzymes related to RNA and amino acid biosynthesis and degradation and to energy production and use (cell metabolism) also are common to all cellular life. A further curious detail that is fully conserved in all cellular life is the symmetry breaking, or homochirality, occurring in its central molecules: all the amino acids used in the cellular proteins are of the same stereoisomer type, which is the left-handed (L)-configuration (see Chapter 2), and the deoxyribose and ribose sugars used in all the nucleotides are of the right-handed (D)-configuration only, meaning that the bond between the carbons in positions 4 and 5 points up, not down (Fig. 4.4).

4.3.1
Containment in a Cell Membrane

As mentioned earlier, all life on Earth is cellular, which means that it is all contained within, and separated from its environment by cell membranes. This containment is important for many reasons. The molecules synthesized in the cells do not escape into the environment, and the concentrations of the intracellular molecules are maintained at desired levels by regulating their synthesis and degradation. Molecules from outside are transported in and out in a regulated manner, and the cellular molecules interact and function in a mutually coordinated way. Further, because the cell membrane physically connects the genomes with their respective gene products, the selection for beneficial products and functions becomes targeted to their corresponding coding sequences, i.e., the genes can evolve based on their products. The physical connection also allows the joint maintenance of different segments of the genomes and allows their coevolution towards synergistic functions. All in all, the cell membrane allows all cellular components to function and evolve as one coordinated entity and unit.

The excellent utility of present-day membranes is related to their highly sophisticated properties. Cell membranes are formed of amphiphilic (i.e., one end is hydrophobic and the other end is hydrophilic) molecules called lipids, which aggregate in water and assemble into bilayers (see Chapter 3). These lipid bilayers are themselves impermeable to water, but they are spanned by several different proteins that serve as transport channels for water-soluble molecules and as receptors for external signals and cues. The cells may also adjust the properties of their membranes by changing their protein components or by changing the saturation level of lipids according to temperature. Cells continuously produce new lipids for the growth of their membranes, and different lipids are produced to construct membranes for different compartments. Thus, membranes in current cells are very dynamic and functional. However, the situation was quite different at the time prior to the existence of the proteins needed for lipid biosynthesis or for membrane transport channels. How could the cells be contained at this stage?

It is possible that some type of membrane was formed in the early world by spontaneous assembly of abiotically formed lipids. It has been shown that different polar hydrocarbon derivatives, containing 10 or more carbons in their hydrocarbon chain, are amphiphilic molecules that may assemble into bilayers, and these again may spontaneously form vesicles in aqueous environments. Such aliphatic (or linear) and aromatic (or circular) hydrocarbons are present, for instance, in the carbonaceous chondrites (see Chapter 3). Hydrocarbon–lipid vesicles might have enclosed the earliest genomes and functioned as compartments for their replication. However, this scenario involves severe problems. Because the hydrophobic lipid membranes separate the organism from the surroundings, they may prevent access to their resources, such as useful molecules available in the surrounding. Further, because the organism is closed inside a "dead" membrane, it may not be able to expand or divide or to allow its progeny to accumulate beyond its own original compartment.

Several membrane chemists, e.g. the research groups of Deamer, Luisi, and Szostak, have attempted to resolve the assembly and the functional parameters of spontaneously assembled membrane vesicles. They have shown that the permeability of the bilayer membranes is related to the length of the aliphatic carbon chains: modified hydrocarbons of less than 14 carbons are permeable for small ionic molecules, such as amino acids and phosphates. Actually, lipid membranes formed of carbon chains of less than 10 carbons are too permeable to maintain any ion gradient. The membrane vesicles formed of longer fatty acids, such as oleic acid, are semipermeable: small molecules, such as nucleotides or amino acids, may diffuse through them while large polymers may not. This property would allow spontaneous uptake of nucleotides into the vesicles and retention of the polymers inside them. This process would increase the osmotic potential inside of the vesicles and cause the growth of the membrane via incorporation of more fatty acids drawn from emptier vesicles or from micelles. Thus, the mere hydrophobicity of fatty acids or of other lipid molecules may have caused the prebiotic assembly and function of primitive, cell-like structures. Indeed, assembly and growth, and even division of such lipid vesicles, have been demonstrated in laboratory conditions. Likewise, nucleotide and amino acid polymerization inside the vesicle has been induced, simulating the assembly of small, artificial cells.

To have an adequate supply of the lipids (or of amphiphilic carbon chains) for membrane assembly, the LUCA would have needed the pathways to synthesize carbon chains containing at least 10 carbons. A significant dilemma related to the origin of the lipid biosynthesis pathways is that in present life two completely different types of lipids are used for cell membranes: the Archaea use isoprenoid-type carbon chains (polyunsaturated hydrocarbons with methyl side chains), and the Bacteria and Eukarya use fatty acid–type (unbranched) carbon chains. These types of carbon chains are produced via completely different pathways, and the pathway utilized in Archaea does not even exist in Eukarya or in Bacteria. Thus, it is difficult to deduce what kind of lipid biosynthesis might have been utilized in the LUCA and how this early membrane synthesis gave rise to two different membrane compositions. It is possible either that several different lipid biosynthesis pathways were established in the LUCA population, or that the cellular membrane synthesis pathways were established after the divergence of the three domains. However, this latter scenario would leave open the question of how, or in what, the LUCA was contained.

It has been suggested by some geochemists that prebiotic reactions and early life would have utilized some other types of compartments, formed of inorganic surfaces. Wächtershäuser (2003) has proposed that prebiotic chemistry might have occurred on surfaces of clays or minerals where the earliest cells might then have assembled. These minerals would have catalyzed the reactions and stabilized the forming polymers (see Chapter 1). Martin and Russell (2002), Koonin and Martin (2005), and Brown propose that the porous, chimney-like structures of metal sulfide (mainly FeS) precipitates, formed at the hydrothermal vents at the bottom of the early ocean, functioned as chemically rich, protective hatcheries that provided containment and suitable chemical conditions for the chemical evolution towards life. The main attractions of these latter ideas are as follows.

- Firstly, those metal sulfide mounts were formed from a multitude of small, interconnected compartments, which might have served as containments for chemicals and as catalytic surfaces for reactions.
- Secondly, the hydrothermal vents produced a continuous flow of warm geothermal fluid, full of several different reduced chemicals, and it maintained a strong reduction gradient between the H^+-rich geothermal fluid and (supposedly) alkaline seawater, driving the reduction chemistry of the molecules.
- Thirdly, the hydrothermal vents produced a temperature gradient, providing convection flows to cycle the macromolecules between double-stranded and single stages (discussed in Section 4.6).

The combination of all these factors might have allowed the evolution of complex molecular communities in the cell-like inorganic compartments, until they were adequately competent to produce lipids and proteins for cellular membranes and structures and might escape from the hatchery to self-sustained life in the outside world.

4.3.2
Genes and Their Expression

One experimental way to study what properties and functions were present in the LUCA is to look at which genes or sequences are still common to all the organisms whose sequences are known today (assumably derived from the common origin). Lazcano and coworkers compared a set of fully sequenced genomes and found 283 genes that are highly conserved in all Eukarya genomes, 24 that are highly conserved in Bacteria, and 145 that are conserved in Archaea. In all domains these genes include putative ATPases (for conversion of proton gradients into adenosine triphosphate [ATP] synthesis and for utilizing the energy of ATP), genes involved in transcription and translation, as well as RNA helicase and enolase, phosphoribosyl pyrophosphate synthase, and thioredoxin genes, which are involved in nucleotide biosynthesis and RNA degradation. Such genes, if they were present in the LUCA, indicate that the organism possessed means for maintaining its genetic information (at least in RNA form) and a translation system. Because the ATPases locate in the membranes and produce ATP (as energy compound) by utilizing proton gradients formed across the membranes, this indicates that LUCA had a cell membrane and produced and used energy in the form of ATP molecules.

Another approach to model primitive life is to analyze what would be the minimal set of genes required by the simplest cells today. Several groups of bioinformatists have worked on such questions. Gill and coworkers have searched the genomes of the smallest existing cells (bacterial and archaeal species, which live as endoparasites inside their host cells) for the set of genes that seem mandatory for them all. The living environments of these species are chemically very rich, and the cells may obtain a large part of all their required biochemical precursors from their hosts. Therefore, they have lost the corresponding genes from their own genomes and have evolved, by reduction, towards ultimately small genomes. Comparison of the genomes of such species revealed that they have 206 genes in common. All species have additional genes that allow them to survive in their specific environment (host), but these 206 genes seem to comprise the obligatory set of genetic information required in all these species. Thus, it can be reasoned that a comparable set of genes is essential for any cellular life, even when living in the most protected and nutrient-rich conditions. Thus, if the LUCA was a well-established cellular species, sustained in its own (probably chemically rich, easy, and protected) environment, it probably had to possess at least this number of gene different functions.

As the early organisms evolved to more complex, the initial sequences were converted via mutation and recombination to code for new functions. The genomes grew larger by duplicating, recombining, and modifying what they already had: new genes were often produced from previously existing ones, by recombining and modifying functional protein domains (3-dimensional structures). Recent protein structure analysis by Doolittle and coworkers has identified a set of 49 protein-folding domains (or folding super-families) that are present in all analyzed genomes of the three domains of life, suggesting that these protein-folding structures were already present in the LUCA. Again, these protein structures occur in the proteins related to translation and to metabolic and glycolytic enzymes.

So far, it has been reasoned that LUCA was a fairly complex organism. But how did it diversify into the three domains of life that we know today? The three domains differ from each other in such ways that their direct descent from one another, in a fixed order, is difficult. Each pair of domains shares some features that are lacking from the third domain (Fig. 4.6), so it seems that none of them is directly derived from the others.

Because the deviation of the three domains from the LUCA is difficult to trace back, the question of the order of the establishment of the three domains remains open, i.e.:

- whether the prokaryotes differentiated as two branches from the LUCA and then formed the eukaryotes later by mutual fusion;
- whether the eukaryotes were derived directly from the LUCA and later gave rise to the prokaryotes (e.g., by plasmid escape and reduction processes); or
- whether the three domains diverged separately from a mixed LUCA population.

When this question has been addressed by phylogenetic studies, different results have been obtained from different genes. For instance, the classical phylogenetic tree based on 16S-rRNA sequences, first produced by Woese and coworkers in 1990, shows the early deviation of the Bacteria and Archaea domains, with the thermophilic Bacteria and Archaea at the base of the different branches, and the later deviation of the Eukarya domain from the Archaea (Fig. 4.7A). However, some authors, e.g., Poole and Forterre, suggest that the phylogenetic analysis may not accurately resolve the most ancient sequences: after very long (or very fast) evolution, sequences may become so saturated with mutations that they appear as more conserved in the phylogenetic analysis.

When different protein gene sequences are analyzed, diverse phylogenetic trees are produced for the different sequences. This may indicate that the domains have not evolved as separate lines but rather that the genes have been mixed by strong lateral gene transfer after the separation of the domains.

It is now widely discussed (see reviews by Pennisi and Doolittle) that the intensive lateral gene transfer between species, and even between domains, may have mixed the initial lineages so effectively that it may be difficult to trace back any original sets of genes associated with the early phylogenetic lines. Such lateral gene transfer would have been possible because the newly separated lineages were still very simple, similar, flexible (nonspecific), and rudimentary in their functions. Therefore, they might utilize the same novel functions and structures, even if those

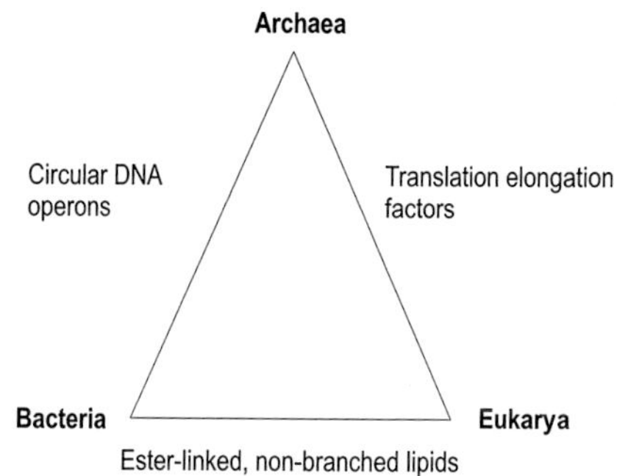

Fig. 4.6 Distribution of some key conserved features among the three domains of life: each of them is shared by two domains only.

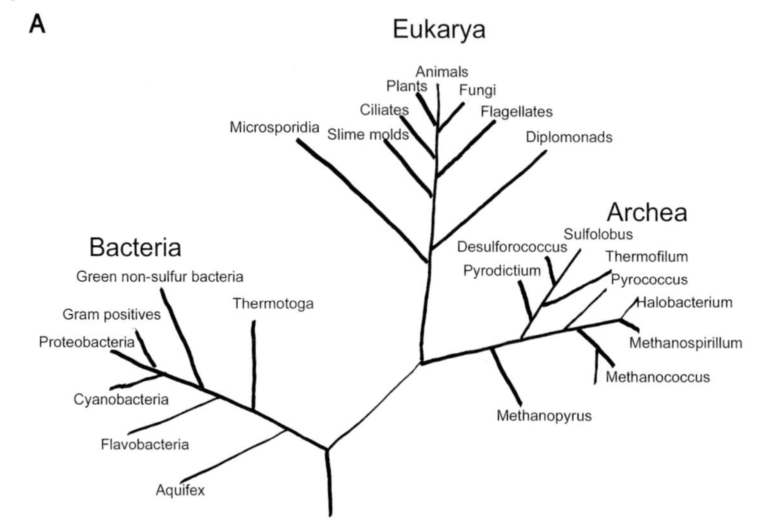

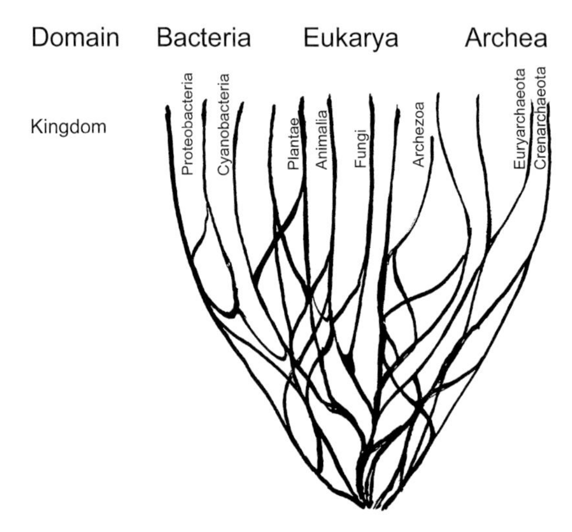

Fig. 4.7 Phylogenetic lineage of the three domains of life determined by different genetic sequences. (A) 16S ribosomal RNA sequence (adapted from Woese et al. 1990); (B) several different sequences (adapted from Doolittle 1999).

were transferred across the domain borders. The DNA exchange could have been mediated via cell fusions or via direct DNA uptake (gene transfer or cellular conjugation) – or via third parties, such as viruses or plasmids (gene transduction or transformation).

Particularly in eukaryotes, the uptake of foreign (prokaryotic) DNA may happen via eating up (endocytosis) of foreign cells. This process is still a distinct function of some eukaryotic cells. Mitochondria and chloroplasts are well-known examples, indicating that endocytosis of other species has occurred in the history of different eukaryotes. In these instances the engulfed prokaryotes were maintained intact in the cytoplasm and established as beneficial endosymbionts, or organelles, of the cells. Considering the strong lateral gene transfer, the tree of life at its earlier stages looks rather like a bush (Fig. 4.7B).

4.3.3
Hypothetical Structure of the LUCA Genome

The dilemma of the differentiation of the three domains relates to the question of the type and structure of the LUCA genome. It is commonly thought that LUCA should have been similar to the simplest life forms existing today, the prokaryotes. Accordingly, it should have been driven by a small, circular DNA genome. However, Poole and Forterre and their coworkers argue that the circular, supercoiled prokaryotic genomes with a single origin of replication might represent a later adaptation to very fast and efficient replication strategies. The most primitive DNA genomes may have been linear, like the RNA genomes (see Section 4.4), which are thought to have preceded the DNA genomes. One argument supporting the idea that the eukaryote-like linear genome structure is the most original one is that specific catalytic RNAs – apparent relicts of the earlier RNA genomes – are still used to process different RNA products in the eukaryotes (Table 4.3), while many of them have been replaced by more efficient (and advanced) protein catalysts in the prokaryotes. Likewise, tRNA-like components are still utilized in the enzymes (telomerases) that maintain the termini (telomeres) of the linear eukaryote chromosomes, suggesting that these terminal structures are derived from the genomes of RNA-based life.

The primitive RNA genomes may have influenced the ploidy level (the copy number) and the fragmentation (chromosome number) of the later-arriving DNA genomes. The early RNA genomes were very fragile and prone to degradation, and the inaccurate replication processes accumulated multiple mistakes in their sequences. For these reasons, the genomic components did not grow very large. In order to accumulate more coding capacity, the genomes should have been composed of multiple short components. To avoid the loss of genetic information due to high mutation rates, the cells should have maintained more than one copy of each genomic component, and thus diploid (or maybe multiploid) rather than haploid (prokaryote-like) genomes should have been used (although one has to bear in mind that prokaryotes also normally maintain several copies of their genome in each cell). The earlier RNA-genome strategy may have been maintained after conversion to the DNA form, suggesting that the earliest cellular genomes should have been composed of multiple, short, linear components, each presented in more than a single copy.

It is not clear whether the LUCA had evolved, as yet, to the DNA stage, or whether the transition from RNA to DNA genomes occurred separately in the three different lineages after their separation from the LUCA. Later conversion from RNA to DNA genomes may be supported by the fact that different types of DNA

Table 4.3 Examples of present-day RNA catalysts and other functional RNAs.

RNA	Function	Distribution	Putative occurrence in the RNA or RNP world
Ribosomal RNAs	Translation: structural and catalytic	Ubiquitous	Yes
mRNAs	Carriers of the genetic code	Ubiquitous	Yes
tRNAs	Translation: Amino acid transport	Ubiquitous	Yes
tRNA	Amino acid metabolism	Ubiquitous	Yes
tRNA	Replication primer	Retrotransposons, retroviruses, pararetroviruses	Possibly
Telomerase RNA	tRNA-like primer for chromosome end synthesis	Eukaryotes	Yes
RNase P	tRNA maturation	Ubiquitous	Yes
snoRNAs	rRNA processing	Eukaryotes, some Archaea	Yes
RNaseMRP	rRNA processing	Eukaryotes	Related to Rnase P
Group I self-splicing introns	mRNA splicing	Eukaryotes, some Prokaryotes	Possibly
Group II self-splicing introns	mRNA splicing	Eukaryotes, some Prokaryotes	Possibly
srpRNA	Protein translocation	Eukaryotes, some Archaea	Yes
vaultRNA	Not known – Protein translocation?	Eukaryotes	Possibly
G8 RNA	Maintenance of ribosomal apparatus at elevated temperatures	Eukaryotes	Probably
Hammerhead and hairpin ribozymes	Self-cleavage and ligation	Viroids and satRNAs	Probably

Source: Modified from Jeffares et al. 1998.

polymerases are utilized by Bacteria and by Archaea and Eukarya. Also, if the RNA genomes were stabilized via methylation of the 2' hydroxyl groups of the ribose molecules, as suggested by Poole, it is conceivable that the RNA genomes might have developed to be fairly large, gaining significant coding capacity. Such RNA genomes may have been adequate to support self-sustained cellular life and to produce codes for the ribonucleotide reductases, thymidylate synthases, and reverse transcription enzymes eventually needed to convert the ribonucleotides into deoxyribonucleotides and RNA sequences into DNA. Forterre suggests that those DNA-producing enzymes and, consequently, DNA genomes were invented several times to establish DNA-based viral genomes colonizing the RNA cells and that they were transferred from those to the RNA-based host cells, in three separate transfer events, to establish the three DNA-based domains of life.

Woese has suggested that at the earliest stage of evolution, when the genomes were still very small and unstable, the fully functional genetic information could not exist in individual discrete organisms or genomes. He suggests that these early genomes instead formed a diverse community of organisms that dynamically interacted with each other, exchanging, testing, storing, and evolving alternative genomic sequences. Woese calls these early life forms "progenotes," which describes the evolutionary stage preceding the fixed genomic life forms, or the "genotes." Again, it is not clear whether such progenotes would have had a DNA or an RNA genome; maybe they could have been characteristic both before and after the transition from RNA to DNA genomes.

4.4 "Life" in the RNA–Protein World: Issues and Possible Solutions

LUCA appears to have been a fairly complex and well-established organism – but from where did this complexity arise? Information flow from DNA into mRNA and then into proteins (Fig. 4.5) demands multiple functions in between; therefore, it is commonly thought that the whole genetic machinery had to evolve from a simpler system. It is supposed that the genetic system preceding the DNA genomes were RNA genomes, which could have functioned as information storage media and as templates for translation. It is thought that the protein synthesis became established at the time of the RNA genomes and that the appearance of the protein catalysts then allowed the "invention" and establishment of the DNA genomes. Thus, present-day DNA–RNA–protein life would have been preceded by life composed of RNAs and their encoded protein products – or by RNA–protein (RNP) life.

At the earliest stages, when the genetic code was translated into protein products, or the genetically encoded proteins were "invented," both the codes and their expression system were certainly very primitive. The codes had not yet evolved to code for specific functions, and also the translation system was very inaccurate. Maybe only a few of the currently known amino acids were used. The system was able to produce only a few short, nearly random polypeptides. The polypeptides had not been adapted to specific functions, but maybe some of them could bind to RNA and function as scaffolding or structural proteins, similar to present-day chaperones. Such structural proteins may still have been quite useful in stabilizing the RNA genomes and in holding them in suitable three-dimensional structures to enhance their catalytic properties. Such structural proteins probably assisted in the most essential functions of the RNP world, i.e., in the replication of the RNAs. The high mutation rate produced more functional gene products, and in due course some proteins appeared that actually catalyzed the replication of RNA genomes.

4.4.1 Evolutionary Solutions

The early enzymes were still very inefficient and prone to errors. Therefore, each time the genomes were replicated, considerable portions of the nucleotides were copied wrong and the genetic information was altered. If the sequences mutated too much, it was changed into a non-functional form, the gene was no longer maintained, and the information was totally lost. This critical level of accumulation of errors has been defined by Manfred Eigen as the "error threshold" and is now known as the "Eigen limit." The number of incorrectly copied nucleotides increased with the genome length; the longer the genomes, the easier they accumulated fatal levels of mistakes. Thus, only fairly short genomes (or genomic components) could be maintained by this very error-prone replication.

Eigen has demonstrated how the evolution of small genomes (e.g., viral genomes) functions in a cyclic fashion, where repeated improvements of different synergistic functions lead to very efficient improvement of fitness and adaptation to new conditions. As described by Poole, this cyclic adaptation and genome-building model seems applicable for the gradual improvement of the RNP organisms: the new sequence variants encoded for improved replicase proteins or new components for the translation machinery. These produced better replicase molecules and thereby improved the accuracy of the replication. This again allowed the maintenance of longer genomes – and the increasing coding capacity allowed for the production of new components for the replication and translation machineries.

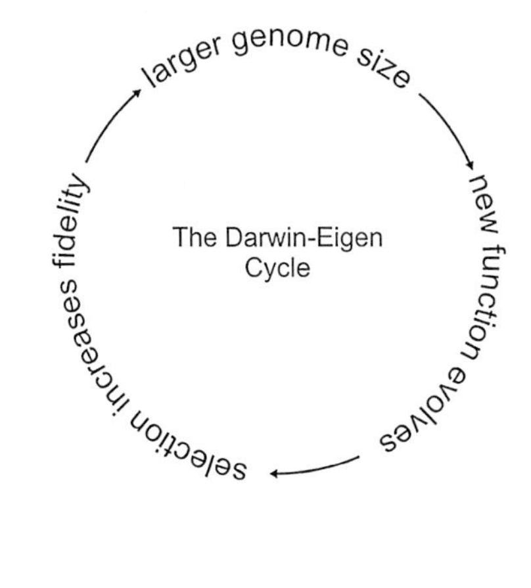

Fig. 4.8 Darwin–Eigen cycle demonstrating that Darwinian selection for improved replication and translation allowed maintenance of larger genomes, and these again allowed for improved replication and gene expression. Thus, the increasing genome size, new gene functions, and improving replication and translation fidelity have mutually enhancing effects on the evolution of complexity (adapted from Poole et al. 1999).

Thus, Darwinian selection for more accurate replication and the corresponding increase of the genome size (determined by the Eigen limit) together enhanced the evolution towards more complex genomes (Fig. 4.8).

4.4.2 Solutions Found in the Viral World

It is of interest that present-day single-stranded (ss) RNA viruses possess multiple features that seem similar to the properties of a hypothetical early RNP organisms. As in the RNP organisms, the viral genomes are composed of short linear RNAs that code only for a few necessary gene products. The total length of RNA virus genomes seems to be limited by both the fragility of the RNAs and the Eigen limit (the avoidance of accumulating too many errors by inaccurate replication); thus, at the present time the longest ss RNA virus genomes – those of the coronaviruses – contain about 30 000 nucleotides. The smallest viral genomes, – those of the leviviruses – are only about 3 500 nucleotides long and code only for three proteins, namely, the replicase protein, the coat protein, and a maturase protein to interact with the host. These seem to make the minimal protein set required for sustainable viral genomes. In many viruses the replicase proteins are divided in two separate components, the helicase and the replicase, apparently because those proteins need to be produced in very different amounts.

The coat protein is required to enclose the RNA and to protect it from degradation. All viruses encode one or more of other gene products to interact with their host cells – mainly in order to suppress the host defense functions – but it appears that in principle the replication and the protective functions are the most important requirements for viral survival. This may also indicate how important these proteins may have been to the early RNA replicators.

Viruses maximize the use of their limited genomic sequences – they use them in the most efficient way. They contain a minimal amount of non-coding sequences, and they do not "invest" any coding capacity for producing regulatory proteins, which are so essential for cellular life. Many viruses regulate their essential functions – initiation of replication and translation – by short and simple sequences at the termini of the RNA genomes and by interactions between RNA loops and sequences. With such RNA interactions, some viruses are capable of dedicating their genomic RNAs to function in either the replication or the translation mode, which avoids collision of the two opposite processes.

Being parasites of the cellular world, RNA viruses make their living in a chemically rich environment: the host cells provide all the building blocks of the macromolecules (the nucleotides and the amino acids) and the translation machinery for use by the viruses. Likewise, the early RNP organisms had to live in a similar environment, because they had no means for synthesizing their own building blocks. Through their evolutionary history, the RNA viruses have become very efficient in their replication and in their adaptation to their own specific host organisms. Indeed, in spite of their ultimately minimal size, the RNA viruses are very successful. There are many more species of viruses than there are cellular species, and each cellular species may be infected by several different viral species. It is possible that the (fully evolved) RNP-based life forms were efficient and successful replicators. Apparently they had to be quite advanced in order to produce adequately large genomic pools to give rise to DNA-based life and cellular organisms. Some of the early RNP-based replicators may have survived to form the base for the massive variation of present-day viruses.

4.5 "Life" Before the Appearance of the Progenote

4.5.1 The Breakthrough Organism and the RNA–Protein World

The first protein-mediated functions initiated evolution towards more and more complex biochemistry. Therefore, the initiation of genetically encoded protein synthesis has been the key step in the development of any more-advanced life. This event has been considered so essential that Poole and coworkers have named the first creature that produced proteins the "breakthrough organism." But how could this hypothetical breakthrough organism break through – or, how might it have come into existence in the first place?

Clearly, it had to evolve from simpler systems through a cascade of improving stages. It was required that it be able to synthesize mRNA-like translatable sequences and to run a primitive translation machinery consisting of ribosomes, tRNAs, and amino acids. Because protein synthesis did not exist prior to this time, the assembly or function of this primitive translation machinery could not have depended on any specific proteins. Thus, the early translation reactions had to be catalyzed by the existing RNAs alone. This is theoretically feasible because in all the present-day life forms, the formation of the peptide bonds between amino acids is specifically mediated by one RNA component of the ribosomes. Different natural or *in vitro*–selected RNA molecules may catalyze many other reactions (for the natural ribozymes, see Table 4.3). For instance, ribozymes have been found that can mediate the amino acylation of tRNAs. Thus, according to experimental data, it may be postulated that mere RNA components were able to mediate the earliest translation process.

4.5.2 Primitive Translation Machinery

A logical requirement for the existence of RNA-based translation machinery is that this machinery itself was initially generated and maintained in a protein-independent manner. Present-day ribosomes are massive complexes, containing either three (in prokaryotes) or four (in eukaryotes) RNAs that vary in size from 120 to 2 904 nucleotides. The genes coding for these RNAs take, together with their intervening sequences, more than 7 500 nucleotides of genomic sequences. It is clear that the earliest RNA components were not as long as they are today. As mentioned in Section 4.4.1, the error-prone replication systems produced only fairly short RNA strands that are estimated not to have exceeded a few hundred nucleotides in length. Thus, it is likely that the proto-ribosome had to be composed of short fragments of RNA that jointly formed one functional complex. This may still be evident in present-day ribosomes, because their interaction with tRNAs may be mimicked by a separate small RNA analogue of the ribosomal RNA sequence and because peptide bond formation is catalyzed by a specific adenine in the core of the ribosomal RNA. Thus, it seems likely that the present-day ribosomal RNAs (rRNAs) developed via a combination of previously existing, small functional subunits.

4.5.3 Origin of Ribosomes

Even if the early ribosomes were formed of small separate RNAs, their maintenance required that they be reproduced fairly accurately. The mere existence of such a functional complex required that it had evolved gradually from some simpler structures – which again had been maintained and selected for their

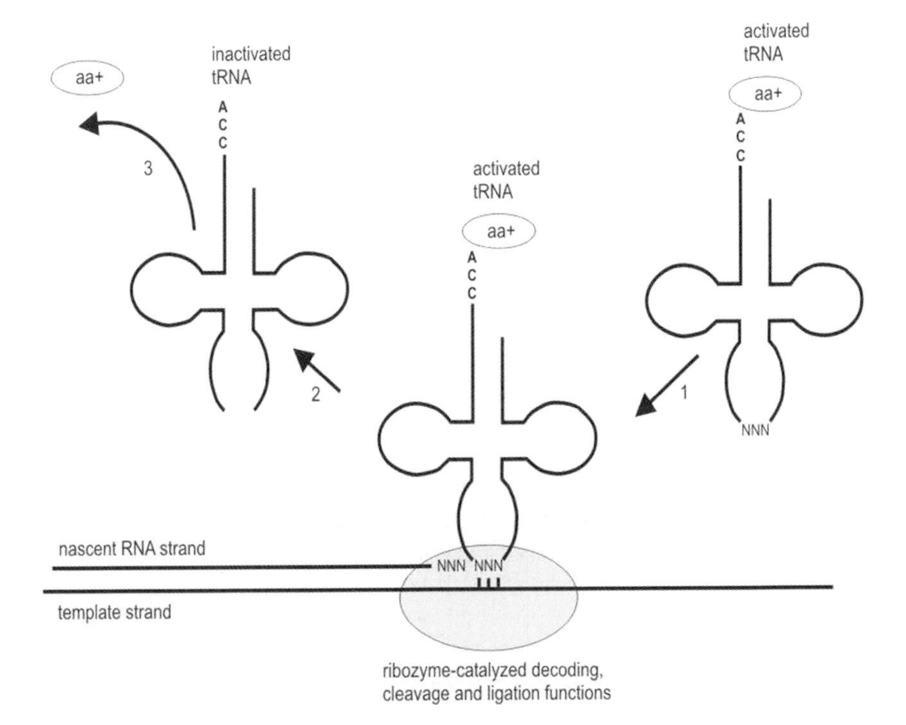

Fig. 4.9 Model of the early function of the proto-translation machinery as RNA replication complex (from Poole et al. 1998, with permission from the publisher).

preexisting function(s). The most important function at this stage would have been the maintenance of the existing sequences, namely, the replication of RNAs. Poole and coworkers have proposed that the original function of the proto-ribosomal complex was the replication of RNAs. According to their model, the early proto-ribosomes extended the complementary RNA strand on the RNA template by ligating nucleotide triplets (corresponding to present-day anticodon loops) brought into the reaction by tRNA molecules or their primitive precursors. This means that early replication was mediated by ligation rather than by polymerization. Incorporation of nucleotide triplets rather than single nucleotides had the advantage of allowing higher affinity and longer-lasting annealing of the incoming nucleotides to the template. This allowed more time for the very slow RNA catalyst to perform the ligation (Fig. 4.9).

According to this model, it is conceivable that amino acid acylation of the incoming tRNAs was a significant step in the early replication process. The covalent binding of the amino acids to the -CCA terminus of the tRNAs might have charged these molecules with energy, which might have been released by the detachment of the amino acid during the polymerization. This process might have been used to drive the polymerization. On the other hand, it is conceivable that the positively charged amino acids allowed the proper confirmation (tight folding) of the primitive (negatively charged) tRNAs and/or allowed their binding to the replicase complex. Release of the amino acids from the tRNAs might then have caused a conformational change allowing for the release of the used tRNA from the complex.

In any of these scenarios, the interaction between template RNA, ribosome, the "triplet-reading" anticodons of the tRNAs, and the amino acids would have already been established at the early RNA replication function, prior to the invention of protein synthesis. Peptide bond formation might have been initiated as a spontaneous reaction between the amino acids brought to close proximity by the incoming tRNAs. However, later conversion of the whole replication machinery into a translation function is difficult to conceive – many things would be have been required to change. For example, the hypothetic cleavage and ligation of the nucleotide triplets from the tRNAs and into the growing chain of RNA needed to be stopped and replaced by peptide bond formation. However, it is conceivable that the introduction of protein synthesis made these changes possible, as it provided protein catalysts for replication and a strong selection for the use of ribosomes for translation. Both processes, translation and replication, had to be made to function on the same template in opposite directions. The function of the same template in both replication and translation seems not to be a major obstacle, as it occurs in all present-day viruses in a coordinated and regulated fashion.

Another model for the early role of tRNAs proposed by Weiner and Maizels is the terminal taging of the replicating RNAs. These terminal tags would have functioned as the selective binding and initiation site of the replicase and thereby would have helped to replicate the genomic ends correctly. This hypothesis is supported by the fact that many present-day viruses have tRNA-like 3' termini and the telomerase enzyme that synthesizes the ends of the eukaryotic chromosomes functions on a tRNA-like template.

4.6
The RNA World

Logically, it is necessary to assume that prior to the invention of protein synthesis, all reactions required for maintenance of macromolecules – and for their evolution towards life – had to be mediated by molecules other than proteins. Present-day RNA chemistry has demonstrated that different RNA molecules (either natural ones or selected *in vitro*) may mediate several different catalytic activities, although the RNA catalysts typically function less efficiently than protein enzymes, i.e., at a much slower rate (Fig. 4.2). Such molecules are called RNA enzymes or ribozymes. Most of the naturally occurring ribozymes mediate cleavage and ligation reactions of their target RNAs (Table 4.3). The smallest ribozymes contain about 50 nucleotides and occur in the RNAs of viroids and viral satellite RNAs, while complex ribozymes capable of many different catalytic functions can be artificially produced by *in vitro* selection.

It must be emphasized that the functionality of the RNA catalysts is based on the sequence information. The information does not code for any protein products, but it dictates the folding of the RNA into secondary structures that can recognize their substrates and cause the desired reactions. Figure 4.10 shows the three-dimensional folding structure, the so-called hammerhead structure, of the smallest natural ribozymes, which are composed of about 50 nucleotides.

Figure 4.11 gives an example of the evolution of an RNA polymerase ribozyme, as mediated by *in vitro* selection steps. The class I ligase ribozyme catalyzes the joining of the 3' end of an RNA strand to the 5' end of the ribozyme (Fig. 4.11 a); the class II ligase ribozyme catalyzes the addition of three nucleotide 5' triphosphates (NTPs) to the 3' end of the primer, directed by an internal template

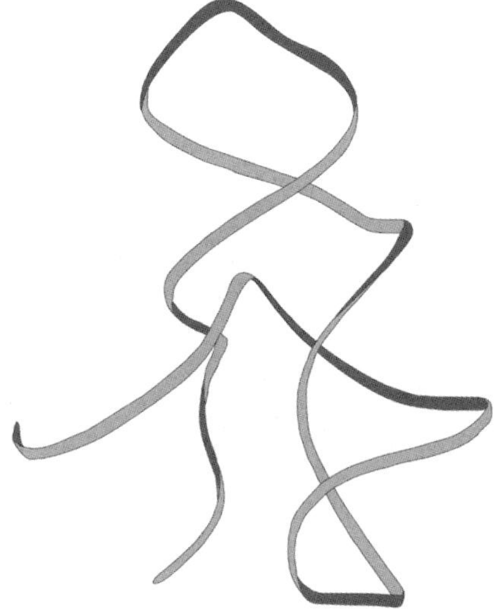

Fig. 4.10 Three-dimensional folding structure (so-called hammerhead structure) of the smallest natural ribozymes. These self-cleaving and -ligating structures occur in several viroid and viral satellite RNAs and are composed of about 50 nucleotides.

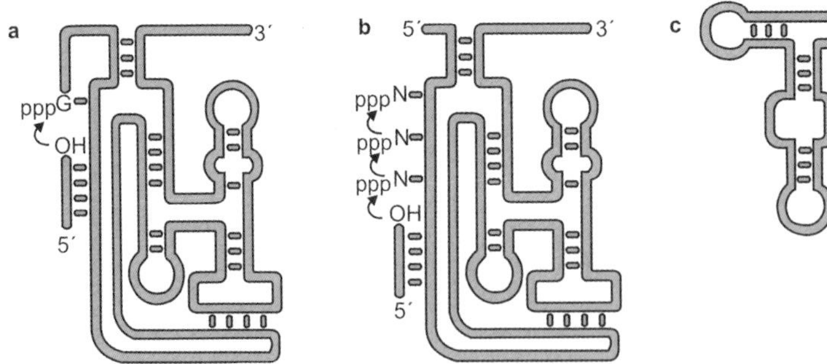

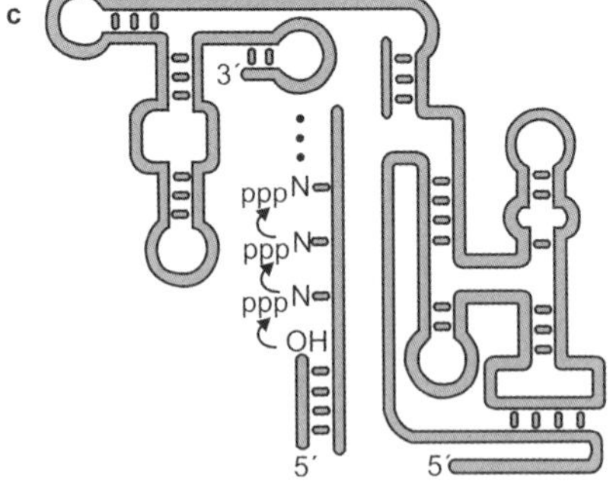

Fig. 4.11 Evolution of an RNA polymerase ribozyme mediated by *in vitro* selection steps. (a) The class I ligase ribozyme catalyzes the joining of the 3' end of an RNA strand (open line) to the 5' end of the ribozyme; (b) the class II ligase catalyzes the addition of three nucleotide 5'-triphosphates (NTPs) to the 3' end of the primer, directed by an internal template; (c) the class I–derived polymerase catalyzes the addition of up to 14 NTPs on an external RNA template (from Joyce 2002, with permission from the publisher).

(Fig. 4.11 b); finally, the class I–derived polymerase catalyzes the addition of up to 14 NTPs on an external RNA template (Fig. 4.11 c).

The hypothetic stage of early evolution, when all biochemistry was mediated by different RNA catalysts, is called the RNA world. Invention of genetically encoded proteins brought out the new, protein-based biochemistry that soon replaced RNA-mediated catalysis. A curious example of this transition to protein-mediated functions is that still today some proteins functioning in the regulation of translation – namely, some translation elongation factors – are folded into a secondary structure similar to that of the tRNAs. Apparently, after the RNA genomes accumulated adequate coding capacity, the protein enzymes were able to convert the RNA genomes to DNA form. Being much more stable than the RNAs, these could evolve into much larger sizes and superior coding capacity.

With these immense improvements in biochemical functions, the original RNA-mediated biochemistry might have disappeared without leaving a trace. However, this did not happen, as we do still have some clear molecular relicts, or molecular fossils, that apparently have been conserved from the RNA era into present-day life. The most convincing molecular fossil from the RNA world is that the primary genetic code is still translatable only from RNA sequence (DNA exists only as an information storage system, comparable to a hard disk, containing all the data but no running program for translating this information). Further, the translation reaction, or the formation of the peptide bond, is still mediated by one RNA component of the ribosomes, and the tRNAs still bring the amino acids into the translation reaction. Several other small functional RNAs exist in present-day cells (Table 4.3) and also may be functional relicts from the early RNA era.

As mentioned in Section 4.3, some of these RNA catalysts – particularly those that function in the maturation of different RNAs, especially small nuclear RNAs and nucleolar RNAs – as well as RNA-containing telomerase enzymes still occur in the eukaryotic organisms (Table 4.3). This suggests that those (slow) RNA-mediated functions have been maintained in eukaryotes from the times of RNA-based life. In prokaryotes, many RNA-processing functions have been replaced by protein-catalyzed functions, apparently due to their adaptation to faster replication and to more efficient RNA processing, particularly in hot environments. For these reasons it might be thought that the eukaryotic genome type – although expanded to huge sizes and complex gene sets over time – would be of a more ancient type than the small and sophisticated prokaryote genomes.

4.6.1 Origin of the RNA World

4.6.1.1 Prebiotic Assembly of Polymers

In order to get the functional RNA world started, preexistence of RNA polymers was required. These RNA polymers supposedly appeared by spontaneous assembly or polymerization of nucleotide monomers. They had to accumulate to significant amounts, and be of considerable length and variation, to ensure that this random collection contained some (or even one) that had the potential of copying itself and, later, of copying even other RNAs.

The prebiotic, i.e., spontaneous, polymerization of RNA nucleotides is difficult to explain with known RNA chemistry, because nucleotide polymerization requires energy and does not happen easily (see Chapter 3). It has been intensively studied – e.g., in the laboratory of James Ferris – and in these optimized conditions, polymers of about 40–50 nucleotides were produced. Such polymerization results have been obtained by allowing activated nucleotides to react in aqueous solution in the presence of clay minerals, which significantly enhanced the polymerization. The enhancement by clays is caused, assumedly, by their fine-layered structure, composed of mineral grains with di- and trivalent positive charges. Those cationic surfaces bind and immobilize the negatively charged nucleotides, placing them in suitable positions to promote their reactions with each other. Further on, binding to the clay surfaces significantly stabilizes the ready-made RNA polymers, which otherwise would be very easily degraded by hydrolysis.

The group of Deamer reports that another environment that promotes RNA polymerization, although to a lesser extent than clays, is the eutectic (very cold) ice solutions; an icy environment is also known to promote the ribozyme-catalyzed ligation of RNAs. The liquid water phase remaining in between ice crystals concentrates the precursors effectively, and the low temperatures help to slow down the reacting components to allow the formation of the phosphodiester bonds between nucleotides.

For building the nucleotide polymers, the nucleotides need to be linked to each other via phosphodiester bond between the 3' and 5' carbons of the adjacent sugars (Fig. 4.3). For this, the nucleosides first have to be phosphorylated at their 5' carbon. In the early world this was problematic because soluble phosphates were not readily available. It is possible that the inorganic phosphate dissolved in low amounts from calcium phosphate (hydroxyapatite) mineral or from volcanically produced linear polyphosphates, or from their breakdown products, and phosphorylated nucleosides to nucleotides. But this reaction is difficult and might occur only in the presence of urea, ammonium chloride, and heat.

Polymerization of the nucleotides requires that they be activated by some high-energy bond at the 5' position to provide energy for the uphill reaction. In present-day life, this binding energy is obtained from the hydrolysis of the triphosphate moiety, but this was not adequate in the prebiotic world because triphosphates alone do not promote (non-enzymatic or non-catalyzed) polymerization. Therefore, it seems that the nucleotides had to be charged with other available molecules, such as amino acids, bound with high-energy bonds at the 5' position of the phosphorylated nucleotides. The most effective activating groups, now commonly used for the polymerization experiments in laboratories, are the different phosphoramidates, such as phophoramidazoles. These compounds are formed from nucleoside 5' polyphosphates and amines or imidazoles; but such complex activating groups might not have been available in the prebiotic environment.

A further complication in the polymerization of ribonucleotides is that in a mixture of monomers, several different reactions take place. To make a functional polymer, the monomers must form the mono-phosphodiester linkage exactly between the 3' and 5' carbons of the adjacent nucleotides. However, the ribose ring has reactive groups in carbons at positions 5', 3', and 2' (Fig. 4.3). In prebiotic conditions all these groups could react with each other, and cyclic compounds could be formed between the 2' and 3' OH groups. Furthermore, the phosphate molecules might have formed different polyphosphate linkages between different carbons. All these varying bonds would have produced dead-end products for further polymerization. It is still unclear what kind of conditions promoted the nucleotides to react in an orderly fashion to produce only the correct, 3'-to-5'-linked RNA polymers (Fig. 4.4).

4.6.1.2 The Building Blocks of the RNA World

For synthesis of nucleosides, the purine bases adenine and guanine (A and G), the pyrimidine bases cytosine and uracil (C and U), and the D-isoform of cyclic ribose sugar are needed. Prebiotic synthesis and assembly of these nucleoside building blocks have been intensively studied by groups of chemists led by Gerald Joyce, Stanley Miller, Leslie Orgel, and Alan Schwartz. The prebiotic synthesis of these components, as well as the prebiotic synthesis of their small organic precursors such as CH_4, NH_3, CH_2O, CNH, CN_2H_4, C_3NH, CH_2O, CN_2H_4O, and SH_2, is covered in Chapter 3.

Provided that ribose sugar and the nucleobases were available, further on the bases had to be covalently linked in β-orientation with the 1' carbon of ribose to form nucleosides. This reaction may be activated by heating and it produces purine nucleosides, although with very low yield. Routes for prebiotic synthesis of pyrimidine nucleosides are not well known, but they may be feasible via multiple sugar phosphate intermediates. Recently, a completely different route has been found, where α-D-cytosine was produced directly from ribose sugar and hydrogen cyanide. This efficient reaction thus produced cytosine, even if of the wrong isoform, and it also converted the existing (unstable) ribose into a stable form.

Prebiotic synthesis of nucleotides would have yielded a racemic mixture of a variety of different nucleotide analogues, both in α and β and in L- and D-isoforms. A spontaneous polymerization reaction of such (randomly) phosphorylated nucleotides would have been very slow and would have led to a wide variety of different linkages, formed between the whole variety of different nucleotide analogues and isoforms. Such randomly linked oligomers or polymers would not have been extendable or functional as templates for replication.

Altogether, prebiotic formation of RNA nucleotides and their polymers, starting from small molecular precursors such as HCN, NH_3, CH_2O, CH_4, PO_4^{3-}, or H_2O, involves so many unlikely or adverse chemical steps that many chemists have speculated that RNA synthesis had to be preceded by simpler chemical polymers – or by nucleic acid analogues whose backbone is composed of non-chiral structures. Polymers proposed as hypothetical predecessors of the RNA world include peptide nucleic acids (PNAs) with a peptide as backbone, threose nucleic acids (TNAs), and pyranosyl-derived nucleic acids (p-RNAs) (see Chapter 1). If any of these polymers was a direct precursor of present genetic material, then it must have been possible to transfer the information by copying or transcription from this molecule into RNA or DNA sequence, which indeed has been experimentally demonstrated for PNAs. However, the role of other polymers in the early steps of molecular evolution does not seem very plausible, because the central molecules of early life (the ribosomes and tRNAs) are based solely on RNA molecules.

To find possible solutions for the appearance of the nucleotide-based (very complex) genetic information, Cairn-Smith (1982) has postulated that instead of organic polymers, entirely different molecules, such as minerals, may have contained the earliest pre-genetic (replicating) information. Still, it is not clear how any specific information would have been coded in the crystal structure of mineral surfaces, or how these would have been converted to be biologically significant. Little is known, as yet, of what the available starting materials were, what the conditions were where the prebiotic chemistry took place, and what chemical and physical reactions and selection processes were driving the increasing complexity of the early polymers (see also Chapter 1).

4.6.1.3 Where Could the RNA World Exist and Function?

The question of the possible location and conditions of the proposed RNA world is very difficult. It is assumed that the initial RNA polymers were formed by random polymerization of nucleotides, but this scenario requires that the nucleotides had to be initially available in the environment. Further on, for the synthesis of nucleotides, their precursors – ribose sugar, purines and pyrimidines, and phosphate molecules – had to be available. For the prebiotic synthesis of those compounds, their precursors or the small organic molecules such as CH_2O, CNH, CN_2H_4, and HNCO had to be available in adequate amounts. To drive repeated polymerization reactions, leading to significant accumulation of the products (adequate for starting a whole new polymerization and replication network), a long-term, high-level supply of nucleotides would be required.

The prebiotic synthesis or at least the accumulation of nucleotides would need to happen in conditions where, further on, the nucleotides could spontaneously polymerize. The conditions also should have been adequately gentle to maintain the polymers intact. The prebiotic synthesis of some of the precursors happens preferably in cold or mildly warm conditions (synthesis of nucleic acid bases), while other steps require heat (nucleoside formation and phosphorylation). So it is conceivable that the locations for the different steps may have been temporally or locally separate.

The polymerization of nucleotides needs to take place in an aqueous environment because the nucleotides are water-soluble and liquid water is needed to carry them into the reaction site. However, water efficiently reacts with the RNA polymers to hydrolyze the phosphodiester bonds, which breaks up the polymers. The hydrolysis reactions occur faster at higher temperatures; therefore, cool, close to freezing temperatures would be more suitable for preserving RNA polymers.

A further complication arises with the templated polymerization – meaning the copying or replication of existing RNA strands. Because the phosphate groups forming the phosphodiester bonds carry a negative charge of −1, the total charge of the polymer is highly negative. In the templated replication, the complementary strand is assembled on the existing template by hydrogen bonding between the complementary nucleotides (A to U and G to C) (Fig. 4.4). For stable assembly of the complementary strand, the negative charges of the phosphates need to be neutralized by an adequate level of soluble cations (positive ions). This neutralization together with the hydrogen bonding of the bases leads to a stable assembly, and the double-stranded product is closed for any further assembly of new strands on the same templates. Thus, the double-stranded structure could be a dead-end product.

To avoid this, some energy input or radical change of environment – e.g., by dilution out of the cationic ions or warming up of the solution – would be needed to open up the double helix. Lathe (2005) has proposed that alternating conditions of low and high salt concentrations (as in tidal pools) would be needed to drive the assembly and separate the daughter strands. Brown and coworkers have experimentally shown that efficient melting (i.e., separation) of double-stranded nucleic acid strands can be obtained by convection-driven cycling of molecules in micro-compartments located in steep temperature gradients, such as the tiny compartments in the FeS precipitate mounts of hydrothermal vents.

Still another putative environment that might have driven prebiotic chemistry and the assembly of macromolecules, possibly including polymers, is the sandy or silty beaches washed by the ocean tides. Bywater and Conde-Frieboes (2005) have proposed that in these locations the ocean water would have brought in the precursors for the reactions, and these might have been bound, by hydrogen bonding or ionic forces, to the cationic minerals in the clay and sand. Small drying ponds might have provided the concentration mechanism for the compounds, and the subsurface layers of sand might have provided protection against UV radiation, as well as size-based fractionation and concentration by chromatographic filtration.

The clay-like minerals would have been suitable catalytic surfaces for the assembly of nucleotide polymers. The repeated washes by new waves or tides might have provided the cyclic dilution effects needed for the separation of the replicating nucleotide strands, as proposed by Lathe.

4.7 Beginning of Life

The above discussion should indicate that the origin of life was by no means simple. Rather, it should have been a procession through multiple stages of spontaneous – but most often uphill – reactions of molecular chemistry and evolution towards cellular life. Within this procession it is difficult to determine when life actually started. It seems that there was a stage where the developing system was not yet clearly alive, and also no longer clearly lifeless, but swaying in between different directions of development in a type of bifurcating stage. Such a threshold level could have been achieved when the first sequences could replicate themselves but when there were no biosynthesis pathways, as yet, for any of the precursors or building blocks for making more of the same molecules. The system was modified and adapted by a selective evolution, but it was still wholly dependent on the supply of complex materials from its environment. It is conceivable that such a threshold stage could have developed in several directions: one, for instance, to produce the semi-living (replicating but not autonomous) viral world and another to produce autonomous cellular life. Parasitic and synergistic interactions also evolved – the viruses ending up as parasites of cellular life – but the early molecular interactions may have been more varied.

The question of the beginning of life is related to how we define life. Life, as we now know it, is very complex and versatile, but the definition should include just the very minimal and first features that were required of a system to become alive. André Brack has defined primitive life as an aqueous chemical system able to transfer its molecular information and to evolve (see Chapter 1). Eigen has described life as phenomena in the multidimensional information space, where the species – or quasi-species – form as their genomic sequences accumulate in the proximity of the optimal or best-adapted sequence, at so-called "fitness peaks." Both these descriptions emphasize the information-driven nature of life, the maintenance of information via replication, and accumulation of new information via natural selection and evolution.

In a reductionist way, it might be claimed that genomic information is the essential part of life and that the cellular structure serves only as a vehicle to maintain and renew the information. But the cellular structure has played a central role in the formation of information: the accumulated information just reflects the needs of the cells, and the optimized information is determined by the adaptation of the cell to its environment.

The very initial functional information appeared by accident, as a random sequence was able to replicate itself. The further accumulation of information took place via improvement of that very function. When the system grew and evolved, new functions were "invented" that allowed the system to survive better in its environment. The selection, or genetic adaptation, was mediated via different interactions. Intracellular molecular interactions were, and still are, of essential importance, as were the interactions of the cell with its surroundings, with the members of its own species, and with the whole surrounding ecosystem. These interactions are the driving force for the accumulation of genetic information.

Even the simplest cellular life is mediated via multiple biochemical reactions driven by energy conversion reactions and, ultimately, by energy drawn from a suitable energy source. With these, life achieves a remarkable organization of matter and energy. The cellular structure is essential for the containment of all these reactions and for the assembly of the organized structures.

Thus, life is indeed based on genetic information and on evolution, but the third essential party in the assembly of life is its cellular form. The cell and its interactions with the environment – and, on the other hand, its separation from the environment –build the entity that has allowed the accumulation of information and that has facilitated evolution, and the self-sustained existence of the organism. Its existence still depends on suitable energy sources and availability of nutrients, i.e., on the life-supporting environment. A cell, adapting in its life-supporting environment, was the last defining requirement for the establishment of life as we know it on Earth.

4.8 Further Reading

4.8.1 Books

Brack, A. *The Molecular Origin of Life. Assembling Pieces of the Puzzle*, Cambridge University Press, Cambridge, 417 pp., 1998.

Cairn-Smith, A. G. *Genetic Takeover and the Mineral Origins of Life*, Cambridge University Press, Cambridge, 1982 (http://originoflife.net/cairns_smith/).

4.8.2 Articles in Journals

Abe, Y. Physical state of the very early earth, *Lithos* **1993**, *30*, 223–235.

Berclaz, N., Muller, M., Walde, P., Luisi, P. L. Growth and transformation of vesicles studied by ferritin labeling and cryotransmission electron microscopy, *J. Phys. Chem.* **2001**, *105, 1056–1064. (http://www.plluisi.org/)*

Biebricher, C. K., Eigen, M. The error threshold, *Virus Res.* **2005**, *107*, 117–127.

Bywater, R. P., Conde-Frieboes, K. Did life begin on the beach? *Astrobiology* **2005**, *5*, 568–574.

Chang, I.-F., Szick-Miranda, K., Pan, S., Bailey-Serres, J. Proteomic characterization of evolutionary conserved and variable proteins of Arabidopsis cytosolic ribosomes, *Plant Physiol.* **2005**, *137*, 848–862.

Chen, I. A., Roberts, R. W., Szostak J. W. The emergence of competition between model protocells, *Science* **2004**, *305*, 1474–1476.

Delaye, D. Becerra, A., Lazcano, A. The last common ancestor: what's in the name, *Orig. Life Evol. Biosph.* **2005** *35*, 537–554.

DeDuve, C. A research proposal on the origin of life, *Orig. Life Evol. Biosph.* **2002**, *33*, 559–574.

Deamer, D., Dworkin, J. P., Sandford, S. A., Bernstein, M. P., Allamandola, L. J. The first cell membranes, *Astrobiology* **2002**, *2*, 371–381.

Doolittle, W. F. Phylogenetic classification and the universal tree, *Science* **1999**, *284*, 2124–2128.

Eigen, M. The origin of genetic information: viruses as models, *Gene*, **1993**, *135*, 37–47.

Ferris, J. P. Montmorillonite catalysis of 30–50 mer oligonucleotides: laboratory demonstration of potential steps in the origin of the RNA world, *Orig. Life Evol. Biosph.* **2002**, *32*, 311–332.

Forterre, P. The two ages of the RNA world, and transition to the DNA world: a story of viruses and cells, *Biochim.* **2005**, *87*, 793–803.

Forterre, P., Confalonieri, F., Charbonnier, F., Duguet, M. Speculation on the origin of life and thermophily: Review of available information on reverse gyrase suggests that hyperthermophilic procaryotes are not so primitive, *Orig. Life Evol. Biosph.* **1995**, *25*, 235–249.

Gil, R., Silva, F. J., Pereto, J., Moya, A. Determination of the core of the minimal bacterial gene set. Microbiol. and molec, *Biol. Rev.* **2004**, *68*, 518–537.

Hanczyc, M. M., Szostak, J. W. Replication of vesicles as models of primitive cell growth and division. Current Opin., *Chem. Biol.* **2004**, *8*, 660–664.

Jeffares, D. C., Poole, A. M., Penny, D. Relicts from the RNA world, *J. Mol. Evol.* **1998**, *46*, 18–36.

Joyce, G. F. The antiquity of RNA-based evolution, *Nature* **2002**, *418*, 214–221.

Kanavarioti, A., Monnard, P.-A., Deamer, D. W. Eutechtic phsces in ice facilitate nonenzymatic nucleic acid synthesis, *Astrobiology* **2001**, *1*, 271–281.

Kapp, L. D., Lorsch, J. R. The molecular mechanics of eukaryotic translation, *Ann. Rev. Biochem.* **2004**, *73*, 657–704.

Koonin, E., Martin, W. On the origin of genomes and cells within inorganic compartments, *Trends Genet.* **2005**, *21*, 647–654.

Lathe, R. Tidal chain reaction and the origin of replicating polymers, *Int. J. Astrobiol.* **2005**, *4*, 19–31.

Lehto, K., Karetnikov, A. Relicts and models of the RNA world, *Int. J. Astrobiol.* **2005**, *4*, 33–41.

Liang, H., Landweber, L. F. Molecular mimicry: quantitative methods to study structural similarity between protein and RNA, *RNA* **2005**, *11*, 1167–1172.

Maizels, N., Weiner, A. M. Phylogeny from function: evidence from the molecular fossil record that tRNA originated in replication, not translation, *Proc. Natl. Acad. Sci. USA* **1994**, *91*, 6729–6734.

Martin, W., Russell, M. J. On the origin of cells: a hypothesis for the evolutionary transformation from abiotic geochemistry to chemoautotrophic prokaryotes, and from prokaryotes to nucleated cells, *Phil. Trans. R. Soc. Lond. B.* **2002**, *358*, 59–85 (also in: www.gla.ac.uk/project/originoflife/html/2001/pdf_files/Martin_&_Russell.pdf).

Mojzsis, S. J., Arrhenius, G., Keegan, K. D., Harrison, T. M., Nutman, A. P., Friend, C. R. L. Evidence for life on Earth before 3,800 million years ago, *Nature* **1996**, *384*, 55–59.

Orgel, L. E. Prebiotic chemistry and the origin of the RNA world, *Crit. Rev. Biochem. Mo.* **2004**, *39*, 99–123.

Pennisi, E. Is it time to uproot the tree of life, *Science* **1999**, *284*, 1305–1307.

Poole, A., Penny, D., Sjöberg, B-M. Methyl-RNA: an evolutionary bridge between RNA and DNA? *Chem. Biol.* **2000**, *7*, R207–R216.

Poole, A., Jeffares, D., Penny, D. Early evolution: prokaryotes, the new kids on the block, *BioEssays* **1999**, *21*, 880–889.

Poole, A. M., Jeffares, D., Penny, D. The path from the RNA world, *J. Mol. Evol.* **1998**, *46*, 1–17.

Sanders, P. G. H. Chirality in the RNA world and beyond, *Int. J. Astrobiol.* **2005**, *4*, 49–61.

Schmidt, G., Christensen, L., Nielsen, P. E., Orgel, L. E. Information transfer from DNA to peptide nucleic acids by template-directed syntheses, *Nucleic Acids Res.* **1997**, *25*, 4793–4796.

Steitz, T. A., Moore, P. B. RNA, the first macromolecular catalyst: the ribosome is a ribozyme, *Trends Biochem. Sci.* **2003**, *28*, 411–418.

Tian, F., Toon, O. B., Pavlov, A. A., De Streck, H. The hydrogen-rich early atmosphere, *Science* **2005**, *308*, 1014–1017.

Vlassov, A. V., Johnston, B. H., Landweber, L. F., Kazakov, S. A. Ligating activity of fragmented ribozymes in frozen solution: implications for the RNA world, *Nucl. Acids Res.* **2004**, *32*, 2966–2974.

Wächtershäuser, G. From pre-cells to Eukarya – a tale of two lipids, *Mol. Microbiol.* **2003**, *47*, 13–22.

Watson, E. B., Harrison, T. M. Zircon thermometer reveals minimum melting conditions on earliest Earth, *Science* **2005**, *308*, 841–844.

Weiner, A. M., Maizels, N. The genomic tag hypothesis: modern viruses as molecular fossils of ancient strategies for genomic replication, and clues regarding the origin of protein synthesis, *Biol. Bull.* **1999**, *196*, 327–330.

Westall, F. Life on the early Earth: A sedimentary view, *Science* **2005**, *308*, 366–367.

Woese, C. R. The universal ancestor, *Proc. Natl. Acad. Sci. USA* **1998**, *95*, 6854–6859.

Woese, C. R., Kandler, O., Wheelis, M. L. Towards a natural system of organisms: Proposal for the domains Archaea, Bacteria and Eucarya, *Proc. Natl. Acad. Science USA* **1990**, *87*, 4576

Yang, S., Doolittle, R. F., Bourne, P. E. Phylogeny determined by protein domain content, *Proc. Natl. Acad.Sci. USA* **2005**, *102*, 373–378.

4.9 Questions for Students

Question 4.1

What are the difficulties in understanding of the origin of life?

Question 4.2

What are the building blocks of life and how do they assemble to make functional cells?

Question 4.3

What are the putative steps required for the establishment of self-sustainable life?

Question 4.4

Explain the RNA world.

Question 4.5

What kind of conditions might have driven the prebiotic molecular evolution towards life?

Question 4.6

What can we say of the LUCA and of the differentiation of the three domains of life from LUCA?

Question 4.7

How do you define the beginning of life?

5
Extremophiles, the Physicochemical Limits of Life (Growth and Survival)

Helga Stan-Lotter

This chapter covers extreme environments, characterized by physical and chemical parameters that had been considered to lie outside the range suitable for life. In recent years, the notion of what constitutes life-limiting conditions in an environment has undergone dramatic changes: microorganisms and occasionally higher organisms have been discovered that not only tolerate but thrive under ranges of temperatures, pH values, pressures, salinity, ionizing radiation, etc., previously thought to destruct biomolecules and organisms. The presence of liquid water is a prerequisite for growth under all kinds of conditions, but survival of resting stages, such as spores or dormant forms, was shown to be possible under vacuum and in space for several months. The detection of numerous types of viable prokaryotes in subterranean locations, such as granite, ancient halite, and sediments, suggests the possibility of even more extensive longevity, namely, over geological time periods, i.e., over millions of years. The several-fold implications of these discoveries for the search for extraterrestrial life are discussed.

5.1
A Brief History of Life on Earth

5.1.1
Early Earth and Microfossils

The Earth is about 4.6 billion years old; while no rocks have yet been found that date back to the origin of the Earth, the most ancient rocks were dated as 3.5–3.86 billion years old. They are located in southern Africa (Swaziland and Barberton Greenstone Belt), western Australia (Warrawoona series and Pilbara formation), and Greenland (Itsaq gneiss complex). Besides volcanic and carbonaceous rocks,

Complete Course in Astrobiology. Edited by Gerda Horneck and Petra Rettberg

ISBN: 978-3-527-40660-9

the sedimentary rocks are of particular interest because their formation required liquid water. Sedimentary rocks of about 3.8 billion years have been discovered (Isua, Greenland), which implies that liquid water must have been present at that time, probably in the form of oceans, and therefore the conditions for life (see Chapter 6) did exist then.

The evidence for early microbial life rests on the fossilized remains of cells in ancient rocks and on an abundance of the lighter carbon isotope (^{12}C), due to its preference over the heavier isotope (^{13}C) by microorganisms. Microfossils found in old sedimentary rocks are similar in shape, size (a few micrometers in length), and arrangements to modern bacteria; the earliest forms were probably cocci (spheres) and short straight or curved rods (Fig. 5.1).

Other somewhat larger microfossils are filamentous structures from 3.5-billion-year-old Australian sediments (Pilbara), which were deemed to be cyanobacteria-like; however, these have recently come under scrutiny because it was suggested that the fossils could perhaps be younger than the surrounding sediments where they were detected.

The environments on the early Earth were different from those of today, and temperatures were likely much higher than now. It is thought that surface temperatures during the first 200 million years exceeded 100°C. The accumulation of water thus occurred only later, when the Earth was cooling down. How much time exactly this process took is not known; if the first life forms originated at the end of the first half-million years – as the microfossils suggest – Earth was still fairly hot, and therefore the first microorganisms were likely thermophiles, or at least thermotolerant. This hypothesis is surprisingly well supported by current knowledge about the evolutionary relationships between organisms (see Section 5.2.4.3).

The atmosphere of the early Earth did not contain substantial amounts of oxygen; the main gases were carbon dioxide, nitrogen, and probably also methane, carbon monoxide, ammonia, hydrogen, and hydrogen sulfide. Only much later was oxygen produced by phototrophic bacteria, and about 2 billion years ago its concentration in the atmosphere rose slowly to reach the present value of 20.95%. Thus, early life forms were anaerobes; even now, there are still many niches where anaerobic microorganisms thrive. Larger unicellular organisms did not appear before about 1 billion years ago, as was deduced from microfossils. Multicellular organisms probably did not appear before about 600–700 million years ago. It is

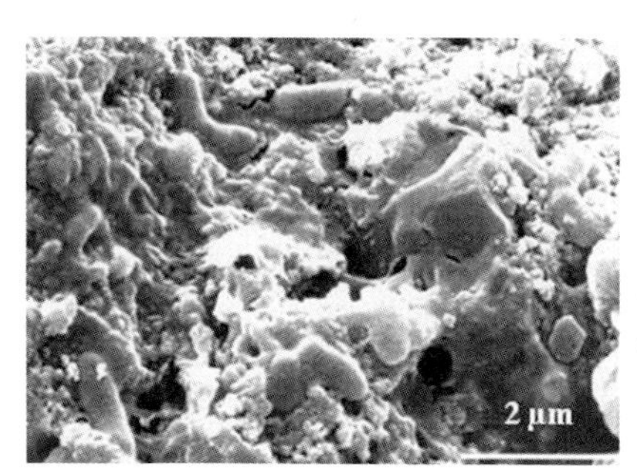

Fig. 5.1 Microfossils from the 3.5-billion-year-old Josefdal Chert in Barberton, South Africa. Scanning electron micrograph showing bacteria-like rods and vibrios (curved rods) that are attached to mineral matter (Courtesy of Dr. Frances Westall, CNRS Orléans).

evident, then, that life has been present on Earth throughout most of its history and that life was only microbial for long periods of geological time.

5.1.2
Prokaryotes, Eukaryotes, and the Tree of Life

Living organisms consist of cells, and much of what is known today about cell structure is due to information from microscopy. Light microscopes can magnify samples at least 1 000-fold, while electron microscopes magnify up to 200 000-fold. The precursor of the light microscope was invented only in the 17th century, and the electron microscope was developed in the 20th century. Cell biology and microbiology are thus very young sciences, compared with astronomy, which may be the oldest science of humankind.

Microscopic examination revealed early on the presence of two fundamental types of cells: eukaryotic and prokaryotic cells, thus named to denote the presence or absence of a cell nucleus, a feature that was easily identifiable. Figure 5.2 shows the general scheme of a eukaryotic and a prokaryotic cell. The nucleus encloses the genetic material of the eukaryotic cell, which is present in the form of chromosomes; prokaryotic cells contain their genetic material as an aggregated mass of DNA, which is called a nucleoid. Other differences between the two types of cells are size differences (eukaryotic cells are on average 10–20 times larger than the typical prokaryotic cell) and the presence of organelles in the eukaryotes. Organelles are small membrane-enclosed structures within cells that have been traced back to bacteria-like precursors. They were incorporated into early eukaryotic cells and became mitochondria and chloroplasts (the latter occur only in photosynthetic cells of plants and algae); they multiply within the host cells and exist in a symbiotic relationship. Prokaryotes are simpler cellular entities; besides the lack of a nucleus and the lack of organelles, their morphology, protein synthesis, cell propagation, and genetic apparatus are less diverse and somewhat more streamlined than the comparative features of eukaryotic cells. On the other hand, prokaryotes display a stunning diversity of metabolic functions; they can exist in a wide range of

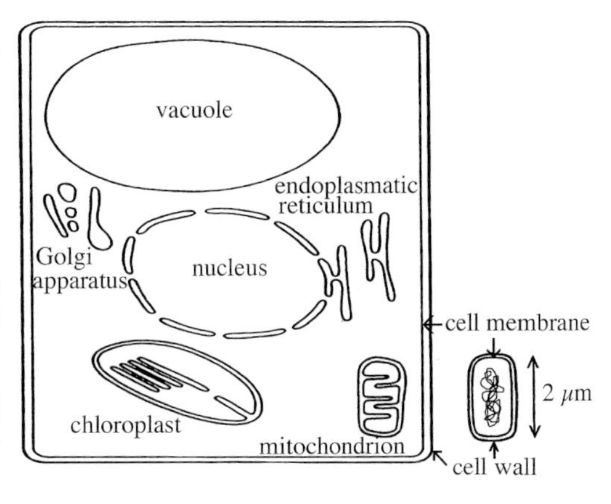

Fig. 5.2 Schemes of a prokaryotic cell (right) and a eukaryotic cell from a plant (left). Differences in sizes are shown, as well as the presence of organelles, nucleus, and inner compartments (Golgi apparatus, endoplasmatic reticulum) in the eukaryotic cell. Mitochondria and chloroplasts are similar in size to prokaryotic cells and possess inner membranes. The DNA of prokaryotes is a coiled structure inside the cell.

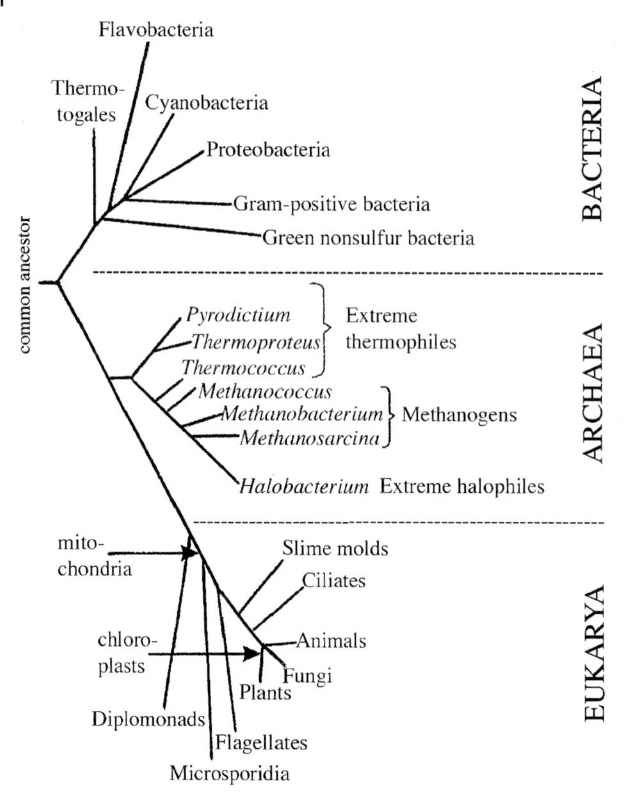

Fig. 5.3 Phylogenetic tree of all organisms based on sequences of small ribosomal RNAs. Prokaryotes form two lineages (Archaea and Bacteria; formerly called archaebacteria and eubacteria) and eukaryotes form one lineage. The length of the branches represents the approximate evolutionary distances between groups, based on the number and positions of different bases in their rRNAs.

inhospitable environments where eukaryotes would not be able to survive. Their influence on all processes of the biosphere is probably still severely underestimated; their biomass alone is thought to be equal to or even greater than the total plant mass on Earth.

Occasionally, eukaryotes are also very tolerant of extreme conditions, in particular, unicellular eukaryotes; however, this chapter will focus on prokaryotes both because they are the record holders for tolerating physicochemical extremes and for evolutionary reasons: as pointed out, the geologically long dominance of prokaryotic life on Earth may also be valid for other planets or celestial bodies.

Relationships between organisms and knowledge about their evolution have now been obtained by the easy availability of nucleic acid sequences. Figure 5.3 shows a so-called phylogenetic tree, which is based on the sequences of an important molecule present in similar form in all organisms, the ribonucleic acid (RNA) of the small subunit of ribosomes. The greater the difference in the sequences between the molecules from two or more organisms, the greater their evolutionary distance is. This is reflected in the length of the "branches" of the tree. Three main lineages of organisms have emerged from these analyses: the eukaryotes (Eukarya), which include all animals, plants, fungi, and many unicellular microorganisms, and two lineages of prokaryotes, the Bacteria (also called eubacteria) and the Archaea (or archaebacteria). The length of the archaeal branch (i.e., the branch connecting the last common ancestor of all Archaea to the point of bifurcation) is shorter than the bacterial and eukaryal branches (Fig. 5.3). Thus, Archaea have more similarity to Bacteria or Eukarya than both of them have to each other, in

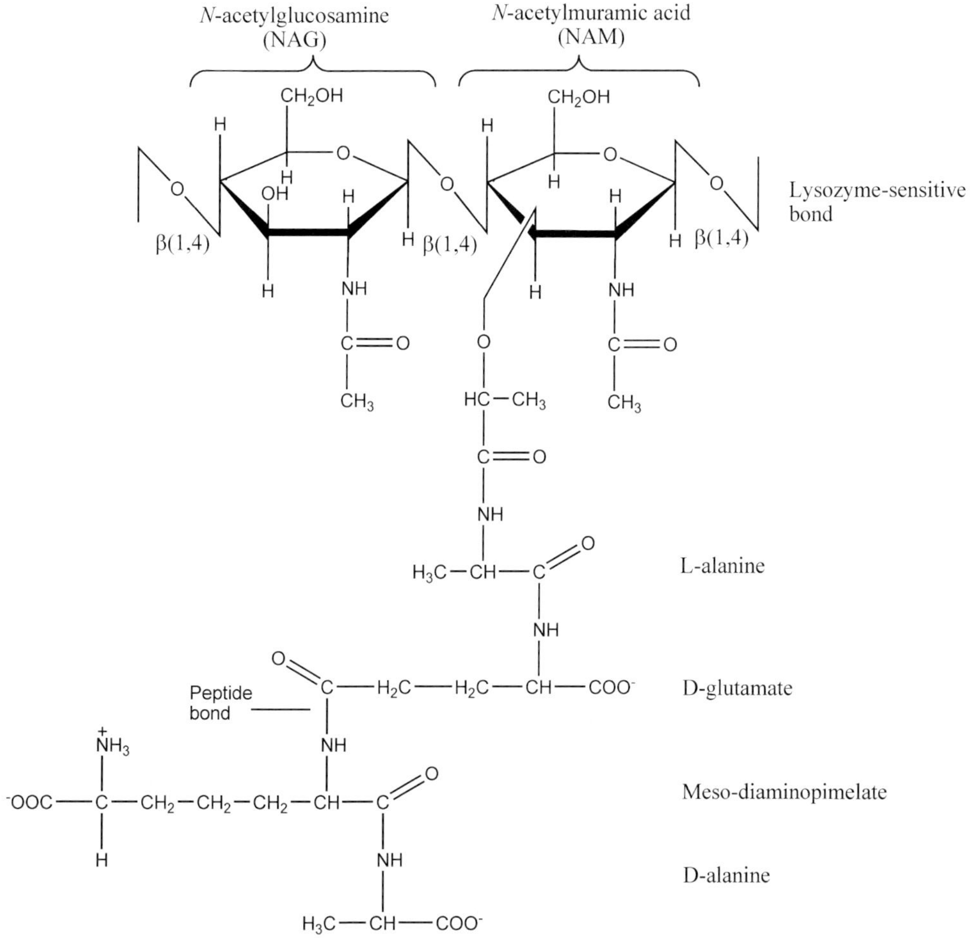

Fig. 5.4 Chemical structure of a repeating unit in peptidoglycan. *N*-acetylglucosamine (NAG) and *N*-acetylmuramic acid (NAM) are connected by the β 1,4 glycosidic bond, which is the target of the bactericide lysozyme. The amino acids (D- and L-isomers) in cross-linking peptides are characteristic of gram-negative bacteria (e.g., *Escherichia coli*); gram-positives replace diaminopimelate with L-lysine and an interbridge of five glycines.

good agreement with the finding that Archaea exhibit a mixture of eukaryotic and bacterial traits at the molecular level (see also Chapter 4).

Nucleic acid sequences also showed clearly the similarity between bacteria and organelles; thus, the uptake of the forerunners of mitochondria is assumed to have occurred about 1.9 billion years ago, while the uptake of photosynthetic endosymbionts, which became modern chloroplasts, occurred somewhat later in Earth's history.

5.1.3
Some Characteristics of Bacteria and Archaea

5.1.3.1 Cell Walls, Envelopes, and Shape

Almost all bacterial cells have walls that contain peptidoglycan, a molecule that is unique for the lineage Bacteria, as the cell wall polymer (Fig. 5.4). This compound consists of repeating units of *N*-acetylmuramic acid (NAM) and *N*-acetylglucosamine (NAG), which are connected by a β 1,4 glycosidic bond. Periodically, the strings of carbohydrates are cross-linked by short peptides, which contain L- and also D-amino acids. This arrangement is the basis for the high mechanical strength of peptidoglycan. NAM is found only in Bacteria; D-isomers of amino acids are rare in other organisms (although higher organisms, including humans, possess some D-amino acids), but in Bacteria they can be present in up to 25% (weight) of their cell walls.

Archaea are considerably more diverse in the composition of their cell envelopes. They lack peptidoglycan, but some species of the order Methanobacteriales contain pseudopeptidoglycan, which has a similar structure: *N*-acetyltalosaminuronic acid replaces *N*-acetylmuramic acid in the backbone of the molecule, and each carbohydrate unit is linked together using β 1,3 glycosidic bonds instead of the β 1,4 gly-

Prokaryotic shapes

Spheric Cylindrical Spiral

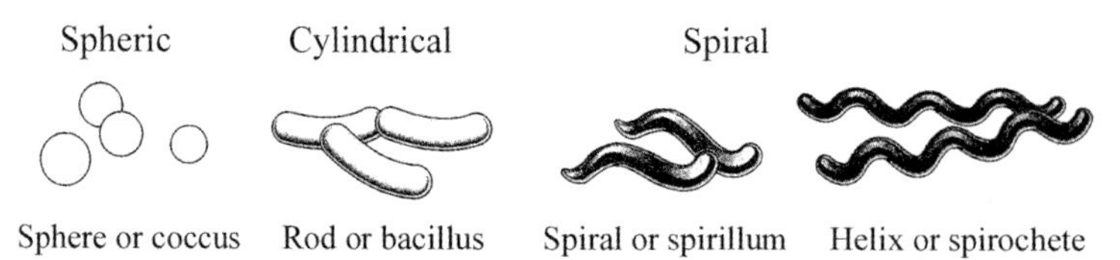

Sphere or coccus Rod or bacillus Spiral or spirillum Helix or spirochete

Prokaryotic arrangements

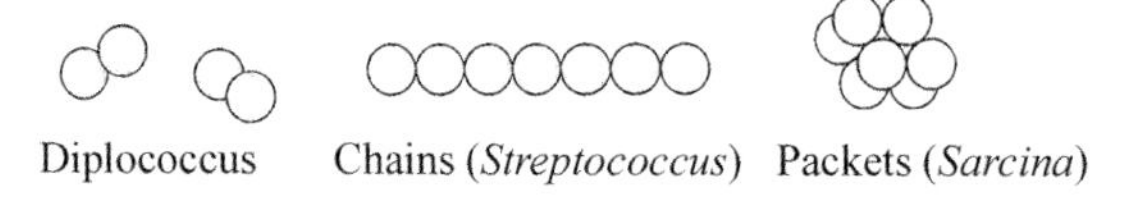

Diplococcus Chains (*Streptococcus*) Packets (*Sarcina*)

Fig. 5.5 Scheme of representative cell shapes and arrangements of prokaryotes. Diameters of cocci are in the range of 0.6–1.4 μm; lengths of rods and spirals are 2–5 μm; filamentous prokaryotes (not shown) may reach lengths of more than 100 μm.

cosidic bonds seen in peptidoglycan. The majority of Archaea does not contain a peptidoglycan molecule in any form, instead covering the outside of the membrane with proteins or glycoproteins, in some cases polysaccharides. These types of cell envelopes are referred to as S-layers (surface layers). The function of the S-layer is similar to that of a cell wall – to contain the cytoplasm and to give the microbe its shape. The most representative shapes of prokaryotes and some typical arrangements of cells are shown in Fig. 5.5; these features are principally the same in Archaea and in Bacteria.

5.1.3.2 Lipids and Membranes

The lipids from Archaea show some significant differences from those of Bacteria and Eukarya. Instead of ester bonds between glycerol and fatty acids (Fig. 5.6C), lipids from Archaea possess ether bonds between glycerol and their hydrocarbons (Fig. 5.6A, B); they do not possess fatty acids but rather isoprenoid chains, which usually consist of either 20 (Fig. 5.6A) or 40 (Fig. 5.6B) carbon atoms.

The general architecture of the cytoplasmic membrane is the same in all organisms. It includes a lipid bilayer, where the inner core is made up of hydrophobic chains, and both outer surfaces are hydrophilic. However, the glycerol tetraether (Fig. 5.6B) forms a monolayer instead of a bilayer because the isoprenoid side chains are covalently linked. Such monolayers are much more resistant towards higher temperatures and pH extremes; they are the predominant membrane structure of hyperthermophilic Archaea.

5.2 Extremophiles and Extreme Environments

The Archaea (Fig. 5.3) consist of three main groups: extreme halophiles, extreme thermophiles, and methanogens. Many Archaea are thus extremophiles, since they live in extreme environments. When attempting to define an extreme environment, it becomes obvious that a strongly anthropocentric component emerges: extreme environments were generally considered "hostile to higher forms of life" and "uninhabitable by other organisms." Thomas D. Brock, a microbiologist who pioneered the study of life in thermal springs, defined extreme habitats as follows: "... they are environments with a restricted species diversity and the absence of some taxonomic groups" (Brock 1986). Why is there a limited diversity, and why are some groups missing?

An examination of the physicochemical parameters characteristic for extreme environments (Table 5.1) shows that they delineate niches that allow inhabitation by only certain groups of organisms; none can be expected to survive the whole range of conditions. A qualitative expression of this fact was stated by the chemist Justus von Liebig in 1842 in his "law of the minimum," albeit developed for plants, which states that (plant) growth will continue as long as all required factors are present (e.g., nitrogen, phosphorus, potassium, and so forth). When one of those

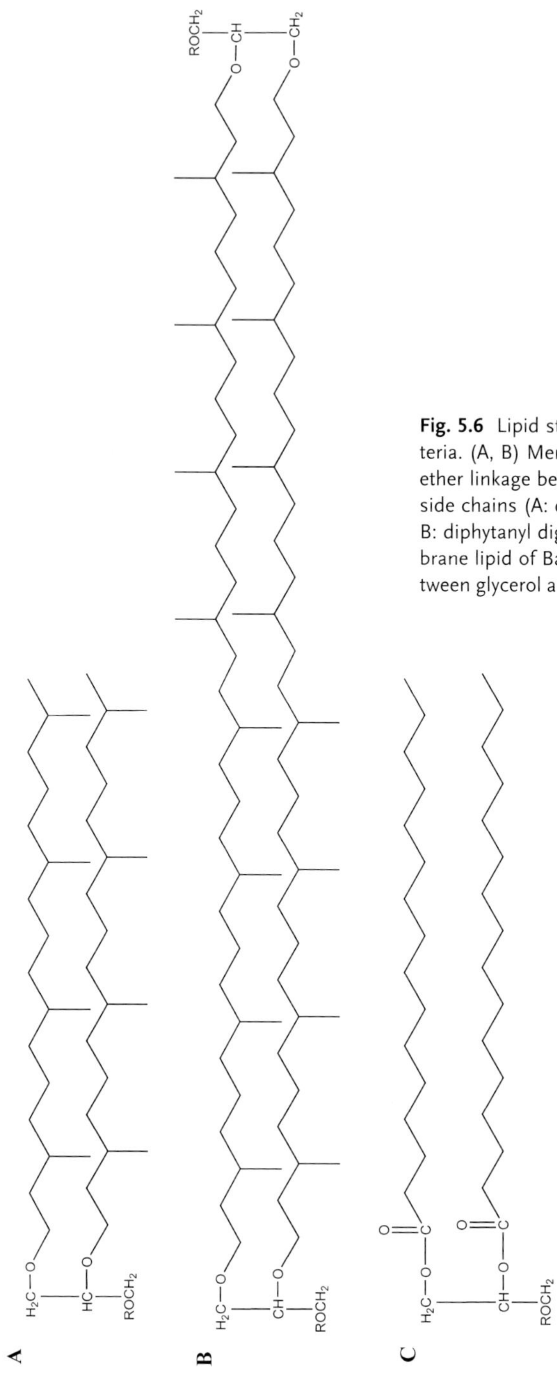

Fig. 5.6 Lipid structures of Archaea and Bacteria. (A, B) Membrane lipids of Archaea with ether linkage between glycerol and isoprenoid side chains (A: diphytanyl glycerol diether; B: diphytanyl diglycerol tetraether) (C) Membrane lipid of Bacteria with ester linkage between glycerol and fatty acids (glycerol diester).

factors is depleted, growth stops. Increasing the amount of the limiting component will allow growth to continue until that component (or another) is depleted.

Table 5.1 Classes and examples of extremophilic prokaryotes.

Physicochemical factor	Descriptive term	Genus/ species	Lineage	Habitat	Minimum	Optimum	Maximum
High temperature	Hyperther-mophile	*Pyrolobus fumarii*	Archaea	Hydrother-mal	90 °C	106 °C	113 °C
High temperature	Hyperther-mophile	Strain 121	Archaea	Black smoker	85 °C	?	121 °C
Low temperature	Psychrophile	*Polaromonas vacuolata*	Bacteria	Sea ice	0 °C	4 °C	12 °C
Low pH	Acidophile	*Picrophilus oshimae*	Archaea	Acidic hot spring	–0.06	0.7	4
High pH	Alkaliphile	*Spirulina* sp.	Bacteria	Alkaline lake	8	9	12
Hydrostatic pressure	Barophile	*Moritella yayanosii*	Bacteria	Ocean sediment	500 atm	700 atm	1 100 atm
Salt (NaCl)	Halophile	*Halobacterium salinarum*	Archaea	Saltern	15%	25%	32%

In 1913, an early ecologist, Victor Shelford, presented in his "law of tolerance" the observation that organisms will usually be limited by abiotic factors. Each particular factor that an organism responds to in an ecological system has what he called limiting effects. The factors function within a range, i.e., a maximum and a minimum value for the factors exists, which Shelford designated "limits of tolerance" (see also Table 5.1). For an organism to succeed in a given environment, each of a complex set of conditions must remain within the tolerance range of that organism, and if any condition exceeds the minimum or maximum tolerance of that organism, it will fail to thrive.

Recent years have seen an unprecedented expansion of knowledge about the physicochemical limits of life. For instance, life beyond the boiling point of water was unthinkable well into the 1970 s; now, numerous microorganisms have been described that flourish at temperatures up to 113 °C and possibly even at 121 °C. Table 5.1 shows some examples of current record holders with respect to growth in extremes of temperatures, salinities, pH values, and pressure.

5.2.1 Growth versus Survival

A distinction is to be made when life in extreme environments and extreme conditions is described: growth of organisms under extreme conditions shall mean the maintenance of an active population that finds its optimum or near-optimum conditions in those environments (see Table 5.1).

In contrast to growth, there is mere survival of environmental extremes for short or even extended periods of time. Such survival is possible by organisms themselves or, more often, by special forms that are produced by organisms when they encounter environmental extremes. For instance, spores or endospores are made by fungi or bacteria, cysts are produced by various eukaryotes, and seeds are produced by plants; all these forms are usually unicellular bodies (seeds are the exception) that can be encased by a thick protective coat, and thus these resting stages are very resistant to environmental extremes such as heat, desiccation, radiation, and so on (see Section 5.3). The survival of extreme conditions, without real growth, is a phenomenon of great importance to astrobiological goals and plans, e.g., when forward contamination of other celestial bodies is considered (see Chapter 13). Table 5.3 contains some current records of microbial survival of extreme conditions, such as ionizing radiation, near vacuum, desiccation, and outer space.

5.2.2
The Search for Life on Mars: The *Viking* Mission

The *Mariner* 9 mission to Mars in 1971 provided convincing evidence that this planet once had flowing rivers and a denser atmosphere (see also Chapter 8). As a result of these data, the *Viking* missions were planned by NASA, with the aim of obtaining high-resolution images of the Martian surface, characterizing its structure and the composition of the atmosphere, and explicitly searching for evidence of life. The mission was composed of two spacecraft, *Viking 1* and *Viking 2*, each consisting of an orbiter and a lander (Fig. 5.7). *Viking 1* was launched on 20 August 1975 and arrived at Mars on 19 June 1976. The first month of orbit was devoted to imaging the surface to find appropriate landing sites for the *Viking* landers. On 20 July 1976, the *Viking 1* lander separated from the orbiter and touched down at Chryse Planitia. *Viking 2* was launched 9 September 1975 and entered Mars orbit on 7 August 1976. The *Viking 2* lander touched down at Utopia Planitia on 3 September 1976. The two landers conducted four experiments (Fig. 5.8) intended to detect the presence of microbiological life on the Martian surface; soil samples were retrieved by the landers' extendible arms.

Fig. 5.7 *Viking* lander on Mars in 1976 (source: http://www.msss.com/mars/pictures/viking_lander/viking_lander.html).

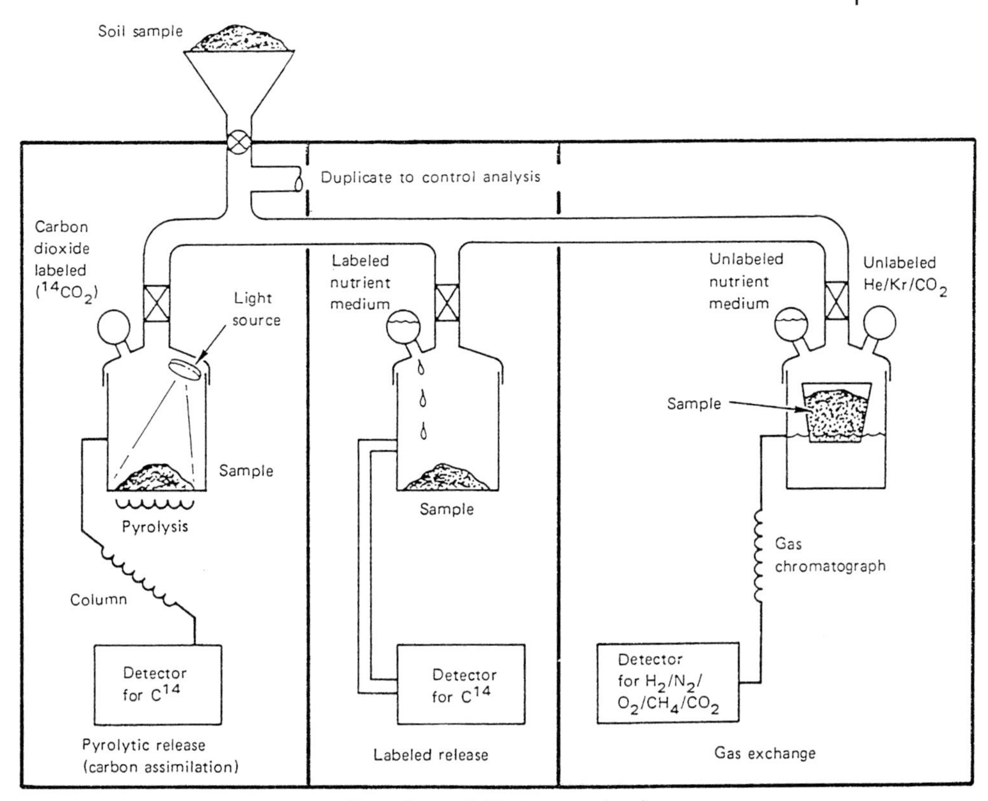

Fig. 5.8 Sketch of the three experiments of the *Viking* landers that searched for metabolic signatures of putative Martian microorganisms in surface samples of Martian regolith (source: *Viking 1,* Early Results, NASA SP-408, 1976).

The Gas Exchange Experiment (GEX) was looking for changes in the makeup of gases in a test chamber, i.e., changes that would indicate biological activity. The results from this test were taken to counter-indicate biology.

The Labeled Release Experiment (LR) was set up to detect the uptake of a radioactively tagged liquid nutrient by microbes. The idea was that gases emitted by any microbes would show the tagging. Initial results were in line with this prediction, but in the end the overall results were deemed inconsistent.

The Pyrolytic Release Experiment (PR) involved heating the soil samples that had been exposed to radioactively tagged carbon dioxide to see whether the chemical had been used by organisms to make organic compounds. Seven of nine experimental runs seemed to show small concentrations indicative of microorganisms, but the results were later discounted.

The Gas Chromatograph–Mass Spectrometer Experiment (GCMS) was looking for organic compounds in the soil but failed to detect any. It did appear that the surface of Mars was a fairly hostile environment, due to the solar ultraviolet radiation.

At a NASA conference to discuss these results, the following summary was provided as the then general consensus:

> *Viking not only found no life on Mars, it showed why there is no life.... reasons are the extreme dryness, the pervasive short-wavelength ultraviolet radiation.... Viking found that Mars is even dryer than had previously been thought.... The dryness alone would suffice to guarantee a lifeless Mars; combined with the planet's radiation flux, Mars becomes almost moon-like in its hostility to life.*

However, there are others who claim that the evidence was not so clear-cut. To this day, Gilbert Levin, who was the principal investigator for the LR experiment, maintains that the experimental results are easier to reconcile with the conclusion that life was detected rather than otherwise. Another consideration is that all the *Viking* experiments used samples from the uppermost Martian surface. But current thinking, from the new knowledge about life in extreme environments on Earth, holds that it is more likely that life would be extant in the deeper subsurface.

Many lessons were learned from the *Viking* experiments that no doubt will be incorporated into future plans for the search for life on Mars and other celestial bodies (see Chapter 12). The vast expansion of knowledge about the limits of microbial life since the *Viking* missions will be considered now.

5.2.3
Temperature Ranges for Microorganisms

Microorganisms can be grouped into broad, although not very precise, categories according to their temperature ranges for growth (Table 5.1):

- Psychrophiles (cold loving) can grow at 0 °C and even as low as –10 °C or –16 °C; their upper limit is often about 20–25 °C.
- Mesophiles grow in the moderate temperature range, from about 20 °C (or lower) to 45 °C.
- Thermophiles are heat loving, with an optimum growth temperature of 50 °C or more, a maximum of up to 70 °C or more, and a minimum of about 20 °C.
- Hyperthermophiles have an optimum above 75–80 °C and thus can grow at the highest temperatures tolerated by any organism. Some will not grow below 60 °C.

The temperature ranges for higher eukaryotes would be different – in accordance with Brock (1986), who defined a thermophile as "... an organism capable of living at temperatures at or near the maximum for the taxonomic group of which it is a part." Thus, a thermophilic vertebrate would be one capable of living near 37 °C, a thermophilic fungus one capable of living near 60 °C, a thermophilic cyanobacterium one capable of living near 70 °C, and so forth.

Temperatures in the range of 50–55 °C are widespread on Earth and are associated with sun-heated habitats, but temperatures higher than 55–60 °C are much rarer and are almost exclusively associated with geothermal habitats. Therefore, a "thermophile boundary" of 55 °C has sometimes been suggested for prokaryotes, beyond which thermophiles grow.

Hot springs and geothermal vents are found in several parts of the world, such as in Yellowstone National Park (Wyoming, USA), Iceland, southern Italy, New Zealand, and Japan and on the floors of the oceans, where spreading zones occur and superheated fluids are emitted.

5.2.4 High-temperature Environments

5.2.4.1 Geography and Isolates

Yellowstone National Park With more than 10 000 geothermal features including hot springs, mudpots, fumaroles, and geysers, Yellowstone National Park provides vast habitats for a wide variety of thermotolerant and thermophilic organisms. Yellowstone's geologic setting serves as the driving force behind its geothermal characteristics. The center of Yellowstone is a caldera (a large crater formed by volcanic explosion or by the collapse of a volcanic cone) that covers a major portion of the central area of the park. Volcanic activity below the caldera and the surrounding area releases tremendous heat to drive the hot springs, geysers, and other features. Groundwater from snowmelt and other precipitation in the greater Yellowstone area is heated by molten rock below the surface. Once heated, the water finds a path through cracks and fissures of lava flows to the surface, where it emerges in geysers, hot springs, mudpots, and fumaroles.

Many hot springs exist in the park; the largest is the Grand Prismatic Spring (Fig. 5.9). It is 90 m across and 50 m deep. In the center of the pool the water is 87 °C, with an acidic pH. In the cooler water along the edges of the pool, colonies of thermophilic cyanobacteria and algae thrive. The colors are due to the presence of photosynthetic (blue, blue-green, brownish) or carotenoid (orange, yellow, and brown) pigments. As a result, the pool displays a spectrum of colors

Fig. 5.9 Aerial view of Grand Prismatic Spring, Yellowstone National Park, Wyoming, USA. The spring is about 90 m across. Colors are due to pigmented thermophilic cyanobacterial, thermophilic bacterial, and archaeal species. Wooden trail for visitors is in the foreground.

from the bright blue of the center to the orange, red, and brown microbial mats along the edges.

Hyperthermophiles were first discovered in the 1960 s in the hot springs in Yellowstone National Park. Since then, more than 50 species have been described. The majority of these are Archaea, although some cyanobacteria and anaerobic photosynthetic Bacteria grow well at 70–75 °C. Members of the genus *Sulfolobus* (Archaea) are among the best-studied hyperthermophiles. They are commonly found in geothermal environments, with a maximum growth temperature of about 85–90 °C, an optimum of about 80 °C, and a minimum of about 60 °C. They also have a low pH optimum (pH 2–3); therefore, they are termed thermoacidophiles. *Sulfolobus* species gain their energy by oxidizing the sulfur granules around hot springs, generating sulfuric acid, and thereby lowering the pH.

Iceland Situated along the mid-ocean ridge of the Atlantic Ocean, Iceland is a geologic "hot spot" with volcanic eruptions, fissure eruptions, shield volcanoes, pillow basalts, and glaciers. Iceland's characteristics are driven by the nature of its geologic setting, a combination of tectonic movements of plates and a hot zone, which supplies molten material for frequent eruptions. Groundwater and sometimes seawater seep into the ground and travel through highly fractured bedrock. The water reaches varying depths, where it comes in contact with heat from volcanic sources. As the water is heated, it ascends through fissures, crevices, and volcanic crust to emerge in hot springs and other geothermal features. This island is also home to a variety of thermophilic life, which inhabits hot springs, mud pots, mud geysers (Fig. 5.10), and fumaroles.

Numerous hyperthermophiles were isolated in Iceland and named appropriately, such as *Pyrobaculum islandicum* and *Thermodesulfovibrio islandicus*. Interestingly, thermophilic viruses that attack the archaeal genera *Sulfolobus* and *Thermoproteus* and are active only at temperatures above 80 °C were first isolated from Icelandic hot springs.

Italy The south of Italy is a zone of intense volcanic activity, with the well-known Mount Etna erupting frequently and producing lava flows of various magnitudes. The island of Vulcano is the southernmost of the Aeolian Islands. It last erupted in

Fig. 5.10 Mud geyser, Iceland (source: Hroarson and Jonsson 1992, with permission).

1888–1890 and presently hosts a vigorous fumarole field on the summit crater, with temperatures inside reaching 400 °C. Lower-temperature fumaroles are found around the beach areas, and soil degassing is prevalent on and around the flanks of the main cone. The gas geochemistry is quite diverse and complex: H_2S, N_2O, H_2O, CH_4, and He are spewed out, and the composition of liquids can be equally rich (boron, sodium, magnesium, vanadium, arsenic, zinc, iodine, antimony, rubidium, and others). Many thermophiles were isolated from the cracks and fissures of the rocks below the seawater level, at temperatures just above the boiling point (103 °C). Hyperthermophilic isolates from these areas include *Pyrococcus furiosus*, which thrives at 100 °C, *Acidianus infernus*, and the genus *Pyrodictium;* its members have a temperature minimum of 82 °C, an optimum of 105 °C, and a growth maximum of 110 °C.

Deep hydrothermal vents In great depths of the deep sea, hydrostatic pressures are high and water does not boil – at a depth of about 2 500 m – until it reaches a temperature of about 450 °C. At certain sites, where the sea floor spreads due to hot basalt and magma, water seeps into the cracked regions and underwater springs develop. These are called hydrothermal vents. Hydrothermal fluids are emitted at temperatures of 270–380 °C; upon cooling, the metal ions from the mineral-rich hot water precipitate and form a chimney. No microorganisms have been found within the superheated fluids, but a temperature gradient forms when mixing with cold ocean water occurs, and the walls of such chimneys – called black smokers, because they emit dark, mineral-containing clouds – are teeming with hyperthermophilic prokaryotes (e.g., species of *Methanopyrus*, *Methanococcus*, and *Pyrodictium*). *Pyrolobus fumarii*, the current classified hyperthermophilic record holder with a growth maximum of 113 °C, was isolated from an Atlantic hydrothermal vent. Recently, an isolate was obtained from a black smoker and termed "Strain 121"; it was able to double its population during 24 hours in an autoclave at 121 °C, hence its name.

Many hyperthermophiles oxidize sulfur, hydrogen sulfide, or thiosulfate and are capable of fixing carbon dioxide. Some vents contain nitrifying bacteria; hydrogen-, iron-, or manganese-oxidizing bacteria; or methylotrophic bacteria. These latter types grow on methane and carbon monoxide, which is emitted from the vents. Black smokers are temporary structures; after several years, they become plugged up, and new smokers form at other locations. Presumably, the prokaryotes follow them and colonize the new chimney walls. Although they require very high temperatures for growth, hyperthermophiles can tolerate cold temperatures remarkably well, and some have even been isolated from Alaskan ocean waters.

Coal refuse piles Self-heating coal refuse piles occur due to spontaneous combustion and can smolder for long times. In these habitats, strains of *Thermoplasma* were found that apparently metabolize organic compounds leached from the coal refuse. The species *Thermoplasma acidophilum* is a thermophile (optimum growth at 55 °C) and also an acidophile (optimum pH 2.0); it became notable because it lacks a cell wall or envelope and is enclosed in a membrane only, similar to the parasitic bacteria of the genus *Mycoplasma*. Mainly for this reason, *Thermoplasma* was considered a likely candidate for the archaeal precursor cell, which would take up smaller bacterial cells as symbionts and evolve into a true eukaryote. However, this idea will have to be abandoned, because in the recently completed genome sequence of *Thermoplasma*, no genes for a pre-nucleus, a cytoskeleton, or a pre-chromatin structure were detected, as would have been expected.

5.2.4.2 **Molecular Properties of Hyperthermophiles**

How can hyperthermophiles tolerate the extreme temperatures that would kill all other organisms instantly? No single cellular feature is responsible for extreme heat resistance; instead, several biomolecular and mechanistic characteristics have been identified that appear to be present in most hyperthermophiles.

In electron microscopic thin sections of hyperthermophiles, a pronounced S-layer of a considerably higher thickness than that of S-layers from mesophiles is visible. *Sulfolobus* cells and the related *Acidianus* possess particularly prominent envelopes, as do members of the Desulfurococcales (genera *Pyrodictium, Pyrolobus, Ignicoccus, Staphylothermus*). Presumably, these envelopes, which consist of a thick protein layer, protect the cells against the acidic environment.

The amino acid composition of proteins from hyperthermophiles is quite similar to that from non-thermophiles and comprises the same 20 amino acids. Features that increase the thermostability of proteins are highly hydrophobic cores, more ionic interactions on the surfaces, and a generally more compact protein structure, which hinders unfolding; small changes in amino acid composition are often sufficient to affect significantly higher cohesion of protein structure.

The ether-linked lipids that all Archaea have in their membranes (Fig. 5.6) are generally more stable towards hydrolysis by heat, acids, or bases than ester-bound lipids are. Hyperthermophiles possess almost exclusively tetraether structures (Fig. 5.6), which result in a monolayer of lipids instead of a bilayer. In addition, five-membered ring structures are found within the tetraether lipids; their number increases with increasing growth temperature. These features confer more rigidity and less fluidity to the membranes of hyperthermophiles.

DNA stability at high temperatures is probably mediated by several mechanisms. The enzyme "reverse DNA gyrase" is produced by all hyperthermophiles; this enzyme introduces positive supercoils into DNA (in contrast to the negative supercoils in the DNA from non-hyperthermophiles), which has a stabilizing effect on DNA towards heat and denaturation. Small DNA-binding proteins have been found in *Sulfolobus* that are involved in maintaining the integrity of duplex DNA at high temperatures. Histone-like proteins (similar to those in eukaryotes), which cause compaction and nucleosome-like structures by interaction with DNA, are present in most Archaea and have been shown to maintain DNA in a double-stranded form at high temperatures.

More difficult to envisage is the heat stability of small essential molecules, such as adenosine triphosphate (ATP) and nicotinamide adenine dinucleotide (NAD); these are known to hydrolyze rapidly (within minutes) at 120 °C, at least in aqueous solutions. It is possible that currently undetected heat-protective agents are present in the cytoplasm of hyperthermophiles.

Although no report has yet been made about life at temperatures above 121 °C, its existence is possible (Strain 121 survived being heated to 130 °C for 2 h but was unable to reproduce until it had been transferred into a fresh growth medium, at a relatively cooler 103 °C). However, it is thought unlikely that microbes could survive at 150 °C or hotter, because the cohesion of DNA and other vital molecules would break down at this point.

5.2.4.3 Early Evolution and Hyperthermophiles

Aquifex and *Thermotoga* species are hyperthermophilic Bacteria that were isolated from the same environments as hyperthermophilic Archaea (mud volcanoes, hot seafloor sediments, some from oil-producing wells). Their optimum growth temperatures are 80–95 °C and they can grow up to 90–95 °C. A phylogenetic tree similar to that in Fig. 5.3, but with an emphasis on the lineages representing hyperthermophilic microorganisms, is shown in Fig. 5.11.

Within such trees, deep branches are evidence for early separation. For example, the separation of the Bacteria from the stem common to Bacteria and Eukarya

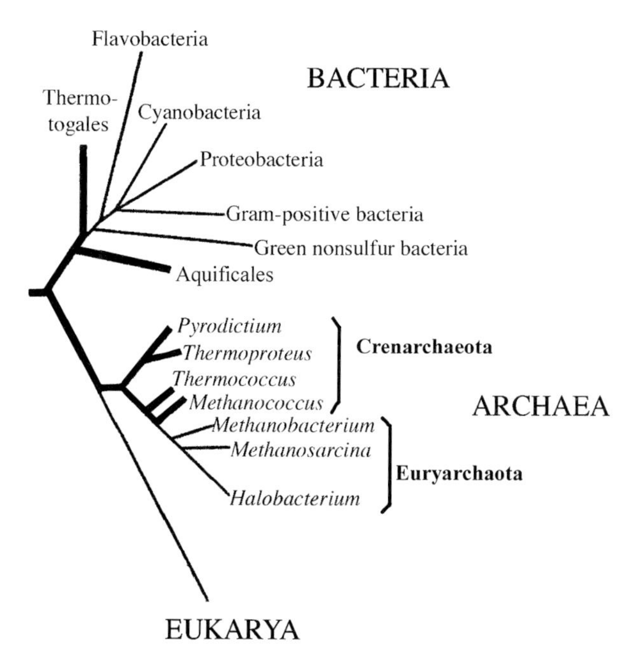

Fig. 5.11 Simplified phylogenetic tree, based on 16S rRNA sequences. Eukaryotes are represented by just one branch. Bold lines represent hyperthermophiles. The order and length of branches suggest a hyperthermophilic common ancestor for all life.

represents the deepest and earliest branching point. Short branches indicate a rather slow rate of evolution. In contrast to the Eukarya, the bacterial and archaeal lineages within this tree exhibit several deep and short branches. Based on the phylogenetic positions of the orders Aquificales and Thermotogales, a thermophilic ancestry of the Bacteria has been suggested. Hyperthermophiles are represented by all short and deep phylogenetic branches, which form a cluster around the phylogenetic root (Fig. 5.11, bold lines). The deepest and shortest phylogenetic branches in this tree belong to *Aquifex* and *Thermotoga* within the Bacteria and *Pyrodictium, Thermoproteus, Thermococcus,* and *Methanococcus* within the Archaea. From this distribution it can be cautiously deduced that the earliest branches of the tree (in bold) all contain hyperthermophiles; therefore, the idea that the first organisms on Earth might have been hyperthermophiles was proposed. This may also apply to life forms on other planets, where volcanism and high heat prevail or have prevailed.

The tree also depicts the introduction of two major groups for the classification of Archaea: "Crenarchaeota" (comprising all thermophiles and hyperthermophiles) and "Euryarchaeota" (methanogens and extreme halophiles).

5.2.4.4 Applications

The study of extreme environments has yielded considerable biotechnological applications. For example, the two thermophilic species *Thermus aquaticus* (a bacterium) and *Thermococcus litoralis* (an archaeon) are used as sources of the enzyme DNA polymerase, for the polymerase chain reaction (PCR) in DNA fingerprinting, and similar applications. The enzymes from these organisms are stable at the relatively high temperatures needed for the PCR process, which involves cycles of heating to break the hydrogen bonds in DNA and leave single strands that can be copied repeatedly. Another thermophilic bacterium, *Bacillus stearothermophilus* (temperature maximum 75 °C), has been grown commercially to obtain its heat-stable proteases and lipases; these enzymes are used in "biological" detergents.

5.2.5 Low-temperature Environments

5.2.5.1 Geography and Isolates

Most habitats on Earth are permanently cold. By volume, 90% of the world's oceans have a temperature of 5 °C or less, supporting both psychrophilic and psychrotolerant microorganisms. When terrestrial habitats are included, over 80% of the Earth's biosphere is permanently cold. Permafrost covers about one-fifth of the world's land surface and is present in most of Alaska, northern Canada, and Siberia, from depths of a few centimeters to 300 m. Even in many frozen materials, small pockets of liquid water are present where microorganisms can grow. Microorganisms that are able to grow at low temperatures are termed "psychrophilic" (sometimes also "obligate psychrophiles") and "psychrotolerant" (or "facultative psychrophiles"). An exact definition is often not possible, because microorganisms are adapted to low temperature to very different degrees and the composition of the environment can have a crucial influence.

Metabolic activity of natural communities of permafrost bacteria below the freezing point was detected recently under laboratory conditions. The uptake of labeled compounds was measured and found to occur at temperatures from 5 °C down to −20 °C. Concomitantly, growth of bacteria was observed at these temperatures; at −20 °C, the doubling time was 160 days.

Although the research with permafrost bacteria and other psychrophiles is often done with microbial communities, increasingly there are strains being isolated that have been taxonomically characterized. So far, psychrophilic bacteria belonging to four phylogenetic groups – the Alpha and Gamma subdivisions of the Proteobacteria, the Flexibacter-Bacteroides-Cytophaga phylum, and the gram-positive branch – have been described; isolated strains include *Arthrobacter* sp., *Psychrobacter* sp., and members of the genera *Halomonas, Pseudomonas, Hyphomonas,* and *Sphingomonas.*

A particularly interesting cold spot is Lake Vostok in Antarctica, which was discovered by the Russian Academy of Sciences, who announced in 1996 that a large lake of liquid water lies beneath the 3 750-m deep glacier at their research station Vostok. The existence of the lake was confirmed by radar imagery and space-borne radar altimetry; its size was estimated to be about 14 000 km^2, which is about as large as Lake Ontario. The water's depth varies from 200 m to 800 m and the total water volume is estimated to be 5 400 km^3. It has not frozen over because the Earth's internal heating warms the bottom and the ice acts like an insulator; the water temperature is about −3 °C. Ice samples from cores drilled close to the top of the lake have been analyzed to be as old as 420 000 years, suggesting that the lake has been sealed under the icecap for between 500 000 and more than one million years. The water is probably different from that of a surface lake, because over the centuries, air bubbles from the ice above have been released into the water, greatly increasing its level of dissolved oxygen. It has been speculated that if the ice were removed, the water would bubble and spout like a bottle containing a carbonated drink, with unforeseeable consequences.

Several microorganisms have already been isolated from the ice sheet, but there has been no breach yet to the surface of the lake. Lake Vostok may be an excellent analogue for extraterrestrial environments of icy worlds in the Solar System, such as the Saturnian moon Enceladus and Jupiter's moon Europa (see Chapter 10). Methods are being developed to ensure that no contamination will occur when drilling down to Lake Vostok is attempted (see Chapter 13); such methods will also be needed for drilling through the presumed ice cover of Europa.

5.2.5.2 Molecular Adaptations

Psychrophilic bacteria have adapted to their cold environments by producing largely unsaturated fatty acids for the lipids in their plasma membranes. Some psychrophiles have been found to contain polyunsaturated fatty acids, which generally do not occur in prokaryotes. Enzymes from psychrophiles continue to function, albeit at a reduced rate, at near-freezing ambient temperatures. The proteins from psychrophiles are, as far as is known, not as rigid as their mesophilic counterparts, because of a high content of α-helix structures and a very low content of β-sheets.

5.2.6 Barophiles or Piezophiles

Organisms that live in the deep oceans are not only psychrophilic – with the exception of the hyperthermophilic vent populations – but also barophilic (or piezophilic). The hydrostatic pressure in the oceans increases by 1 000 hPa (1 atm) for every 10 m of depth. The maximum depth is in the Mariana Trench; with its 10 900-m depth, the pressure is about 1.1×10^6 hPa (1 100 atm). Even from this site, viable microorganisms have been isolated. The extremely barophilic bacterium *Moritella yayanosii* is unable to grow at pressures below 500 000 hPa (500 atm); its optimum growth occurs at 700 000–800 000 hPa (700–800 atm), and it can grow at pressures up to 1.035×10^6 hPa (1 035 atm), which is the pressure of its natural habitat. Other isolates from the ocean floors and sediments of about 4 000–6 000 m depth grow optimally at pressures of 400 000–600 000 hPa (400–600 atm) but have usually retained the capacity for growth at a normal pressure of 1 000 hPa (1 atm).

Proteins do not appear to be so greatly affected by high pressures, because cells that were grown at high or low pressure showed nearly the same protein composition, with some exceptions. When cells were grown at high pressures, the membranes and particularly the lipids, which contained a very high proportion of unsaturated fatty acids, appeared to be more affected. Presumably, this composition prevents gelling of membranes and ensures retainment of fluidity.

5.2.7 High-salt Environments

5.2.7.1 Hypersaline Environments and Isolates

Halophilic Archaea (traditionally called halobacteria; now haloarchaea is preferred) thrive in environments with salt concentrations approaching saturation, such as natural brines, the Dead Sea, alkaline salt lakes, and marine solar salterns; they have also been isolated from rock salt of great geological age (about 250 million years). The halobacteria are a group of microorganisms with many unusual features: besides growth at high-salt concentrations, they possess striking colors of red, orange, or purple because of their pigments; obligately salt-dependent enzymes; and, in some species, the proton pump bacteriorhodopsin, which is driven only by sunlight. Figure 5.12 shows a commercial salt-producing facility near Swakopmund, Namibia, where seawater is pumped into flat basins for evaporation. When the concentration of salt (mostly sodium chloride) increases to more than 12–15 %, only halophilic microorganisms remain alive. The red color of the brines is due to high concentrations of haloarchaea (about 10^7 to 10^8 cells per

Fig. 5.12 Salt-producing facility near Swakopmund, Namibia. In red-colored lagoons, haloarchaea are dominant; green color is due to *Dunaliella* sp. (algae).

milliliter of liquid); in some ponds halophilic microalgae, mostly *Dunaliella* species, dominate, imparting a green color to the lagoons (Fig. 5.12).

The principal morphological types of haloarchaea are rods, cocci, and irregular forms, which are often very flat cells. The first genera that were described are *Halobacterium, Haloarcula, Halococcus,* and *Haloferax.* Alkaliphilic haloarchaea from African soda lakes are represented by the genera *Natronobacterium, Natronomonas, Natrialba,* and *Natronococcus*; now there are a total of 19 haloarchaeal genera. An unusual shape is exhibited by the famous "square bacterium" (*Haloquadratum walsbyi*): almost perfectly quadratic cells are attached to each other like postage stamps and are capable of forming large thin sheets. The intense red, pink, or purple pigmentation of haloarchaea is due to carotenoids in their membranes and to the related compound bacterioruberin, a C_{50} molecule (comprising 50 carbon atoms). These nonpolar lipids provide protection against the strong sunlight that is usually characteristic for the natural environments of haloarchaea, and they may act as membrane reinforcers.

The retinal-containing protein bacteriorhodopsin, which imparts a strong purple color to cells, is present in some species of haloarchaea and has been thoroughly studied because it is a comparatively simple, light-driven proton pump, capable of producing a pH gradient across the membrane, which is used for the production of ATP.

5.2.7.2 **Viable Haloarchaea from Rock Salt**

During several periods in the Earth's history, extensive sedimentation of halite and some other minerals from hypersaline seas took place. An estimated 1.3×10^6 km^3 of salt was deposited in the late Permian and early Triassic periods (about 240–280 million years ago). The continental landmasses at that time were concentrated and formed the supercontinent Pangaea. Salt sediments developed in large basins that were connected to the open oceans by narrow channels. About 100 million years ago, Pangaea started to break up; the continents were displaced to the north, and folding of new mountain ranges took place. As a result of these movements driven by plate tectonics, huge salt deposits are found today mainly in the northern regions of the continents.

Several viable haloarchaea were isolated from Permian rock salt, which was collected in Alpine salt mines following blasting or drilling. Based upon the comparison of numerous taxonomic data, they were recognized as novel species with relationships to known haloarchaea. Figure 5.13 shows a coccoid isolate from Permian salt, *Halococcus dombrowskii*, which grows in small aggregates of four to eight roundish cells.

Another coccoid strain from Alpine salt, *Halococcus salifodinae*, was isolated from a brine in a salt mine in Cheshire, England, and from a salt bore core near Berchtesgaden, Germany; all three isolates proved identical when their cellular properties were compared. Thus, viable haloarchaea that belong to the same species occur in geographically separated evaporites of similar geological age. So far, these haloarchaea have not been found in any hypersaline surface waters or in any location other than salt mines. Analysis of dissolved alpine rock salt with molecular methods by extracting DNA and subsequent amplification and sequencing of 16S rRNA genes resulted in numerous novel sequences that indicated the presence of a very diverse microbial community in ancient rock salt. These findings support the hypothesis that halophilic isolates from subterranean salt deposits may be the remnants of populations that once inhabited ancient hypersaline seas. However, at this time there are no methods available to prove directly a great prokaryotic age. The mass of an average prokaryotic cell is only about 10^{-12} g (picograms); it is composed of about 3 000 different biomolecules that are present at femtogram levels or less; therefore, no current dating procedures can be applied.

5.2.7.3 Molecular Mechanisms

High concentrations of salt, as found in natural or manmade brines and salt lakes, belong to environments that are characterized by reduced availability of water or

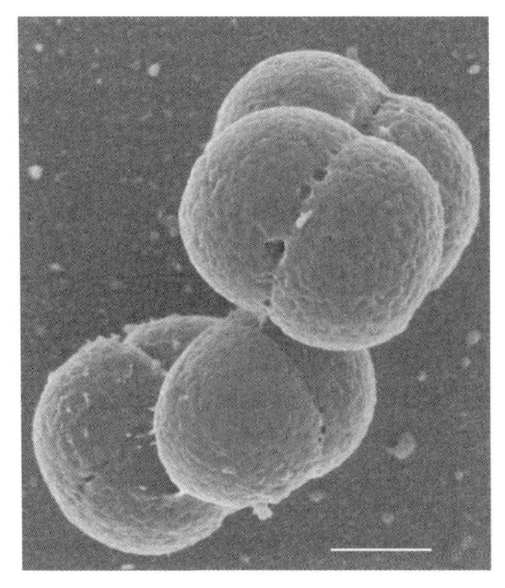

Fig. 5.13 Scanning electron micrograph of cells of *Halococcus dombrowskii* strain H4, an isolate from Permian salt. Bar indicates 0.5 μm (photo by Chris Frethem, with permission).

low water activity (a_w). Microorganisms use basically two strategies to obtain scarce water from their environments: they pump inorganic ions into their cells, or they synthesize small molecules, so-called compatible solutes, inside their cells. Both have the same purpose – to increase osmotic flow of water into the cytoplasm. The compatible solutes must be non-inhibitory and nontoxic to the cell – hence their designation. They are highly soluble compounds, such as sugars, alcohols, amino acids, or derivatives of these molecules. For example, cyanobacteria use sucrose, trehalose, or glycine betaine; they grow at a_w between 0.75 and 0.9. The halophilic alga *Dunaliella*, xerophilic yeasts, and xerophilic fungi all use glycerol as a compatible solute and are capable of growing at a_w from 0.75 down to 0.61. Haloarchaea are exceptional in that they do not use organic compounds but rather the inorganic K^+ ion; the internal concentration of K^+ can be 5.3 M, which allows haloarchaeal growth at a_w of 0.75.

Most enzyme proteins from haloarchaea, with only a few exceptions, are optimally active and stable in the presence of 20% NaCl or KCl. Often, a preference for K^+ instead of Na^+ is observed; this is consistent with the intracellular concentration of K^+ ions in haloarchaea (up to 5.3 M). The stability and solubility of halophilic proteins under conditions where non-halophilic proteins would precipitate are due to a high excess of the acidic amino acids aspartic and glutamic acid, which possess negative charges and – at least partially – compensate for the highly positive charges of the ions K^+ and Na^+.

5.2.7.4 Extraterrestrial Halite

Mars is a planet where the presence of salts has been demonstrated. Evidence for halite was found in the SNC meteorites in 1992 (named for their places of impact: Shergotty, India; Nakhla, Egypt; and Chassigny, France); their Martian origin has been confirmed independently by several researchers (see Chapter 8). Elements from Martian soil and rocks recently determined with the alpha-particle X-ray spectrometer on the Martian rovers include Na, Mg, Cl, and Br. Therefore, saturated salt solutions, which would possess greatly depressed freezing points, could be envisaged on Mars. They may not be present as standing pools but rather could occur in small pore spaces between mineral grains. The apparent longevity of haloarchaeal strains in dry, salty environments is thus of interest for astrobiological studies and the search for life on Mars. On Earth, microorganisms were the first life forms to emerge and were present perhaps as early as 3.8 billion years ago. If Mars and Earth had a similar geological past, microbial life, or the remnants of it, could still be present on Mars.

Carbonaceous meteorites, such as the well-studied Murchison meteorite, also contain traces of salt. A rather spectacular find was the Monahans meteorite (named after the town Monahan, Texas), which fell in 1998 and was immediately recovered and analyzed. It contained large crystals of halite (NaCl) together with sylvite (KCl), some with small fluid inclusions; its age was estimated by Sr–Rb dating as 4.7 ± 0.2 billion years. Analysis of the Zag meteorite, which fell in Morocco in 1998, revealed with even greater sensitivity an ancient origin of the halite and water inclusions of this meteorite (4.57 billion years); hence, they suggest the formation of evaporites as a very early event in the Solar System.

Another example of putative extraterrestrial salt comes from the exploration of the Jovian moon Europa: the *Galileo* spacecraft collected evidence that supports the existence of a liquid ocean on Europa (see Chapter 10). *Galileo*'s onboard magnetometer, which measures magnetic fields, detected fluctuations that are consistent with the magnetic effects of currents flowing in salty brine.

If halophilic prokaryotes on Earth can remain viable for very long periods of time, it is reasonable to consider the possibility that, even if present-day conditions are inimical for life, viable microorganisms may exist in similar salt deposits or salty liquids on other planets or moons.

5.2.8
Subterranean Environments

What seemed rather inconceivable just a few years ago is now being considered more seriously: the traditional scientific concept of an abiological terrestrial subsurface has to be abandoned. Vast numbers of microbes have been detected repeatedly in deep subterranean locations; they may have been there for centuries or even millennia, as they were found almost everywhere at great depths (the current record is 5 278 m for thermophilic anaerobes). The deep terrestrial subsurface is another extreme environment; it is characterized by dryness, few nutrients, and an increase in temperature with increasing depth. Drilling up to 3 000 m has shown that microorganisms are present in subsurface environments such as granite, sediments, gold and uranium mines, and in the tailings (leftovers from ore extraction) of such mines.

The presence of microorganisms in samples from bore holes is frequently ascertained by fluorescence microscopy using specific stains, including phylogenetic stains, that provide an initial classification; both Bacteria and Archaea were detected by this method. However, cultivation of subterranean isolates is more problematic and can be very time-consuming because of their largely unknown metabolic requirements. Several isolates from subterranean sites (besides isolates from rock salt; see Section 5.2.7.2) and subsurface sediments have been formally classified, including: *Bacillus infernus* (from about 2 700 m below the land surface in the Taylorsville Triassic Basin in Virginia, USA); *Bacillus subterraneus* (from a deep aquifer in Australia); *Geobacter bemidjiensis* (from subsurface sediment in Minnesota, USA); *Geobacter metallireducens* (from a mud sediment in the Potomac River, USA); *Geobacter chapellei* (from deep aquifer sediments in South Carolina, USA); *Hydrogenobacter subterraneus* (from deep subsurface geothermal water in Japan); and *Thiobacter subterraneus* (from subsurface geothermal aquifer water in the Hishikari gold mine in Japan). By using 16S rRNA analysis, *Arthrobacter* sp., *Commamonas* sp., *Pseudomonas* sp., and *Sphingomonas* sp. (all from a deep subsurface well in the U. S.) were identified.

Subterranean microorganisms represent a wide and novel area for research. Some central questions to be answered by investigation of such subterranean environments include: Where do these microorganisms come from? How long were they included in the rocks? How did they get in? What is their role deep down there? Could they be models for life prospering far below the Martian surface, where it would be sheltered from radiation and the harsh surface environment?

5.2.9 Radiation

Vegetative (actively growing) strains of *Deinococcus radiodurans* are more resistant to ionizing radiation and strong UV light than even bacterial endospores (Fig. 5.14). *Deinococcus radiodurans* is a coccoid microorganism that can withstand a dose up to 30 000 Gy (Gy stands for Gray, which is the unit that denotes the energy deposited into biological tissue or cells; 1 Gy equals 1 kJ kg^{-1}; the former unit is rad, with 100 rad = 1 Gy), whereas less than 5–10 Gy is fatal for humans. Radiation damage consists mostly of double-strand breaks and of chemical modification of the DNA molecule; only a few breaks are usually lethal for a cell. Why is *Deinococcus radiodurans* so resistant? One reason is that it possesses a very efficient DNA repair system, which is able to repair hundreds of double-strand breaks. The system consists of an enzyme complex that can excise misincorporated bases and replace them with the correct ones. Secondly, *Deinococcus* can apparently reassemble its DNA chromosome in a short time when it has been fragmented by radiation. But these properties do not really explain resistance to a radiation dose of a magnitude that does not occur naturally – at least not on Earth. By analysis of various mutations of *Deinococcus radiodurans*, it was discovered that

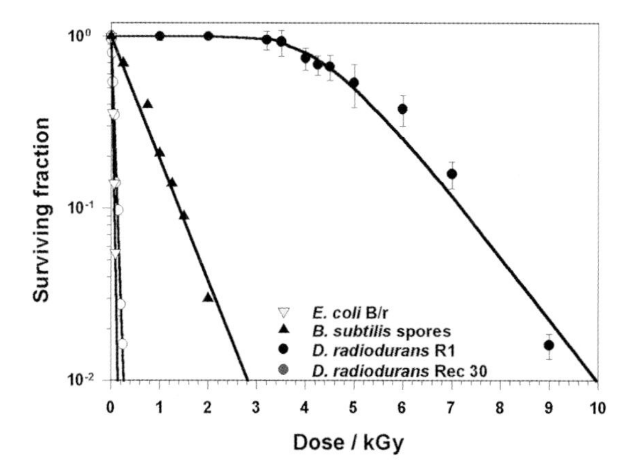

Fig. 5.14 Representative survival curves for *Deinococcus radiodurans* R1 and its recombination-deficient mutant Rec 30, compared to survival curves for spores of *Bacillus subtilis* and cells of *Escherichia coli* B/r following exposure to X-rays (from Horneck and Baumstark-Khan 2002).

loss of radiation resistance went along with a loss of viability following desiccation; thus, it may be concluded that this microorganism evolved cellular responses to survive long periods of dehydration and that the resistance to radiation is only incidental to the discovery and development of radiation-emitting technology in modern times. It not only survives exposure to radiation but also can grow in the presence of large doses of it, as it has been found living in the cooling water of nuclear power plants. In addition to tolerance of radiation and desiccation, there is also a marked resistance to DNA damage from ultraviolet light compared with other bacteria.

Deinococcus radiodurans was the first highly radiation-resistant bacterium to be discovered and is still the most thoroughly investigated one; however, other microorganisms are being described with similar radiation resistance. For example, *Deinococcus deserti* was isolated recently from the sands of the Sahara desert and was found to tolerate the same doses of gamma radiation as *Deinococcus radiodurans*, and even higher doses of ultraviolet light (de Groot et al., to be published). Table 5.2 contains radiation doses that produce 90% killing of radiation-resistant species and non-resistant bacterial strains.

Table 5.2 Survival of microorganisms following irradiation with gamma rays.

Bacterial strains	Radiation dose yielding 10% CFU[a] survival
Deinococcus radiodurans	16 000 Gy[b]
Deinococcus mumbaiensis	16 400 Gy
Deinococcus geothermalis	10 000 Gy
Deinococcus grandis	7 400 Gy
Enterococcus faecium	2 000 Gy
Escherichia coli	700 Gy
Pseudomonas putida	250 Gy

a CFU: colony-forming units; CFU survival: CFU determined following radiation and compared with CFU of non-irradiated cells.
b 1 Gy = 1 kJ per kg of cells.

5.3 Microbial Survival of Extreme Conditions

Table 5.3 shows several examples of microorganisms that survived exposure to extremes of physicochemical factors. The time of exposure varies greatly and can be quite brief. For example, *Deinococcus radiodurans* is listed here as, depending on the source of gamma radiation, surviving a dose of about 20 000 Gy obtained by a pulse lasting a few hours.

Table 5.3 Microbial survival in extreme conditions.

Microorganism	Physicochemical parameter	Time of exposure	Other information
Deinococcus radiodurans	Ionizing radiation	5–20 hours	~20 000 Gy
Streptococcus mitis	Surface of the Moon	2.5 years	In a camera
Numerous microorganisms	−20 °C	~10^6 years	Permafrost
Numerous microorganisms[a]	−193 °C	>10 years	Liquid N_2
Numerous microorganisms[a]	a_w <0.75	>10 years	Vacuum
Halococcus salifodinae	NaCl >30%	>10^6 years (?)	In salt crystals
Endospores (*Bacillus, Clostridium*)	Heat, chemicals	>3 000 years	Mummies, sediments
Endospores (*Bacillus*)	Outer space	6 years	Surface of a space probe[b]

a With protective substances (e.g., 25–40% glycerol).
b Shielded against solar UV radiation.

Other examples are of importance precisely because of the aspect of a long or even very extended exposure to an extreme environment. The survival of the bacterium *Streptococcus mitis*, which traveled to the Moon during the *Apollo* missions and apparently survived there for 2.5 years in a camera, has, unfortunately, not been rigorously documented; in some descriptions, survival of 20–500 cells is mentioned, without reference to the method of determination. Still, the fact of its apparent resistance to space conditions over several years is of profound importance. However, it cannot be ruled out that this bacterium came from contamination of the sample during the collection by the astronauts or during the return trip. The main reason is that *S. mitis* was identified as part of the microflora of the astronauts of *Apollo 12* (see Chapter 13).

Careful documentation of survival was carried out with *Bacillus* endospores that were fully exposed to space conditions in the LDEF (Long Duration Exposure Facility) mission for six years, from which a significant fraction was recovered in a viable state. From samples of the same mission that were shielded against UV light with metal foils, about 70% survived the six years in space (see Chapter 11).

Permafrost is an environment in which many viable microorganisms have been found and recultivated; some stem from sediments that are 1–2 million years old. Because of the frozen state, their metabolism is extremely slow or perhaps nonexistent. Some microorganisms in permafrost soils apparently form resting stages with thick coatings, but these are not endospores. Freezing will prevent microbial growth, but it does not necessarily kill microbes. Culture collections use freezing in the presence of protective substances for long-term preservation of microbial strains and other biological materials. Prerequisites are the addition of water-miscible liquids such as glycerol or dimethyl sulfoxide (DMSO) in concentrations of 20% or more; they will penetrate the cells and protect them mainly by hindering crystal formation upon freezing. Temperatures around that of liquid nitrogen (−193 °C) are survived by strains for decades or longer.

Endospores are a survival form for certain bacteria (mostly from the genera *Bacillus* and *Clostridium*); they are formed when they find themselves in an unfavorable environment, such as low nutrients or high salt concentrations. Endospores are perhaps the most resilient life form on Earth. They are resistant to extreme temperatures (an autoclave operating at 121 °C will kill endospores of most species, but exceptional endospores survive up to 150 °C) and to most disinfectants, radiation, and drying. They can survive for thousands of years in this dormant state, as they have been found in Egyptian mummies and lake sediments of known age. They may survive even longer, because it was reported that *Bacillus* species were revived from bees encased in 25- to 30-million-year-old amber; however, because phylogenetic analysis showed that the isolates from ancient amber were very similar to extant microbes, some caution is advised and an independent confirmation of these results would be desirable.

Still-older microbial isolates have been reported from salt sediments and from other deep subterranean sources (bore cores, mines) that are of great geological ages (up to 250 million years; see Section 5.2.7.2). Further research is required to determine whether any metabolic maintenance activities of these microorganisms – including potential repair of damaged DNA, as would be expected – are occurring.

The survival of endospores in adverse conditions, including the space environment (see above and Chapter 11), raises the possibility that bacterial endospores could travel to Mars on the surface of a spacecraft and survive on or in Martian soil. This could seriously compromise future efforts to establish whether there is, or has been, life on Mars, as it would be difficult for researchers to know whether any endospores found originated from Earth or Mars. Similarly, survival of non-spore-forming microorganisms (*Streptococcus mitis*, permafrost bacteria, many subterranean bacteria, haloarchaea) under Martian or other space conditions should be explored in detail.

There are regulations already agreed upon by all space-faring nations, notably by COSPAR (the International Committee on Space Research), with the goal of preventing contamination by spacecrafts (see Chapter 13); the rules will have to be updated and revised, taking into account the potential of extreme survival of microorganisms.

5.4 Conclusions

Microorganisms thrive in extreme environments that once were believed to be entirely hostile to life. Extreme environments are usually characterized by one or several physicochemical parameters. Species diversity in extreme environments may be lower than in moderate environments; however, improved methods led to the identification of numerous novel phylogenetic groups. Extremophilic microorganisms survive exposure to space conditions, desiccation, high and low temperatures, and low nutrient environments, perhaps for millions of years. Therefore, they may also survive space travel and be able to live in extraterrestrial environments. This suggests that the old concept of panspermia (see Chapter 11) might be an acceptable idea for the distribution of life in the Universe.

5.5 Further Reading

5.5.1 Books or Articles in Books

Amy, P. S., Haldeman, D. L. (Eds.) *The Microbiology of the Terrestrial Deep Subsurface*, CRC Press, Boca Raton, 1997.

Brock, T. D. (Ed.) *Thermophiles. General, Molecular and Applied Microbiology*, John Wiley & Sons, New York, 1986.

Horneck. G., Baumstark-Khan, C. (Eds.) *Astrobiology – The Quest for the Conditions of Life*, Springer, Berlin, Heidelberg, New York, 2002; especially the following chapters therein:

- Horneck, G., Mileikowsky, C., Melosh, H. J., Wilson, J. W., Cucinotta, F. A., Gladman, B., Viable Transfer of Microorganisms in the Solar System and Beyond. pp. 57–76.
- Gilichinsky, D. A., Permafrost Model of Extraterrestrial Habitat, pp. 125–142.
- Kunte, H. J., Trüper, H.g., Stan-Lotter H., Halophilic Microorganisms, pp. 185–200.
- Stetter, K. O., Hyperthermophilic Microorganisms, pp. 169–184.

Hroarson, B., Jonsson, S. S., *Geysers and Hot Springs in Iceland*, Mal og Menning, Reykjavik, 1992.

Madigan, M. T., Martinko, J. M., *Brock Biology of Microorganisms*, International Edition. 11th Ed. Pearson Prentice Hall, Upper Saddle River, New Jersey, 2006.

Stan-Lotter, H., Radax, C., McGenity, T. J., Legat, A., Pfaffenhuemer, M., Wieland, H., Gruber, C., Denner, E. B. M., From Intraterrestrials to Extraterrestrials – Viable Haloarchaea in Ancient Salt Deposits, in *Halophilic Microorganisms*, Ventosa, A. (Ed.), Springer Verlag, Berlin, Heidelberg, New York, pp. 89–102, 2004.

5.5.2 Articles in Journals

McCord, T. B., Mansen, G. B., Fanale, F. P., Carlson, R. W., Matson, D. L., Johnson, T. V., Smythe, W. D., Crowley, J. K., Martin, P. D., Ocampo, A., Hibbitts, C. A., Granahan, J. C. Salts on Europa's surface detected by *Galileo*'s near Infrared Mapping Spectrometer. The NIMS team, *Science* **1998**, *280*, 1242–1245.

McGenity, T. J., Gemmell, R. T., Grant, W. D., Stan-Lotter, H. Origins of halophilic microorganisms in ancient salt deposits (MiniReview), *Environ Microbiol* **2000**, *2, 243–250.*

Pedersen, K. Exploration of deep intraterrestrial microbial life: current perspectives, *FEMS Microbio. Lett.* **2000**, *185*, 9–16.

Rieder, R., Gellert, R., Anderson, R. C., Bruckner, J., Clark, B. C., Dreibus, G., Economou, T., Klingelhöfer, G., Lugmair, G. W., Ming, D. W., Squyres, S. W., d'Uston, C., Wänke, H., Yen, A., Zipfel, J. Chemistry of rocks and soils at Meridiani Planum from the alpha particle X-ray spectrometer, *Science* **2004**, *306*, 1746–1749.

Westall, F., De Wit, M. J., Dann, J., Van Der Gaast, S., De Ronde, C., Gerneke, D. Early Archaean fossil bacteria and biofilms in hydrothermally influenced, shallow water sediments, Barberton Greenstone Belt, South Africa, *Precambrian Res.* **2001**, *106*, 91–112.

Zolensky, M. E., Bodnar, R. J., Gibson, E. K., Nyquist, L. E., Reese, Y., Shih, C. Y., Wiesman, H. Asteroidal water within fluid inclusion-bearing halite in an H5 chondrite, Monahans (1998), *Science*, **1999**, *285*, 1377–1379.

5.5.3 Web Sites

Life on Mars: Viking and the Biology Experiments: http://www.resa.net/nasa/mars_life_viking.htm

5.6 Questions for Students

Question 5.1

The search for extraterrestrial life is one of the scientific goals of the 21st century. But the term "life" is not clearly and unequivocally defined. What, in your opinion, should be searched for? (Make some suggestions.)

Question 5.2

Hypothesis: Mars is already contaminated with terrestrial microorganisms and there is no chance of detecting potential Martian forms of life. Would you agree or disagree? (Provide some reasons for your decision.)

Question 5.3

Which molecular mechanisms enable microbial life at extreme temperatures?

Question 5.4

Which molecular mechanisms enable microbial life at extreme salt concentrations?

6
Habitability

Charles S. Cockell

The term "habitability" means many different things depending on the organisms that are considered, as it defines the chemical and physical envelope for a given organism. Because there is currently no direct evidence for life on another planet, habitability is necessarily constrained by our knowledge of life on Earth. We use our knowledge of the extremes of life on Earth to assess extraterrestrial environments and the plausibility that they can support life. In this chapter the use and limitations of the concept of habitability are discussed and examples are provided of how the habitability of other worlds is assessed using extreme environments on Earth. The chapter starts with a brief discussion of the history of habitability, including some historical watersheds in this concept. Then, some of the factors and mechanisms that determine the habitability of other planets are discussed. Finally, a postulate for habitability is given, which can be used to guide the way we think about the characteristics of other environments that we wish to assess as habitable or uninhabitable.

6.1
A Brief History of the Assessment of Habitability

The assessment of the "habitability" of other planetary bodies has been a pervasive intellectual problem throughout the history of planetary sciences and biology. Indeed, erroneous assertions of habitability, borne of optimism about the prospects for life, have clouded attempts to make astrobiology a serious field of scientific endeavor. In any field of science where technical or logistical constraints limit the amount of data that can be acquired, it is necessary to proceed with caution. Ultimately, the way in which we define habitability, the factors that determine it, and the way we use instrumentation to assess the habitability of other planets underpin the entire astrobiological enterprise. As efforts are developed to search

Complete Course in Astrobiology. Edited by Gerda Horneck and Petra Rettberg

ISBN: 978-3-527-40660-9

for "Earthlike" planets around distant stars (extrasolar planets), understanding what constitutes habitability has become an area of great relevance.

Prior to the invention of the telescope, the only planetary body that provided any detailed information on its surface characteristics was the Moon. Mars was a red anomaly in the night sky, but this color, spread across a tiny disk, did not reveal details. The pre-telescope era and the view of habitability it engendered was an era that I call the Birth of Optimism. This period of optimism was one dominated by theoretical discussions on the plurality of worlds. The debate was infused by religious considerations (i.e., that humans were special and unique) and contrasting considerations more at home in natural sciences (i.e., the notion of "plenitude") that nature makes use of all possibilities and, therefore, other worlds would be inhabited like ours.

The Birth of Optimism is marked by some remarkable comments and assertions that would not be considered out of place in today's debates on habitability. Among the most notable is the comment by the Greek philosopher Metrodorus of Chios, a student of Epicurus, who in the fourth century B. C. stated: "It would be strange if a single ear of corn grew in a large plain, or there were only one world in the infinite" (citation in Cornford 1934). The Roman poet Lucretius similarly stated: "It is in the highest degree unlikely that this Earth and sky is the only one to have been created."

These general, but important, thoughts were also surrounded by an entourage of more-specific claims and beliefs concerning the inhabitants of the Moon. Many Greeks and Romans believed that the Moon was inhabited by intelligent beings.

One would have thought that the invention of the telescope in the 17th century would have provided a means to assess the habitability of other planets more accurately, thus leading to a reduction in speculation as worlds yielded their secrets to empirical observation. However, the opposite occurred. Early telescopes allowed the discs of new worlds to be resolved, but their power was not sufficient to be able to tell anything about their surfaces or their environmental conditions. The 17th and 18th centuries were thus a time when optimism ran riot. This was a period when we knew of the existence of countless worlds, from Mars to the Jovian and Saturnian moons, but we knew little about the physical conditions in the Solar System. I call this period, which runs right up to the 1960 s, the Age of Optimism.

It is misleading to summarize such a long period of history with just a few quotes, but to give some flavor of the time and the way in which the telescope made possible observations of other planetary bodies without an ability to assess their habitability, a few quotes are worthwhile, which are taken from an extensive review by Crowe (1999). French author Bernard le Bovier de Fontenelle (1657–1757) stated of the Venusians that they were "little black people, scorched with the Sun, witty, full of fire." British astronomer William Herschel (1738–1822), whose observational astronomy was par excellence, informs us: "By reflecting a little on this subject I am almost convinced that those numberless small Circuses we see on the moon are the works of the Lunarians and may be called their Towns." And as recently as 1909, in his book *Mars as the Abode of Life*, Percival Lowell asserted: "Every opposition has added to the assurance that the canals are artificial; both by disclosing their peculiarities better and better and by removing generic doubts as to the planet's habitability."

The Age of Optimism is a fascinating age for the discussion of the habitability of other worlds, but it is also a warning from the past. Enveloped in an era where many scientists had optimism about the habitability of other planets, none of the comments listed above seemed absurd to their authors, despite their lack of empirical data to support their claims. To them it was just obvious that other planets were inhabited. It was a case of culture driving science. Today, we are similarly living through a period of optimism that, although more constrained by empirical observations from spacecraft and telescopes, is nevertheless strongly driven by an undercurrent of new-found optimism in the likelihood of life elsewhere.

The 1960 s witnessed the beginning of the space age and, with it, the possibility of visiting other planets with spacecraft. I call this age of the discussion of habitability, which runs until the early 1990 s, the Age of Retreat. The failure to find canals on Mars, culturally sophisticated Venusians, or fortresses on the Moon, irrefutably supported by landers and orbiters, forced a retreat in our view of habitability. We realized that many of these worlds we had viewed through rose-tinted biological spectacles were in fact cold or hot, barren wastelands with little or no prospects for any life at all. The *Viking* landers did nothing to dispel the vision of Mars as a barren, dead planet (see Chapter 5). Despite the somewhat negative view of habitability, this period of space exploration was nevertheless remarkable, leading to new and astonishingly detailed insights into the conditions on other planetary surfaces. Images sent back from the far reaches of the Solar System by the *Voyager* robots, for instance, truly ushered in an extraordinary period of exploration. Of course, habitability, at least in the case of the moon, was also directly assessed by human explorers during this period.

Despite the biologically disappointing data sent back from other planets, there was still room for maneuvering. The period from the 1990 s onwards I refer to as the Rebirth of Optimism. Two developments came together to generate renewed optimism in the possibility of life in our Solar System. Firstly, spacecraft instruments became of better quality and had higher resolution. Where previously we saw barren Martian deserts with some evidence of past liquid water, we now saw extant ice sheets and even putative gullies formed by liquid water (see Chapter 8). Rocky moons around Jupiter yielded images of ice sheets, and studies of the magnetic fields around them suggested liquid water oceans that are conducive to life (see Chapter 10). The array and sophistication of instruments on spacecraft have provided many different imaging methods, which together have provided insights into environments on other planets that may be conducive to life.

Secondly, studies of microorganisms on Earth expanded the envelope of life to such an extent that many environments on Earth, previously thought to be uninhabitable, were found to harbor life (see Chapter 5). Some of these environments overlap with extraterrestrial planetary environments in some physical and chemical conditions. Because they contain life on Earth, renewed optimism about their capacity to sustain life elsewhere has been encouraged. Such environments include

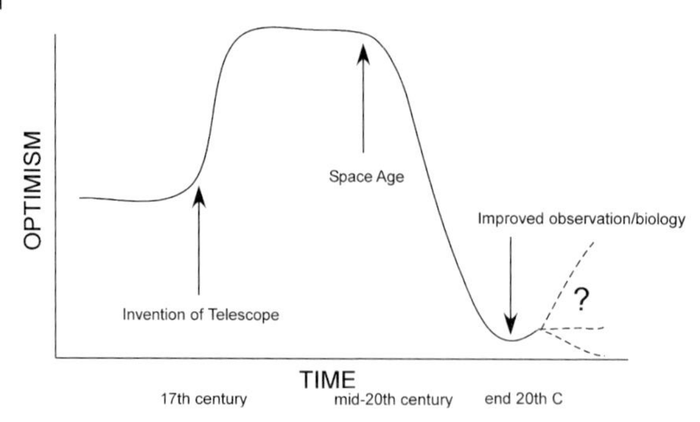

Fig. 6.1 Qualitative graph showing the change in our views of habitability over time.

permafrost in Earth's polar regions, which may be analogous to permafrost regions on Mars, and hydrothermal vents on the ocean floors and ocean-spreading axes, which have been suggested to provide insights into the biological potential of the bottom of Europa's oceans (see Chapter 10).

In Fig. 6.1 I have drawn a graph showing, qualitatively, the change in optimism about habitability over time. The current challenge in astrobiology is, of course, to determine the direction in which we should extrapolate this graph and whether we can transgress from "optimism" to the positive identification of life on particular planetary bodies.

This brief review of the history of habitability is lacking in many details, as there is insufficient space to review adequately the many diverse discussions since the time of the ancient Greeks. However, it should provide some understanding of the major periods of history in the assessment of habitability, from which more detailed research can be undertaken. It should also place in context where the student of astrobiology stands today in this long history, and how we have arrived at where we are.

6.2 What Determines Habitability?

The habitability of an environment is firstly, and quite obviously, constrained by the organisms we wish to contemplate in the environment under consideration. The two end members of habitability considerations are extremophilic micro-organisms, which define the outer boundaries of life, and humans and other complex multicellular organisms, which require much more narrow ranges of environmental conditions. In between them is a diversity of prokaryotes and multicellular organisms that may be adapted to specific extreme environments and have different environmental tolerances. Table 6.1 shows some environmental tolerances of humans and microbes to illustrate this difference in ranges (see also Chapter 5).

Table 6.1 Some physical and chemical factors important for defining whether an environment is habitable and the ranges tolerated by microorganisms and humans.

Parameter	Microorganisms	Humans[a]
Temperature	< 0–121 °C	0–30 °C
Pressure	<50 hPa–>10 000 hPa[b]	700 hPa–5 000 hPa
Radiation	4 000 Gy[c]	1–3 Gy[c]
pH	0–13	Near neutral
Oxygen	0–~100%	15–25%
Carbon dioxide	0–100%	<1%

Source: Some of these values are taken from McKay et al. 1991.

a Some of these values are approximations. For example, humans can tolerate a far wider temperature range than shown here by using technology, and a wide range of pHs can be tolerated by humans over short time periods.

b 1 hPa = 1 mbar

c Sublethal dose; 1 Gy (Gray) = 100 rad

Given the obvious need to define the types of organisms we are considering, what are the fundamental requirements for any life forms based on what we know about life on Earth? All life must have a solvent to carry out chemical reactions (and to act as a reactant itself). At least based on life on Earth, this solvent is water, and thus any planetary environment we postulate to be habitable must have liquid water (see Chapter 1). The presence of liquid water on a planetary surface has often been taken, alone, to be the definition of "habitable." The "habitable zone" around a star is defined as the region around a star where liquid water is stable on a planetary surface (see Section 6.4.1). It should be noted that this classic definition of a habitable zone is based on liquid water made possible as a result of direct solar heating from the parent star. In comparison, the Jovian moon Europa lies far outside the classic habitable zone but has a liquid water ocean on account of the tidal heating of the moon.

There must be an energy supply to drive the reactions that life requires to grow, reproduce, and repair itself. Energy supplies can range from chemolithotrophic metabolism (using inorganic redox couples such as reduced iron and nitrate to drive energy acquisition) to heterotrophic modes of metabolism (using organic molecules to provide energy) to photosynthesis (using photons from the parent star to drive energy acquisition) (see Chapter 5).

There must be a source of nutrients in an accessible form, with which life can construct itself. On Earth, we know these to be carbon, hydrogen, oxygen, nitrogen, sulfur, and phosphorus (C, H, O, N, S, P). Many organisms use more than this complement of elements. Elements as diverse as iron, zinc, and molybdenum can be required, and many anaerobic microorganisms on Earth are known to require tungsten and selenium in their culture medium. However, whether many of these are absolutely required for life (as opposed to being an evolutionary quirk of enzymes and other biochemical systems on Earth) is not understood. At least as a minimum we suppose that life requires those six elements as summarized in CHONSP.

A planetary environment that possesses all three of these requirements, i.e., water, energy, and basic nutrients, can be presumed to be a place that is potentially habitable. Although one can postulate entirely novel biochemistries that might use completely different elements and solvents (such as silicon-based life or life using ammonia as a solvent), because no such life has yet been found, it does not fit within the boundaries of science. It goes without saying that the discovery of a novel biochemistry using other types of elements and solvents on another planet could simply be used to expand the definition of habitability. At the present time we use life on Earth to define its boundaries. In Section 6.4 two of these factors that determine the boundaries, i.e., temperature and energy, are considered.

6.3 Uninhabited Habitable Worlds

It is important to understand that the presence of habitable conditions does not imply that an environment is inhabited. On Earth, where the conditions described above are met, microorganisms are invariably found, because there is a plenitude of microorganisms inhabiting many diverse environments. For example, $4–6 \times 10^{30}$ microorganisms are estimated to exist on Earth. The appearance of a habitable environment offering conditions for life usually leads to its rapid colonization. However, on young planets or planets where, for whatever reasons, life has not originated, one could imagine habitable environments where life would not be around to take advantage of the conditions, particularly if these environments were transient, such as a warm hydrothermal system in an impact crater on an otherwise frozen, dead world. Such environments are "uninhabited habitable habitats," and determining their distribution throughout the Universe is a challenge of astrobiology.

6.4 Factors Determining Habitability

6.4.1 Habitability and Temperature

The temperature of the surface of a planetary body is, in principle, simply determined by the heat it receives from its star in the form of visible and near-infrared (IR) light and the energy it releases in the form of thermal IR radiation. The solution of these simple inputs and outputs yields a "black body" temperature for a planetary surface. Although this simple formulation works quite well for a planetary body without an atmosphere, an atmosphere can cause a warming effect by trapping IR radiation, the so-called greenhouse effect. In the case of the Earth, water and carbon dioxide provide the greatest contribution to warming. More recently, anthropogenic greenhouse gases, such as chlorofluorohydrocarbons (CFCs), have contributed to the global warming. The greenhouse effect is itself greatly complicated by the quantity of dust in an atmosphere, the cloudiness, and the possible presence of chemical smog, all of which will influence the magnitude of the effect. In the case of the Earth, the greenhouse effect is sufficiently large to bring our planetary surface temperature to above freezing (about 15 °C average), whereas otherwise it would be below freezing, demonstrating quite effectively the role of greenhouse gases in influencing surface habitability.

The temperature of a planetary surface will be influenced by many feedback processes, both positive and negative. Snow cover, for instance, will increase the albedo (reflectivity) of a planetary surface, thus causing greater emissions of IR radiation back into space, cooling the planet further.

This positive feedback is suggested to be responsible for the onset of "snowball Earth" conditions, whereby runaway growth of ice sheets, which increase the albedo and drive further cooling, ultimately cause a freeze-over of all, or most, of the planetary surface. These snowball Earth episodes are postulated to have occurred during the Proterozoic (2.5–0.54 billion years ago). One of these events, which occurred in the period between 2.45 and 2.2 billion years ago, is thought to be linked to the rise in atmospheric oxygen at that time; the oxygen may have reacted with methane (another greenhouse gas) in the atmosphere, reducing the greenhouse effect and leading to glaciations. There is evidence for snowball glaciations in the late Proterozoic as well. The existence of widespread evidence for glaciers during these periods is also supported by modeling studies. These glaciations provide dramatic empirical evidence for the importance of the ice and snow albedo feedback.

Another feedback effect is controlled by water vapor. Because water vapor contributes to greenhouse warming, it too can cause positive feedback. As the planet warms, more water evaporates from the oceans into the atmosphere and causes yet more warming. Conversely, cooling of the surface reduces evaporation and reduces the greenhouse effect caused by water vapor.

The extent to which these feedback processes would be important on other planets obviously depends on their mean temperature, which will influence the extent of ice and snow on their surfaces. It also depends on their inventory of water and thus the size of their oceans and the extent to which water evaporation drove the greenhouse effect. Therefore, we can see that although the principles are quite easy to grasp, applying habitability criteria to other planets would require a specific knowledge of each body being assessed.

The ice/snow and water vapor feedback processes function over quite short timescales (almost instantly), but there are other feedback processes that operate over millions of years and are profoundly important for determining the habitability of a planet. One of the most important of these is the carbonate–silicate cycle.

The carbonate–silicate cycle regulates the amount of CO_2 in the atmosphere and thus the contribution of this gas to the greenhouse effect. The CO_2 in the atmosphere is there mainly as a gas, but some of it is dissolved in atmospheric water as carbonic acid (H_2CO_3). When it rains, this weak acid washes over silicate rocks and dissolves them. The product of the weathering process is bicarbonate (HCO_3^-), which itself is washed into the oceans. Bicarbonate is used by diverse organisms for photosynthesis and to make calcium carbonate shells, e.g., those of the foraminifera. This CO_2 is thus locked up in these carbonates and effectively buried and removed from the atmosphere. The continuous operation of this process would, of course, remove all the CO_2 from the atmosphere and cool the planet, but the CO_2 fixed as carbonates is eventually returned to the atmosphere. Deep in the crust, plate tectonics pushes the carbonates into high-pressure and high-temperature regions where it reacts with silicate rocks, releasing the CO_2 back into the atmosphere through volcanism, and hence the cycle is complete.

With this in mind, we can now understand how the carbonate–silicate cycle works to regulate the surface temperature of a planet. If the planet slowly heats up (e.g., by the slowly increasing solar luminosity over time), the weathering rate is increased at the planetary surface (because chemical reactions occur faster at higher temperatures), and thus more CO_2 is drawn down from the atmosphere in these reactions. The greenhouse effect is thus reduced and the temperature drops. Conversely, cooling reduces the weathering rate and less CO_2 is drawn down. Because volcanism is still returning the existing CO_2 pool back into the atmosphere at the same rate, the atmospheric CO_2 concentration rises and the greenhouse effect is increased, thus warming the planet. Thus, we can see how the carbonate–silicate cycle provides a negative-feedback control of planetary temperature.

The importance of the carbonate–silicate cycle for habitability can be easily appreciated by observing Mars and Venus (see also Chapter 8). Mars long ago ceased to have plate tectonics. The presence of remnant magnetic regions on Mars suggests that it may have had some plate tectonics, but the small size of the planet caused it to rapidly cool. The shutdown of widespread volcanism would have prevented CO_2 from being rereleased into the atmosphere, thus rendering the carbonate–silicate cycle inactive. Combined with this problem is the problem of keeping Mars warm at the further distance it occupies from the Sun, which makes the CO_2 greenhouse less effective to the point where sufficient CO_2 to keep the planet warm would cause it to condense out (although other solutions such as a hydrocarbon smog and CO_2 ice clouds might solve this problem). Thus, even if any putative carbonates could be returned by volcanism, the greenhouse effect has serious problems in operating on Mars. The result is a planet that today has a mean temperature of –60 °C, although the subsurface may have environments conducive to life.

By contrast, Venus has a very different problem. Volcanism is still evident on the planet; therefore, unlike on Mars, the return of CO_2 to the atmosphere does not seem to be a problem. However, the high temperature of Venus long ago caused the complete evaporation of its water inventory. The water was eventually lost into space from the top of the atmosphere, thus shutting down weathering. With no way for the CO_2 to react in surface weathering reactions with silicate rocks mediated by water, all of the CO_2 remained in the atmosphere, leading to the 464 °C mean temperature now found at its surface. The atmospheric CO_2 content would itself have contributed to greenhouse warming, thus accelerating the loss of water.

Although there is necessarily some simplification in these accounts, we can see how the carbonate–silicate cycle can be used to understand some quite fundamental reasons for the conditions of habitability on Venus and Mars. Venus provides us with an inner bound at which the carbonate–silicate cycle becomes ineffective and a planet is transformed into a scorched desert. Mars can be seen as a proxy for the outer bound of habitability. These bounds can therefore be conveniently used to define the region around a star where the liquid water is stable on a planetary surface within temperature conditions conducive to life.

Some of the simplifications in this account include the role of plate tectonics. For example, if Mars were larger than it is, it is possible that it would retain a functioning carbonate–silicate cycle on account of the presence of plate tectonics, leading to a cycle of carbon and a less extreme loss of its atmosphere in its early history. Might a larger Mars have resulted in a second habitable world in our own Solar System? Thus, the bounds of habitability within a star system must be treated as somewhat of an approximation.

In addition, we should bear in mind that mechanisms other than the carbonate–silicate cycle might influence the stability of liquid water. The tidal forces exerted by Jupiter on its moon Europa, for example, allow for the stability of liquid water under a thick ice sheet, independent of the carbonate–silicate cycle's temperature regulation (see Chapter 10). Europa lies well outside the traditional habitable zone defined by solar luminosity. Nevertheless, the habitable zone provides a convenient method to help define regions around other stars that might support liquid water, and it helps us to understand how conditions on a planetary surface, down to the scale of microorganisms, can be defined by feedback processes operating over long time periods.

Figure 6.2 shows a diagram of where the habitable zone lies around different stars. This zone is much the same size when the distance from the star is plotted on the logarithmic scale.

Stars cooler than our own Sun, such as M stars, have habitable zones that are closer to their star, because planets need to be closer in for the greenhouse effect to work sufficiently. Hotter stars, such as F stars, have habitable zones that are further away. For very cool stars, a complication arises in that if the planet is too close to the star it becomes tidally locked, meaning that only one side is facing towards the star. Although this might cause the dark side of the planet to freeze out, causing the atmosphere to collapse, if the atmospheric transport of heat is sufficient, this may not occur. It raises the intriguing possibility of life on planets where photosynthesis is possible only on one side and life forms on the other side live in perpetual darkness.

In our own Solar System the inner region of the habitable zone is thought to lie at about 0.95 AU and the outer region somewhere between 2.37 and 2.4 AU (AU = astronomical unit = average distance of the center of the Earth from the center of the Sun = approximately 149.6×10^6 km) (Fig. 6.2).

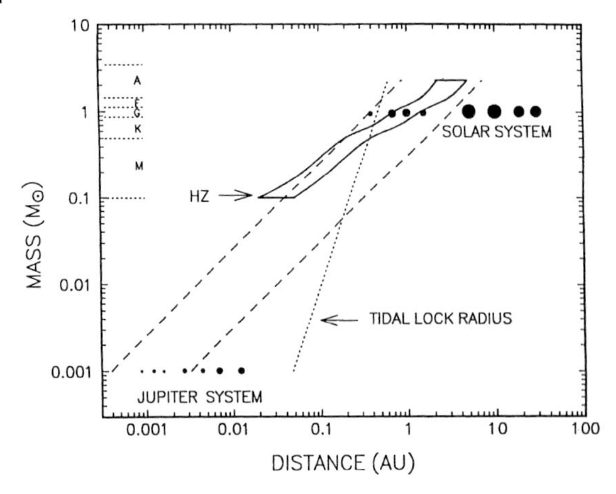

Fig. 6.2 Habitable zone (boxed area) around other stars as a function of the mass of the star and the distance from the star. $M_\odot$ = Solar mass; the long dashed lines delineate the probable terrestrial planet accretion zone (figure is taken from Kasting et al. 1993).

A further augmentation of the habitable zone is the "continuously habitable zone" (CHZ). This is the region around a star where conditions for habitability are met over some defined period of time. For example, because we know that it has taken 3.5 billion years to go from microorganisms to intelligent life on Earth (see Chapter 1), a search for planets around other stars might focus on regions where the habitable zone exists for at least this length of time. This is especially necessary if one wants to search for radio signals from other civilizations, e.g., the project Search for Extraterrestrial Intelligence. Shorter lengths of time for the existence of the CHZ might be sufficient for the rise of oxygen and thus the possible detection of oxygen in the atmosphere of extrasolar planets.

6.4.2
Habitability and Energy

For a planetary surface to be habitable, it must also have a supply of energy. At least for much of life on Earth, this energy is derived from the flow of electrons through a biologically mediated electron transport chain. The fundamental chemical process that drives energy acquisition is called the "chemiosmotic theory." Electron transport that occurs during redox reactions expels protons from the cell membrane. This translocation process creates a proton gradient across the cell membrane. The protons can move back across the cell membrane to do useful work, e.g., adenosine triphosphate (ATP) synthesis (see Chapter 4). Thus, to make energy for growth, reproduction, and repair, most life must have access to electron donors, which supply electrons and electron acceptors, which take up electrons in order to drive the energy-yielding reactions. Examples of these electron transfer reactions are shown in Fig. 6.3.

Inorganic redox couples perhaps represent the most widespread form of energy availability. Many of these elements or compounds are formed in supernovae and exist in stars. Figure 6.4 shows oxygen and iron in the Sun, which can be used by iron-oxidizing microorganisms as a source of energy (they require reduced iron). The Sun is obviously not habitable because it has no water and is too hot, but this graph nicely illustrates the universally widespread distribution of energy sources for life.

Many microorganisms can use reduced iron as an electron donor and compounds such as nitrate and oxygen as electron acceptors. Because electron donors and acceptors are inclined to react through abiotic processes (e.g., reduced iron "rusts" in the presence of oxygen), having a supply of electron donors and acceptors requires some type of geochemical disequilibrium where both electron donors and acceptors can be continuously produced, coexist, and be available for life. Active volcanism, plate tectonics, and asteroid and comet impacts are examples of processes that can cause turnover and movement in the planetary surface and subsurface, generating geochemical gradients that life can take advantage of.

Furthermore, some inorganic redox reactions occur very slowly because of a high activation energy; in this case, life can overcome these activation energies and increase the rate of the energy-yielding reaction.

The range of energy-yielding processes on Earth is immense. As well as those reactions shown in Fig. 6.3, which are used by chemolithoautotrophs (literally "rock eaters"), other organisms rely on heterotrophy (using organic carbon as a source of energy – including direct ingestion of prey and fermentation) and still others, i.e., photoautotrophs, use sunlight as a source of energy and derive electrons from water.

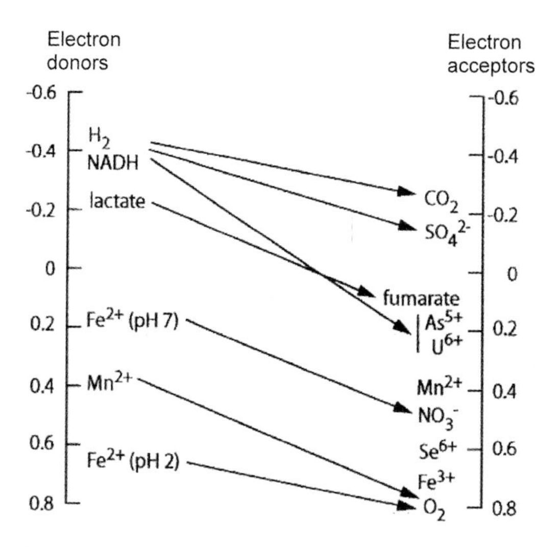

Fig. 6.3 Examples of energetically favorable reactions for the acquisition of energy by microorganisms. The downward-pointing arrows indicate thermodynamically favorable reactions.

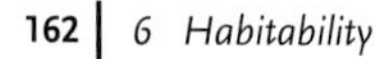

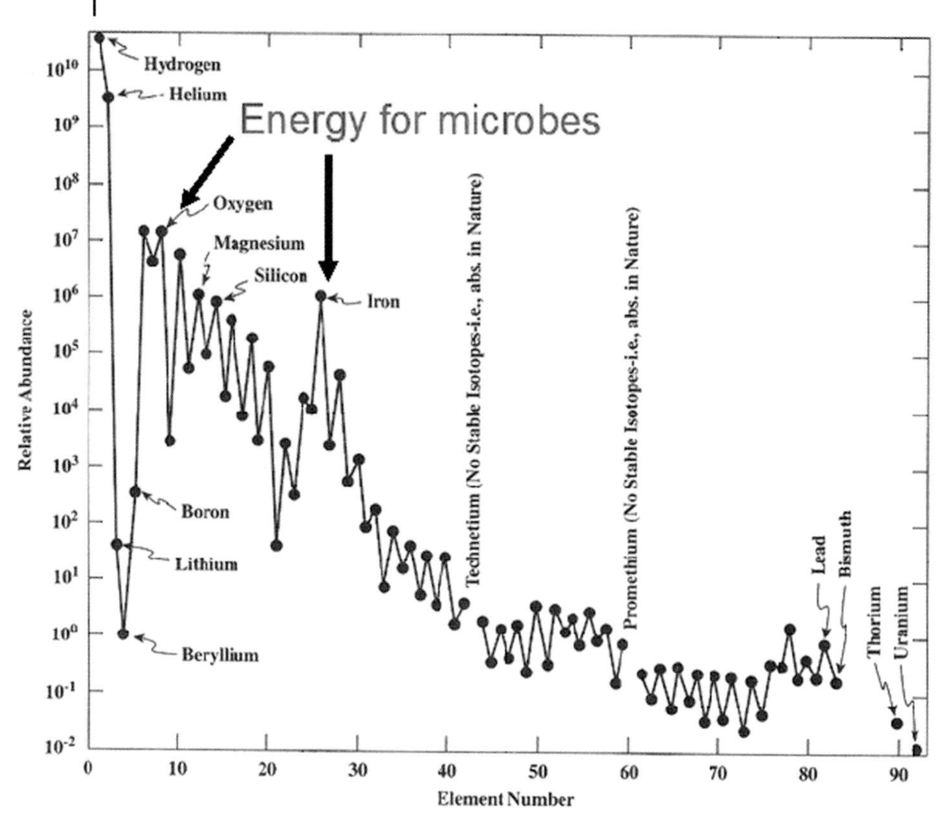

Fig. 6.4 Relative abundance of chemical elements in the atmosphere of our Sun. The graph indicates that redox couples for energy production are widespread throughout the Universe.

In the case of fermentation, the production of energy relies on the breakdown of organic molecules without an electron acceptor being involved – the molecules are essentially dismutated with a release of energy and no overall change in the redox levels of the substrate and products. In that sense, fermentation differs from other types of energy-yielding processes.

The photoautotrophs produce sugars from the reaction of carbon dioxide and water according to the classic reaction in Eq. (6.1). Here, light energy is used to activate chlorophyll or bacteriochlorophyll, which then provides an electron to the electron transport chain.

$$6\ CO_2 + 6\ H_2O + \text{light energy} \rightarrow C_6H_{12}O_6 + 6\ O_2 \tag{6.1}$$

The sugars produced in this reaction are then available for other organisms to respire (burn) in oxygen again, thus yielding the energy locked up in the sugars.

To a large extent, the biosphere we know of on Earth is made possible by photoautotrophs. Without them the biosphere would probably have a productivity much less than about 2% of its current productivity, and the planet would be relegated to chemolithoautotrophs living off rocks and heterotrophs living off the organics they produce. The existence of "complex" multicellular life is dependent on photosynthesis, because photoautotrophs provide the biologically bound carbon and oxygen that make possible the sorts of energy-yielding reactions required for large organisms to operate. The rise in atmospheric oxygen, which seems to have occurred about 2.4 billion years ago (Fig. 6.5), was the result of photosynthesis,

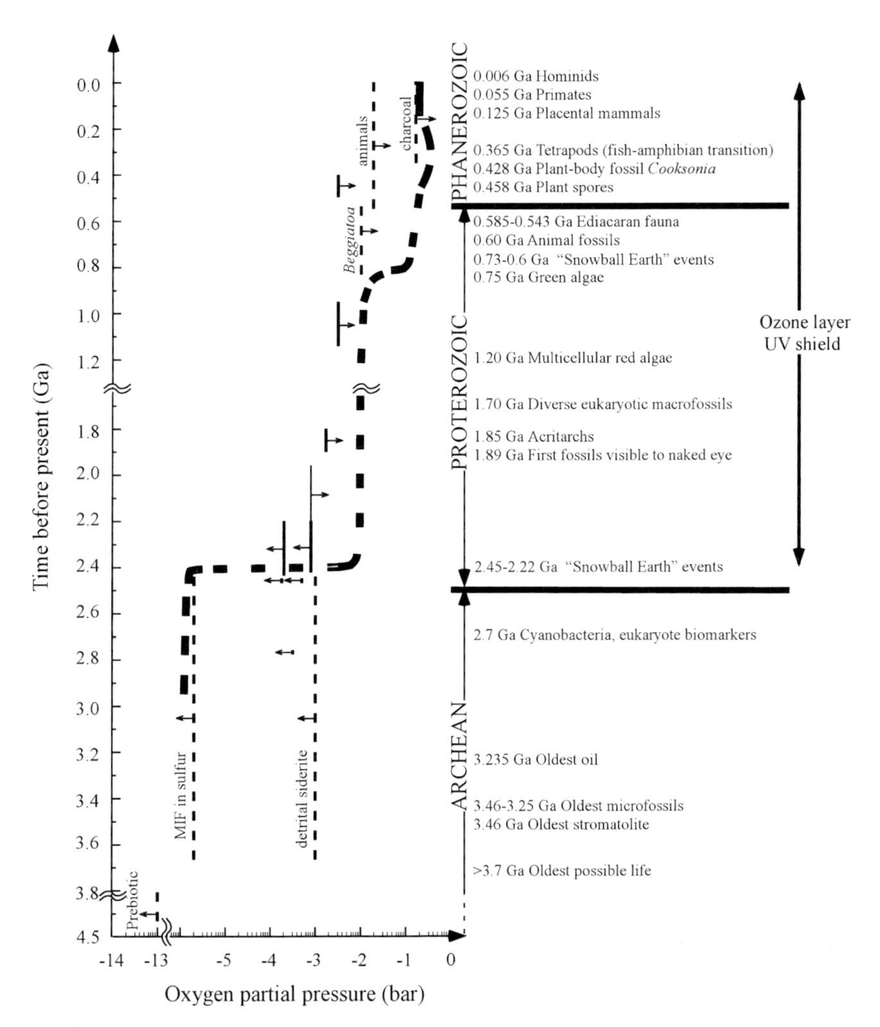

Fig. 6.5 Bounds of oxygen concentrations during the history of the Earth. The innovation of using sunlight as a source of energy and water as an electron donor resulted in the formation of oxygen. Eventually, levels of the gas were sufficient to change the atmospheric composition of the planet (1 bar = 1000 hPa; Ga = billion years) (from Catling with permission)

allowing for the widespread use of this oxidant. Thus, it is important to remember that habitability can also be influenced, even determined, by organisms themselves.

Each successive group of organisms inhabiting the Earth may change the conditions in such a way that it influences the environmental conditions and thereby the nature of the organisms that follow them. In that sense, habitability is not merely a function of planetary chemical and physical conditions but is also predicated on the course of biological evolution that preceded any given point in the planet's history. This course of evolution, at least on Earth, has been strongly coupled to life's mechanisms for gaining energy, photosynthesis having had the most profound effect on planetary habitability.

The minimum energy that life can use on Earth is set by the energy needed to make ATP, which is essentially the energy currency molecule used by life. ATP is a chemical form of energy. The minimum energy that life on Earth seems capable of exploiting is 20 kJ mol^{-1}, although studies on hydrogen production and consumption in sediments suggest that much lower energy availabilities might be exploited by life. The degree to which energy-yielding reactions used by life on Earth represent the "universal" set of energy-yielding possibilities is not known. Nevertheless, life on Earth has mastered a vast array of reactions, and almost every available redox reaction known, at least amongst chemolithoautotrophs, has been shown to have a corresponding organism that uses it. Any planet that has sources of energy in the form of organic molecules, inorganic redox couples, or plentiful availability of light and carbon dioxide can be presumed to be habitable with respect to energy availability.

6.4.3 Other Factors that Determine Habitability

Still other factors may be important for determining habitability and its characteristics over the lifetime of a planet. Many of these other factors are contentious and open to contrary interpretations. For example, the Moon is thought to stabilize the Earth's obliquity. Thus, without it the Earth's obliquity would swing wildly over tens millions of years from 0° to 85°. One might argue that this would make the Earth less habitable by causing dramatic climate changes. One could equally argue that by causing frequent climate change, the lack of a moon would be a selection pressure for generalist organisms that were not so sensitive to rarer climate changes and were more likely to have longer species lifetimes because of their ability to withstand a greater range of environmental changes. Perhaps extinctions would be less common.

Other factors that influence habitability include the impact flux and the magnitude of objects that hit a planet of a given size, which will determine the extent to which surface habitats and the atmosphere are perturbed. Further, the ratio of oceans to land might influence the availability of oceanic and land-based environments for the origin and evolution of life.

Determining which factors are essential for habitability and which are not is subject to anthropocentric interpretations colored by our own evolutionary path. Thus, care must be taken. However, the assertion that the availability of a solvent, energy, and nutrients is the minimum requirement for an environment to be habitable seems to be a safe proposition.

6.5 A Postulate for Habitability

As the brief review of the history of our perception of habitability has shown, a pervasive problem in biology is assessing the habitability of planets for which there is a limited dataset. The expense and logistical difficulty of launching missions to other planets and moons necessitates drawing biological conclusions based on vastly fewer data than are available for habitats on Earth.

On Earth, at least, we can approach the question of whether habitats are habitable or uninhabitable by two means. Firstly, we can directly explore the region in question. Because most regions of the Earth are accessible multiple times within the budgets that can be acquired through research grants, it is possible for scientists to assess the habitability of environments directly. Initial datasets can be reviewed by returning to the site of interest. Secondly, environments can be assessed by using as analogies environments that have been explored previously and that are known to possess very similar physical and chemical characteristics. Thus, a comparison of certain unexplored habitats to similar explored habitats on Earth allows an extrapolation of habitability criteria (surfaces for growth, liquid water, nutrients, light, etc.) to other habitats.

The assessment of actual or past habitats on other planets poses quite different problems for a number of reasons:

- It is, for the time being, logistically difficult for scientists to directly visit extraterrestrial sites of interest themselves.
- The data that do exist are limited and may have been acquired by just a few spacecraft measurements or by observations from Earth, thus greatly limiting the empirical basis of assessment of habitability.
- The data gathered do not necessarily have a biological focus, because that might not have been the focus of the missions in the first place. For example, the concentration of the major macronutrient nitrogen (in its various biologically accessible states) in the Martian regolith is unknown, despite the numerous landers that have been to Mars and the obvious biological importance of this measurement.
- Extraterrestrial environmental conditions are often very different from conditions known to support life on the present Earth, frustrating any analogue parallels that can be used to assess the possibility for life. An obvious example is surface

atmospheric pressure and composition, which is different for Mars, Venus, and Earth (see Chapter 8).

We can use the present-day Earth as our point of reference. However, the Earth itself evolved through time and so did its habitability. In this respect, the early Earth can be regarded as a different planet, and the problems associated with the assessment of its habitability are similar to those that challenge us regarding the other terrestrial planets. These problems are related to the fact that although we have unlimited access to the locations where the oldest well-preserved rocks occur (about 3.5 billion years for evidence of life), the dataset is limited because

- there are only two localities where these ancient rocks are exposed (the greenstone belts of Barberton in South Africa and the Pilbara in Australia);
- these ancient terrains are already relatively young, having formed a billion years after the formation of the Earth; and
- the evidence for potential life in these rocks is disputed and, consequently, so is the habitability of the Earth at that time (see Chapter 1).

The scarcity of the data points is related to the dynamics of a tectonically active planet where the process of lateral plate tectonics has caused the destruction of the oldest rocks.

In view of the inherent complexities in making comparisons between present-day Earth and other planets, including early Earth, one useful approach to the problem is to define a postulate that can be used to guide attempts to make comparisons and to arrive at conclusions on habitability and the likelihood of life.

Owing to the different environmental conditions occurring on other planets compared to the Earth, and to the limited number of datasets available from spacecraft and Earth observations, it is necessary to create a bridge between what we know about habitats on Earth at present and the environment under consideration (Fig. 6.6).

In the case of early Earth the "astrobiological gap" in the assessment of habitability can be filled by visiting the remnants of early Earth or in the case of extraterrestrial environments, by visiting the planet under study. Thus, the questions that remain about the habitat being studied can be directly addressed. However, if the database is poor, as in the situation of the early Earth, or if it is

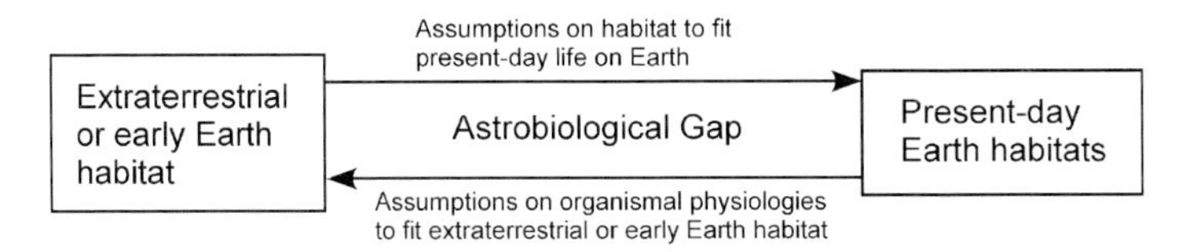

Fig. 6.6 Astrobiological gap in knowledge between present-day Earth habitats and extraterrestrial or early Earth habitats.

not logistically easy to visit the planet in question, the gap must be filled by using the existing data.

Historically, in the literature the gap between extraterrestrial or early Earth habitats and present-day habitats has been bridged in two ways: assumptions about the habitat or assumptions about life.

6.5.1 Assumptions about the Habitat

Assumptions can be made about the physical and chemical characteristics of the early Earth or the environment on the planet under study. The problem with this approach is that it can, without conscious effort, err on the side of biological optimism, especially with respect to extraterrestrial environments.

As an example, we can use nitrogen as a limiting nutrient. Nitrogen is essential for the biosynthesis of proteins and nucleic acids. We know that nitrogen existed in the environment of the early Earth, but we do not know in what biologically available form it occurred. Likewise for Mars, although the concentration of biologically available nitrogen in the Martian regolith is not known, there is nitrogen in the atmosphere. One might assume that this nitrogen could have been abiotically fixed into biologically available nitrogen compounds during the past history of Mars. However, without direct observation of nitrogen in the regolith, we empirically do not know whether it exists, and therefore we cannot empirically assess the habitability of Mars at present. As a result of these uncertainties there is a temptation to succumb to "biological optimism" (see Section 6.1), making extrapolations from the particular environment being investigated that are most suited to proving life.

6.5.2 Assumptions on Life

Theoretical life forms can be matched with the known or inferred conditions on the early Earth or in the extraterrestrial environment. This method of bridging the astrobiological gap is the most pervasive in the literature. In the case of extraterrestrial habitats, its inherent problem is the temptation to resort to "invented physiologies."

Because of a desire to show that the habitat can support life, new physiologies with no counterparts on Earth are invented to show that life might be possible with a new innovation. The problem with this line of reasoning is that it requires a transgression from science into science fiction.

6.5.3 Attempt to Formulate a Habitability Postulate

The purpose of a postulate for habitability is to bridge the gap between the known and the hypothesized while attempting to reduce the chances of biological optimism. The principles are self-evident, but they need to be enunciated. One attempt to formulate a habitability postulate might be that derived of the following postulate: "The proposition that a planet is or was habitable requires that the physiological requirements of microorganisms known at the time of assessment match the empirically determined combined physical and chemical conditions in the environment being assessed" (Cockell and Westall 2004).

How does this simple postulate help to reduce biological optimism? First, we will take the case of habitat assumptions. The second part of the postulate – "the empirically determined combined physical and chemical conditions in the environment being assessed" – requires us to confine ourselves purely to direct observations that have been made by spacecraft or from Earth. Thus, in the absence of a measurement of biologically available nitrogen on the Martian surface, we must conclude that we do not know whether Mars is presently habitable and we cannot know this until we know the concentration of biologically available nitrogen. Thus, until such measurements are made, any hypothesis concerning the presence of active life on Mars must remain a biological speculation. The same is true for measurements of all other known macro- and micronutrients required for life. Thus, the assessment of an extraterrestrial habitat as being conducive – or not, as the case may be – for life requires the empirical measurement of *all* factors that are currently known to be critical to the kind of life being considered as a potential inhabitant.

The requirement to measure all the currently known nutrients in an extraterrestrial or early Earth environment may be a little too stringent. In some particular cases, models might provide realistic insights. However, in the absence of total knowledge of all physical and chemical parameters known to influence life, we are left with gaps. For example, models used to assess the ultraviolet radiation environment on Mars are an excellent example of where habitability has, at the time of writing, been assessed using models rather than direct field observations. In this case light penetration through a CO_2 atmosphere is simple enough for us to be confident that these models give realistic insights into the UV flux at the Martian surface or on the early Earth, but even they must be treated with caution, as we do not know precisely the absorption properties of the ubiquitous Martian dust.

How does the postulate help us to reduce biological optimism in the assumptions about life? The most parsimonious way to prevent biological optimism is not to make assumptions regarding the biochemistry and physiologies of extraterrestrial life and to use only those physiologies currently demonstrated to exist on Earth. This is required in the first part of the postulate – "requires that the physiological requirements of microorganisms known at the time of assessment." This seems intuitive, but in the case of Venusian exobiology, biological optimism has been a pervasive problem, as we shall see in Section 6.6. Even if a postulated physiology appears to be thermodynamically possible, it does not necessarily mean that it is biologically likely. There are costs and benefits to different modes of living. Thermodynamically plausible modes of existence in energy acquisition, motility, UV protection, etc., may not occur if there are some unknown biological or biochemical reasons why such a physiology cannot exist. A possible example of such organisms is the hypothesis of thermosynthetic organisms that use temperature differences at the micron scale as a source of energy. These organisms seem thermodynamically possible, but they have never been observed on Earth. Their existence is therefore, by definition, science fiction at this time.

Thus, using the above-suggested constraints, it is possible to bridge the astrobiological gap in assessing the possibility of life on other planets and in habitats on early Earth while at the same time minimizing the chances of falling prey to biological optimism. If the postulate does not hold, the planet or the microenvironment is uninhabitable at the time of assessment.

For the environment to become habitable, either

- new information about the extraterrestrial or early Earth environment must be obtained that brings the previously uninhabitable environment into the physical and chemical hyperspace for life, or
- new organisms must be discovered on Earth that could potentially grow in the previously uninhabitable environment, i.e., new organisms must be discovered that fall within the physical and chemical hyperspace defined by the extraterrestrial or early Earth environment under consideration.

6.6 Some Test Cases for Habitability

It might be valuable to briefly look at two test cases of habitability. Aside from applying the postulate described above, they will also provide some additional information on habitability.

6.6.1 Test Case One: Life on Venus

The habitability of the clouds of Venus has been a matter of discussion in the exobiology community since the 1960 s. The clouds of Venus provide an example of the way in which exobiological thinking develops in favor of life, and they illustrate the potential value of a postulate on habitability. The life on Venus debate will be briefly reviewed and the postulate of habitability will be applied as a test case.

The habitability of Venus has been a matter of speculation since the turn of the 20th century. Arrhenius (1918) speculated that the planet was covered in microbially dominated swamps. Seckbach and Libby (1970) speculated that acidophilic algae lived in hot Venusian oceans near the poles. Ichinose and Folsome (1973) also considered the habitability of Venus.

These early speculations were afflicted by a lack of knowledge about the environment of Venus at that time and limited knowledge about the range of microbial physiologies found on Earth. In the strictest sense, they do not conform to the second part of the postulate discussed here, because at the time of assessment it was known that microbes required nitrogen, phosphorus, liquid water, and a range of other macro- and micronutrients. Because there was no empirical knowledge of the abundance of such elements and compounds on Venus at the time of assessment, it was not possible to determine whether the physical and chemical environment fitted the physiologies of organisms known on Earth at that time. Thus, "we do not know if the planet is habitable, we require more measurement" would have been a stance consistent with the postulate discussed here.

The conclusions of these scientists can be defended on the basis that models of the environmental conditions on Venus at that time did support the potential for Earthlike organisms (the greenhouse effect on Venus had not been revealed), and so it was reasonable for them to infer that liquid water existed, and perhaps, knowing that Venus was rocky, other elements for life as well. These early authors did stay consistent with the first part of the postulate in that they used only organisms empirically known on Earth. Arrhenius considered swamps, while Seckbach and Libby were more specific and assessed Venus using the acidophilic alga *Cyanidium caldarium*, concluding that it could grow. In that sense they did not succumb to biological optimism in their assumptions about life.

As models and observations of Venus improved, it became apparent that the surface is not conducive to life. The temperature on the surface is almost globally isothermal at 464 °C (see Chapter 8). This is at least 340 °C higher than the currently known upper temperature limit for life (Table 6.1, see also Chapter 5), and there is no liquid water on the surface. Because of this apparent hostility, the clouds of Venus then became a focus of astrobiological studies and have remained so to this day.

Morowitz and Sagan (1967) speculated on the existence of hydrogen-evolving float bladders in the atmosphere of Venus. They proposed that these organisms would float through the Venusian cloud layer and would be 75 μm to 4 cm in size. This paper was the first suggestion of a life form for the Venusian clouds. The problem with the hydrogen-evolving float bladders is that they do not exist in the cloud deck of Earth, although there are marine organisms, e.g., most fishes and the nautiloids, that use buoyancy to regulate their position in the oceans of Earth and cyanobacteria that use gas vesicles for buoyancy. Hydrogen-evolving float bladders might seem thermodynamically plausible, but the lack of them on Earth forces us, under the conditions of the postulate proposed here, to reject them on Venus. The paper, although presenting a beautiful idea, tried to bridge the astrobiological gap by conceiving of entirely new organisms to fit the extraterrestrial environment being considered, and thus it falls foul of the first part of the postulate.

Since these early papers were written, a great deal of knowledge about the Venusian cloud layer has been gathered. It contains sulfuric acid of concentrations between and 81% and 98%. These concentrations present an extreme problem for osmoregulation in any putative cloud biota. Between 48 km and 55 km, the temperature and pressure within the clouds are similar to ground conditions on Earth; however, Venus, with a maximum distance of 0.723 AU from the Sun, does not lie in the habitable zone of our Solar System (Fig. 6.2). The UV flux at the top of the cloud layer, because of Venus's closer proximity to the Sun, is higher than on the surface of the Earth, although deeper in the clouds it is likely to be significantly reduced by CO_2 scattering and the presence of UV absorbers in the atmosphere.

There are a number of ways around the sulfuric acid problem. One could hypothesize the existence of microorganisms that can tolerate this high level of sulfuric acid. However, the assumption of an entirely new physiology that has not been demonstrated on Earth requires a transgression into science fiction and is not tenable under the first part of the postulate.

Another approach might be to assume that there are regions of the Venusian clouds that do not have sulfuric acid and are long-lived. However, this has not been empirically demonstrated at the time of this writing. Thus, this assumption would also require a transgression into science fiction at the time of this writing, and it is not tenable according to the second part of the postulate.

Recently, the cloud deck was discussed as an abode for life (Cockell 1999). The author concluded in a synthetic review that anaerobic sulfate-reducing chemoautotrophic organisms fitted the habitat most closely on account of the presence of sulfate as an electron acceptor, the presence of hydrogen as an electron donor, and the abundance of carbon dioxide and other elements such as phosphorus and nitrogen in an essentially oxygen-free environment. But the paper did not suggest that the organisms were actually there.

More recently, there have been new proposals for organisms in the clouds of Venus that consider new types of physiologies for growth and survival in the cloud layer (e.g., Schulze-Makuch and Irwin 2002). Thus, the debate about life on Venus continues, and it illustrates the breadth of discussions in assessing the environments of other planets and the potential value of a postulate.

At the time of this writing (2006), what we empirically know about the cloud deck of Venus is that there is sulfuric acid at concentrations between 81% and 98% suspended as droplets. We know there is very little water in the clouds. One positive observation for life that we can make is that temperatures and pressures are favorable in certain regions. Because we do not know at the time of assessment of any microorganism on Earth that naturally grows under these combined conditions in the aerosol state permanently in clouds, we should conclude, if we use the postulate formulated above, that Venus is a dead world.

This is not to say that Venus *is* a dead world but rather that at the time of assessment – in other words, at the time this chapter was written – we must conclude that it is a dead world.

6.6.2
Test Case Two: Life on the Early Earth

The use of modern life on Earth as an analogy for ancient life has both positive and negative aspects. Obviously, dinosaurs and the weird, but wonderful, Ediacaran fauna have no modern analogues. But what about single-celled organisms, such as bacteria?

Using the theory of uniformity, i.e., "the present is the key to the past," propounded by the 18th century Scottish geologist James Hutton, the first observations of microfossils in ancient rocks described microorganisms in the 1.9-billion-year-old Gunflint Formation of Canada that resembled contemporary microbial mat and stromatolite-forming, oxygenic photosynthetic cyanobacteria. Since then, a plethora of investigations reported a wide variety of microfossils (often identified down to the species level), from increasingly older rocks up to the highly metamorphosed, 3.8-billion-year-old Isua Greenstone Belt in southwest Greenland. In the latter case, cyanobacteria-like and even yeast-like species were described. The well-preserved 3.5-billion-year-old chert horizons of the Barberton (South Africa) and Pilbara (northwest Australia) Greenstone Belts have yielded many species of cyanobacteria, as interpreted by Schopf, Walter, and others (see also Chapter 1). The identification of these organisms stimulated a cycle of further studies of modern cyanobacterial mats and the continued search for such organisms in the ancient rocks.

The result of nearly 50 years of such investigations is the creation of a framework of understanding and search methodology that is rigidly locked into the identification of fossil cyanobacteria. Although cyanobacteria are wide ranging in morphology, they tend to be much larger than most other types of bacteria and exhibit a type of cell differentiation that does not occur in other prokaryote groups. This makes them relatively easy to observe and identify with simple optical microscopic methods. Moreover, their use of a more efficient energy-producing metabolism led to

- a vast increase in biomass and
- an important influence on the precipitation of calcium carbonate.

The prominent stromatolites so typical of the later Precambrian are a result of these two factors.

Microbial mats formed by non-cyanobacterial microorganisms are, on the other hand, subtler in terms of size, morphology, and volume importance. They, and the microorganisms that form them, also require more sophisticated methods of observation, such as electron microscopes.

The existence of cyanobacteria and yeasts in these ancient rocks would imply very rapid evolution of life, because oxygenic photosynthesis is a sophisticated method of obtaining energy, while the yeasts belong to the eukaryote group. This produced a dichotomy regarding the environment of the early Earth, because modeling by the group of Kasting, as well as the geochemical analyses of the rocks used by Rye and coworkers (1995), suggested that the early environment was very poor in oxygen. (On the contrary, Ohmoto [1999] contended that oxygen levels were as high as they are today).

The dichotomy can, however, be resolved very simply. It was shown by Westall and Folk (2003), for example, that the yeast-like and cyanobacterial microfossils from the Isua rocks are actually recent windblown or endolithic contaminants in the cherts. With respect to the 3.5-billion-year-old cherts from Barberton and the Pilbara, it has been suggested by Westall and coworkers and Brasier and coworkers (2002) that the fossil cyanobacteria are most likely artifacts. Westall (2003) demonstrated that, although life was relatively abundant in the Pilbara and Barberton cherts, it consisted of small microorganisms that formed mats in non-oxygenic environments. These microfossils, with extremely rare exceptions, are individually not visible with the optical microscope, although the mats that they form are.

Was the identification of cyanobacteria in the very ancient cherts biological optimism? In fairness to the original interpretations, little was known about the environment of the early Earth, and thus the interpretation of oxygenic photosynthesizers could be defended (the existence of sophisticated eukaryotes, such as yeasts, was rather surprising, however). The problems involved in studying the most ancient habitats on Earth have obvious relevance to the study of actual and ancient habitats on other planets, such as Mars.

6.6.3
Résumé of the Two Test Cases

These two test cases show that the assessment of environments on other planets or the early Earth as being habitable or uninhabitable needs a postulate that allows the environments and potential life to be evaluated with the minimum chance of biological optimism. Such a postulate was discussed in Section 6.5.

The postulate suggests firstly that the physical and chemical conditions determined by direct or indirect analysis of rocks, in the case of ancient terrestrial environments, and by Earth or spacecraft observations, in the case of other planets, be used as a basis to evaluate the environment as an abode for life. Secondly, the physiologies of known organisms should be used as a basis to assess the plausibility of a potential physiology capable of tolerating those conditions. When applied to Venus, this postulate suggests that the planet is lifeless. With regard to the early Earth, it argues for new approaches to the interpretation of early life and ancient microfossils.

In view of the increasing number of worlds coming under scrutiny and the possibility that even the environmental conditions on the surface of Earthlike extrasolar planets might one day be empirically determined, the debate about the definition of habitable and uninhabitable, and if habitable, by what type of life, has become a useful one for standardizing approaches for the biological evaluation of other planets.

6.7
Conclusions

According to the postulate of habitability, the following items need to be considered:

- Habitability is dependent upon the biochemical system under consideration. However, assuming that life is carbon-based and requires liquid water, sets of conditions required for life can be broadly defined and we can use these conditions to assess the habitability of other planets.

- Conditions for habitability favor microorganisms because they have a broader range of environmental tolerances than multicellular life and therefore may be more widespread, or at least the potential for their existence may be more widespread.
- Defining habitability and searching the Universe for its presence (or demonstrating its absence) constitutes one of the great challenges of biology.

6.8 Further Reading

6.8.1 Books and Articles in Books

Arrhenius, S. (Ed.) *The Destinies of the Stars*, Putman, New York, 1918.

Cockell, C. S. *Impossible Extinction – Natural Catastrophes and the Supremacy of the Microbial World*, Cambridge University Press, Cambridge, 2003.

Crowe, M. J. *The Extraterrestrial Life Debate, 1750–1900*, Dover Publications, 1999.

Dole, S. H. *Habitable Planets for Man*, Blaisdell, New York, 1964.

Haynie, D. T. *Biological Thermodynamics*, Cambridge University Press, Cambridge, 2001.

Lowell, P. *Mars as the abode of Life*, 1909 (reprinted 2005, Adamant Media Corporation).

Lucretius *The Nature of the Universe* (Baltimore), 1951.

Marov, M., Grinspoon, D. H. *The Planet Venus*, Yale University Press, New Haven, 1998.

Robbins, E. I. Appelella Ferrifera, a Possible New Iron-Coated Microfossil in the Isua Iron-Formation, Southwestern Greenland, in: Appel, P. W. U., LaBerge, G. L. (Eds.) *Precambrian Iron Formations*, Theophrastes, Athens, 1987, pp. 141–154.

Schopf, J. W. Tracing the Roots of the Universal Tree of Life, in: A. Brack (Ed.) *The Molecular Origins of Life*, Cambridge Univ. Press, Cambridge, 1998, pp. 336–362.

Schopf, J. W. Earth's Earliest Biosphere: Status of the Hunt, in: Eriksson, P. G. et al. (Eds.) *The Precambrian Earth: Tempos and Events*, Elsevier, Amsterdam, 2003, pp. 516–538.

Schopf, J. W., Walter, M. R. Archean Microfossils: New Evidence of Ancient Microbes, in: Schopf, J. W. (Ed.) *Earth's Earliest Biosphere*, Princeton Univ. Press, Princeton, 1983, pp. 214–239.

Westall, F. Precambrian geology and exobiology, in: Eriksson, P. G. et al. (Eds.), *The Precambrian Earth: Tempos and Events*, Elsevier, Amsterdam, 2003, pp. 575–586.

Wharton, D. *Life at the Limits: Organisms in Extreme Environments*, Cambridge University Press, 2002.

6.8.2 Articles in Journals

Barghoorn, E. S., Tyler, S. A. Microorganisms from the Gunflint Chert, *Science* **1965**, *147*, 563–577.

Beaumont, V., Robert, F. Nitrogen isotope ratios of kerogens in Precambrian cherts: a record of the evolution of atmospheric chemistry? *Precambrian Res.* **1999**, *96*, 63–82.

Brasier, M. D., Green, O. R., Jephcoat, A. P., Kleppe, A. K., van Kranendonk, M., Lindsay, J. F., Steele, A., Grassineau, N. Questioning the evidence for Earth's oldest fossils, *Nature* **2002**, *416*, 76–81.

Caldeira, K. Continental-pelagic carbonate partitioning and the global carbonate-silicate cycle, *Geology* **1991**, *19*, 204–206.

Catling, D. C., Glein, C. R., Zahnle K. J., McKay C. P. Why O2 is required by complex life on habitable planets and the concept of "oxygenation time", Astrobiology, **2005**, *5*, 415–438.

Cockell, C. S. Life on Venus, *Planet Space Sci.* **1999**, *47*, 1487–1501.

Cockell, C. S. Astrobiology and the ethics of new science, *Interdiscipl. Sci. Rev.* **2001**, *26*, 90–96.

Cockell, C. S., Westall, F. A postulate to assess "habitability", *Internat. J. Astrobio.* **2004**, *3*, 157–163.

Cockell, C. S., Catling, D., Davis, W. L., Kepner, R. N., Lee, P. C., Snook, K., McKay, C. P. The ultraviolet environment of Mars: biological implications past, present and future, *Icarus* **2000**, *146*, 343–359.

Cornford, F. M. Innumerable worlds in presocratic philosophy, *Classic Quarterly* **1934**, *28*, 1–16.

Gilichinsky, D., Rivkina, E., Shcherbakova, V., Laurinavichuis, K., Tiedje, J. Supercooled water brines within permafrost – An unknown ecological niche for microorganisms: A model for astrobiology, *Astrobiology* **2003**, 3, 331–341.

Hoehler, T. M., Alperin, M. J., Albert, D. B., Martens, C. S. Apparent minimum free energy requirements for methanogenic Archaea and sulfate-reducing bacteria in an anoxic marine sediment, *FEMS Microbiol. Ecol.* **2001**, *38*, 33–41.

Hoffman, P. F., Kaufman, A. J., Halverson, G. P., Schrag, D. P. A Neoproterozoic snowball earth, *Science* **1998**, *281*, 1342–1346.

Ichinose, N. K., Folsome, C. E. Lack of bacterial survival under cytherean-oriented conditions, *Space Life Sciences* **1973**, *4*, 332–334.

Joshi, M. M., Haberle, R. M., Reynolds, R. T. Simulations of the atmospheres of synchronously rotating terrestrial planets orbiting M dwarfs: Conditions for atmospheric collapse and the implications for habitability, *Icarus* **1997**, *129*, 450–465.

Kashefi, K., Lovley, D. R. Extending the upper temperature limit for life, *Science* **2003**, *301*, 934.

Kasting, J. F. Earth's early atmosphere, *Science* **1993**, *259*, 920–926.

Kasting, J. F., Catling, D. Evolution of a habitable planet, *Ann. Re. Astron. Astrophys.* **2003**, 41, 429–463.

Kasting, J. F., Whitmire, D. P.,Reynolds, R. T. Habitable zones around main sequence stars, *Icarus* **1993**, *101*, 108–128.

Kopp, R. E., Kirschvink, J. L., Hilburn, I. A., Nash, C. Z. The paleoproterozoic snowball Earth: A climate disaster triggered by the evolution of oxygenic photosynthesis, *Proc. Natl. Acad. Sci. USA* **2005**, *102*, 11131–11136.

Kuhn, W. R., Atreya, S. K. Solar radiation incident on the Martian surface, *J. Mol. Evol.* **1979**, *14*, 57–64.

Laskar, J., Joutel, F., Robutel, P. Stabilization of the Earth's obliquity by the moon, *Nature* **1993**, *361*, 615–617.

Lowell, R. P., DuBose, M. Hydrothermal systems on Europa, *Geophys. Res. Lett.* **2005**, *32, Art. No. L05202.*

Malin, M. C., Edgett, K. S. Evidence for recent groundwater seepage and surface runoff on Mars, *Science* **2000**, *288*, 2330–2335.

Mancinelli, R. L. The search for nitrogen compounds on the surface of Mars, *Adv. Space Res.* **1996**, *18*, 241–248.

McKay, C. P., Toon, O. B., Kasting, J. F. Making Mars habitable, *Nature* **1991**, *352*, 489–496.

Melezhik, V. A. Multiple causes of Earth's earliest global glaciation, *Terra Nova* **2006**, *18, 130–137.*

Morowitz, H., Sagan, C. Life in the clouds of Venus, *Nature* **1967**, *215*, 1259–1260.

Muller, A. W. J. Finding extraterrestrial organisms living on thermosynthesis, *Astrobiology* **2003**, *3*, 555–564.

Oder, R. K., Forsythe, J., Webber, D. M., Wells, J & Wells M. J. Activity levels of nautilus in the wild, *Nature* **1993**, *362*, 626–628.

Ohmoto, H. Redox state of the Archean atmosphere: Evidence from detrital heavy minerals in ca. 3250–2750 Ma sandstones from the Pilbara craton, Australia: Comment, *Geology* **1999**, *27*, 1151–1152.

Orlando, T. M., McCord, T. B., Grieves, G. A. The chemical nature of Europa surface material and the relation to a subsurface ocean, *Icarus* **2005**, *177*, 528–533.

Patel, M. R., Berces, A., Kerekgyarto, T., Ronto, G., Lammer, H., Zarnecki, J. C. Annual solar exposure and biological effective dose rates on the Martian surface, *Adv. Space Res.* **2004**, *33*, 1247–1252.

Pimenov, N. V., Savvichev, A. S., Rusanov, I. I., Lein, A. Y., Ivanov, M. V. Microbiological processes of the carbon and sulphur cycles at cold methane seeps of the North Atlantic, *Microbiology* **2000**, *69*, 709–720.

Pinti, D. L., Hashizume, K. ^{15}N-depleted nitrogen in Early Archean kerogens: clues on ancient marine chemosynthetic-based ecosystems? *Precamb. Res.* **2001**, *105*, 85–88.

Pflug, H. D. Archean fossil finds resembling yeasts, *Geol. Palaeontol.* **1979**, *13*, 1–8.

Pflug, H. D., Jaeschke-Boyer, H. Combined structural and chemical analysis of 3,800-Myr-old microfossils, *Nature* **1979**, *280*, 483–486.

Rye, R., Kuo, P. H., Holland, H. D. Atmospheric carbon dioxide concentrations before 2.2 billion years ago, *Nature* **1995**, *378*, 603–605.

Schink, B. Energetics of syntrophic cooperation in methanogenic degradation, *Microbiol. Mol. Biol. Rev.* **1997**, *61*, 262–280.

Schopf, J. W. Microfossils of the Early Archean Apex Chert: new evidence of the antiquity of life, *Science* **1993**, *260*, 640–646.

Schopf, J. W., Packer, B. M. Early Archean (3.3 billion to 3.5 billion-year-old) microfossils from Warawoona Group, Australia, *Science* **1987**, *237*, 70–73.

Schulze-Makuch, D., Irwin, L. N. Reassessing the possibility of life on Venus: proposal for an astrobiology mission, *Astrobiology* **2002**, *2*, 197–202.

Schulze-Makuch, D., Grinspoon, D. H., Abbas, O., Irwin, L. N., Bullock, M. A. A sulphur-based survival strategy for putative phototrophic life in the Venusian atmosphere, *Astrobiology* **2004**, *4*, 11–18.

Seckbach, J., Libby, W. F. Vegetative life on Venus? Or investigations with algae which grow under pure CO_2 in hot acid media at elevated pressures, *Space Life Sciences* **1970**, *2*, 121–143.

Takami, H., Inoue, A., Fuji, F., Horikoshi, K. Microbial flora in the deepest sea mud of the Mariana Trench, *FEMS Microbiol. Lett.* **1997**, *152*, 279–285.

Walsby, A. E. Gas vesicles, *Microbiol. Rev.* **1994**, *58*, 94–144.

Walsh, M. M. Microfossils and possible microfossils from the Early Archean Onverwacht Group, Barberton Mountain Land, South Africa, *Precambrian Res.* **1992**, *54*, 271–293.

Westall, F. The nature of fossil bacteria, *J. Geophys. Res.* **1999**, *104*, 16437–16451.

Westall, F. Gould S. J. Les procaryotes et leur évolution dans le contexte géologique, *Palévol.* **2003**, *2*, 485–501.

Westall, F., Folk., R. L. Exogenous carbonaceous microstructures in Early Archaean cherts and BIFs from the Isua greenstone belt: Implications for the search for life in ancient rocks, *Precambrian Res.* **2003**, *126*, 313–330.

Westall, F., Steele, A., Toporski, J. Walsh, M., Allen, C., Guidry, S., Gibson, E., Mckay, D., Chafetz, H. Polymeric substances and biofilms as biomarkers in terrestrial materials: Implications for extraterrestrial materials, *J. Geophys. Res.* **2000**, *105*, 24511–24527.

Westall, F., De Wit, M. J., Dann, J., Van Der Gaast, S., De Ronde, C., Gerneke, D. Early Archaean fossil bacteria and biofilms in hydrothermally-influenced, shallow water sediments, Barberton greenstone belt, South Africa, *Precambrian Res.* **2001**, *106, 93–116.*

Whitman, W. B., Coleman, D. C., Wiebe, W. J. Prokaryotes: The unseen majority, *Proc. Natl. Acad. Sci. USA* **1998**, *95*, 6578–6583.

6.9 Questions for Students

Question 6.1

Describe two factors that might be important for the habitability of a planet. Why are they important?

Question 6.2

Why is Jupiter's moon Europa such a focus for astrobiologists? What factors about this moon make it biologically interesting?

Question 6.3

Describe how the carbonate–silicate cycle is involved in regulating planetary temperature.

Question 6.4

Discuss the assertion, “The circular objects on the moon are the fortresses that lunarians have built to defend their cities.”

7
Astrodynamics and Technological Aspects of Astrobiology Missions in Our Solar System

Stefanos Fasoulas and Tino Schmiel

The chapter gives an introduction into some basic technological aspects related to space missions in general and to astrobiology missions to other planets, asteroids, comets, and/or moons in particular. We focus on explaining the payload mass limits and the specific constraints given by orbital mechanics.

7.1
Introduction

Space missions are associated with many different applications, e.g., telecommunication, navigation, Earth observation, planetary exploration, and, very often, manned spaceflight. In the past few years, another discipline, namely, astrobiology, has become more and more important. Today, many scientific space missions are especially dedicated to tackling astrobiological questions. The opportunities of space missions in low Earth orbit (LEO), by using space-based telescopes or by exploring *in situ* the planets (see Chapter 12), moons, and other bodies in our Solar System, provide unique possibilities for all scientists seeking a deeper understanding of the origin and nature of life. Thus, the community of scientists from other disciplines that formerly were not necessarily linked to the space sciences is continuously increasing.

In the meanwhile, astrobiology missions are much diversified, using different experiments in LEO, space-based telescopes, different landing probes, satellites orbiting other planets, and, possibly in the very near future, sample return missions from different locations in our Solar System. All these missions have their specific constraints, but they also have some common boundary conditions. Therefore, this chapter will concentrate on some selected basic boundary conditions for space missions, especially for the reader who is not familiar with space engineering. One main focus is to understand what space engineers mean when they talk about a required velocity change Δv to realize a particular mission, or about payload

Complete Course in Astrobiology. Edited by Gerda Horneck and Petra Rettberg

ISBN: 978-3-527-40660-9

and structure fractions, and why they are always worried about each gram of a mission.

Probably everyone has the feeling that the mass budget of a space mission is critical, but why is this so? This shall be explained by the basic rocket equation. Then some basics of orbital mechanics will be discussed, followed by a brief introduction into orbital maneuvers. Finally, in order to illustrate our general findings, missions to Mars will be briefly addressed.

7.2 The Rocket Equation

7.2.1 Single-staged Rockets

The original derivation of the rocket equation is attributed to Konstantin Tsiolkovsky (1857–1935), a Russian schoolteacher (Fig. 7.1). The Tsiolkovsky rocket equation, which is simply the result of the momentum conservation equation for a single-staged rocket, states that the characteristic velocity change Δv_{ch} of a rocket is proportional to the velocity of the exhaust gases c_e multiplied with the logarithm of the initial total mass m_0 divided by the final, end-of-burn mass m_b:

$$\Delta v_{ch} = c_e \ln\left(\frac{m_0}{m_b}\right) \quad (7.1),$$

where $$m_0 = m_{Structure} + m_{Payload} + m_{Fuel} \quad (7.2),$$

and $$m_b = m_{Structure} + m_{Payload} \quad (7.3).$$

The structure mass ($m_{Structure}$) comprises the mass of all needed subsystems of a rocket, e.g. the tanks, thrusters, pumps, pipes, sensors, and other subsystems. The payload mass ($m_{Payload}$) essentially comprises the scientific instruments of a mission plus all equipment supporting these instruments. Now, by defining a structure mass ratio σ and a payload mass ratio μ according to

$$\sigma = \frac{m_{Structure}}{m_0} \quad (7.4),$$

and $$\mu = \frac{m_{Payload}}{m_0} \quad (7.5),$$

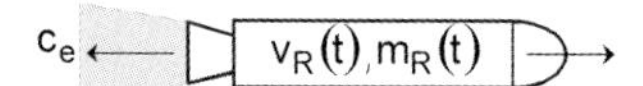

Fig. 7.1 Scheme for the derivation of the rocket equation; c_e = velocity of the exhaust gas; $v_R(t)$ = velocity of the rocket depending on time; $m_R(t)$ mass of the rocket depending on time.

Eq. (1) can be rewritten as

$$\Delta v_{ch} = c_e \ln\left(\frac{1}{\sigma + \mu}\right) \qquad (7.6).$$

Table 7.1 shows that typical values for the structure fractions of chemical rockets are below 10%. Based on this relation, Fig. 7.2 shows the resulting payload fractions for different structure fractions as a function of $\Delta v_{ch} / c_e$. Now, in order to illustrate the meaning of a structure fraction σ = 5–10%, let us consider an example from biology. What is the structure fraction of an egg, which means, what is the shell mass related to the total egg mass?

Table 7.1 Initial total mass m_0, structure fraction σ for the first stage, and launch acceleration of three different rocket types (g_0 = acceleration of gravity).

Rocket type	Total mass m_0 (kg)	Launch acceleration	Structure fraction σ for the first stage
Scout	18 100	2.53 g_0	8.9%
Ariane	202 510	1.22 g_0	6.6%
Saturn V	2 850 000	1.19 g_0	4.8%

In a non-representative experiment the author performed for a lecture in Dresden, it turned out that the average structure fraction of six free-range eggs was about 11% of the total mass. Certainly, an eggshell is not made of the strongest

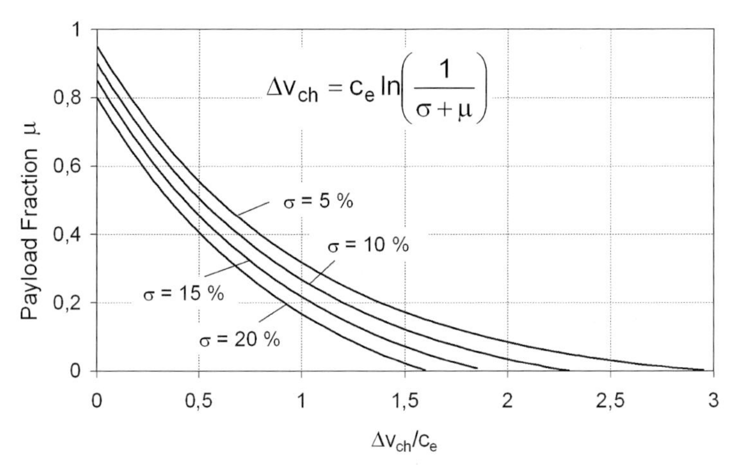

Fig. 7.2 Payload fraction as a function of $\Delta v_{ch} / c_e$; see Eq. (6) (σ = structure fraction; μ = payload fraction).

material that could be provided by nature, but, on the other hand, it usually does not have to withstand accelerations of up to several g's, extreme vibrations, high internal temperatures and pressures, aggressive and explosive fuel compounds, and so forth. This figurative example of an egg may give an impression of how thin and fragile rockets must be, having only about half the structure fraction of an egg.

Indeed, some rocket structures are so thin that they would buckle under their own weight as soon as they are put upright. How is this done, then? First the fuel is put into the tanks under a higher pressure and then the rocket is put upright. The tanks serve simultaneously as the skin structure of the rocket, and therefore the internal fuel pressure stabilizes the rocket.

As a matter of fact, the results of Eq. (6) as depicted in Fig. 7.2 show that, even for a very ambitious structure fraction of about 5%, the characteristic velocity change would be lower than 2.8 times the effective exhaust velocity if some payload, let's say of about 1% of the total mass, is assumed:

$$\Delta v_{ch,max} \leq 2.8 \cdot c_e \text{ for } \sigma = 0.05 \text{ and } \mu = 0.01 \qquad (7.7).$$

The situation is of course worse for a higher structure fraction. On the other hand, the effective exhaust velocity c_e depends on the total energy in the combustion chamber of a rocket, which is given for chemical rocket propellants mainly by the heat of formation Δh_R, e.g., for liquid hydrogen (LH_2) and liquid oxygen (LOX) as propellant, by

$$LH_2 + LOX \Rightarrow H_2O + \Delta h_R \qquad (7.8)$$

$$c_{e,max,ideal} = \sqrt{2\Delta h_R} \qquad (7.9).$$

Subsequently, the heat of formation must be chosen as high as possible in order to achieve a high exhaust velocity. Some values for chemical compounds with the highest known heat of formation are given in Table 7.2. With the exception of the compound BN, these compounds are oxides or fluorides of metals, hydrogen, or carbon. The burning of metals in a combustion chamber is technically not an easy task, as they often build impermeable oxide layers that could prevent a complete burn or could block the nozzle throat. Some of them are extremely toxic and highly reactive and thus are very difficult to store. Nevertheless, some metals as powders, mainly aluminum, are given as energy-increasing additions to solid rocket propellants. Thus, practically all chemical propellants used in rocketry are based on hydrogen and its compounds with carbon and nitrogen. Fluorine has never reached an operational status because of its high reactivity. Thus, only oxygen and its compounds are used as oxidizers.

Table 7.2 Heat of formation and maximum velocity of the exhaust gases [$c_{e\,(max,\,ideal)}$] for some chemical compounds.

Compound	Heat of formation ($MJ\ kg^{-1}$)	Melting point (°C)	Boiling point (°C)	$c_{e,\,max,\,ideal}$ ($m\ s^{-1}$)
BeO	29.670	2 550	3 850	6 921
LiF	23.606	848	1 767	6 869
BeF_2	21.495	547	1 170	6 554
Li_2O	19.986	1 427	2 997	6 320
B_2O_3	18.352	450	2 217	6 056
AlF_3	17.766	–	1 279	5 959
MgF_2	17.640	1 263	1 357	5 937
BF_3	16.676	–129	–101	5 773
Al_2O_3	16.425	2 045	2 700	5 729
SiF_4	15.000	–90	–	5 475
MgO	14.958	2 642	2 800	5 468
SiO_2	14.623	1 610	2 727	5 406
HF	14.192	–85	19	5 326
NaF	13.576	992	1 704	5 209
H_2O	13.442	0	100	5 183
CF_4	10.391	–184	–128	4 557
BN	9.805	–	(2 327)	4 427
CO_2	8.967	–58	–79[a]	4 233

a Sublimation point

Of course, the ideal value for the velocity of the exhaust gases c_e cannot be reached because of internal losses in the thrusters. When expanding the propellants through a nozzle into a vacuum, for the hydrogen–oxygen combination a value of about 4–4.5 km s^{-1} can be achieved, corresponding to an efficiency of about 75 %. Additionally, when launching from the ground, c_e is further limited to about 3–3.4 km s^{-1}. Thus, one can calculate the Δv limit for single-staged rockets to about 9–10 km s^{-1}. It will be shown later that this is the minimum value needed to get a satellite into a low Earth orbit.

7.2.2 Multiple-staged Rockets

The solution to obtaining a higher Δv, the staging concept, is based on the idea of continuously jettisoning structure mass that is not needed anymore, e.g. thrusters, tanks, boosters, and payload fairing. The describing equation is then

$$\Delta v_{ch,total} = c_{e1} \ln\left(\frac{m_{01}}{m_{b1}}\right) + c_{e2} \ln\left(\frac{m_{02}}{m_{b2}}\right) + \ldots = \sum_{i=1}^{n} c_{ei} \ln\left(\frac{m_{0i}}{m_{bi}}\right) \quad (7.10),$$

$$\text{where} \quad m_{bi} \neq m_{0(i+1)} \quad (7.11).$$

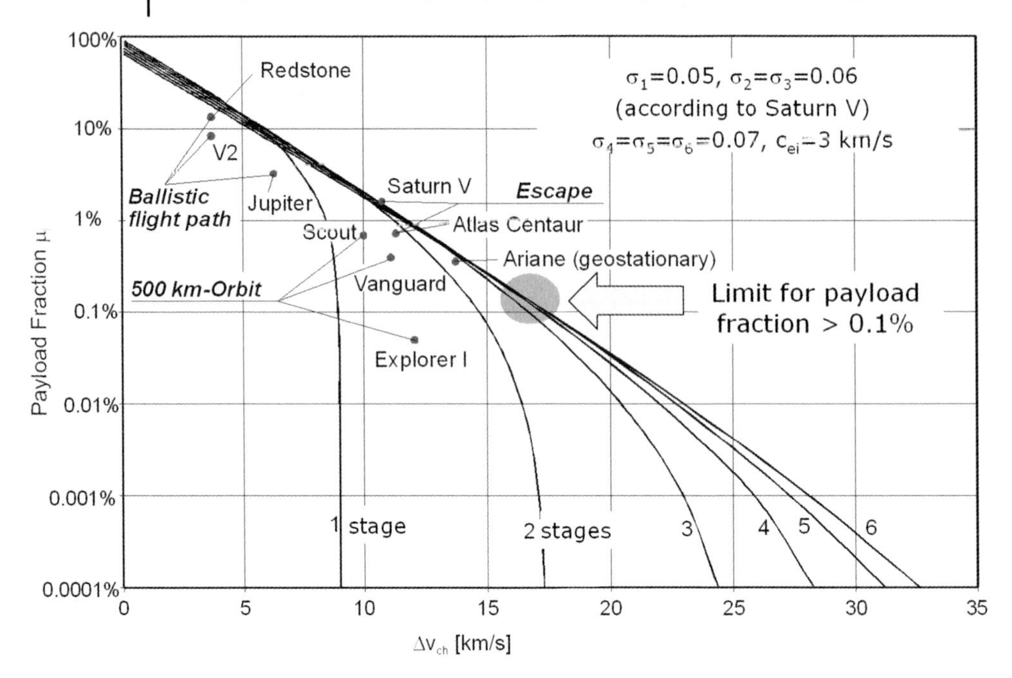

Fig. 7.3 Payload fraction as a function of Δv_{ch} for single- and multiple-staged rockets; index in σ_i gives the number of stages, e.g., σ_1 for 1-stage rocket, σ_2 for 2-stage rocket, etc. The names give examples of different rockets launched.

Thus, the overall Δv can be increased as depicted in Fig. 7.3. Indeed, Δv could be increased almost linearly with increasing number of stages. However, in this case the payload fraction is decreased exponentially (notice that the payload fraction is depicted on a logarithmic scale in Fig. 7.3). Thus, for multiple-staged rockets Δv is limited if a payload fraction of at least 0.1% is assumed.

7.3
Orbital Mechanics and Astrodynamics

7.3.1
Some Historical Notes

The description of planetary motion has challenged scientists for many centuries. The early Greeks believed that celestial bodies moved in circles around the Earth. At that time this idea was in excellent agreement with observations for all distant stars. However, some of the observed bodies had a somewhat strange motion. Sometimes they moved backward and forward in the celestial sphere reference. Thus, they were called planets, from the Greek word for "wanderer" or "nomad." This geocentric universe model probably reached its peak of refinement around the year 140 with Ptolemaeus (85–165), who calculated the orbits of the planets using various combinations of circles known as epicycles.

It took almost 1 500 years before Nicolaus Copernicus (1473–1543) placed the Sun at the center of the Solar System, with the Earth rotating on its axis once a day while moving around the Sun once a year. However, Copernicus still believed that a planetary orbit should be circular, and his geometry was therefore as complex as that of Ptolemaeus. It was Johannes Kepler (1571–1630) who, based on the precise observations of Tycho Brahe (1546–1601), discovered and formulated the laws of planetary motion that are known today as Kepler's laws (Fig. 7.4).

Kepler's First Law: *The planets orbit the Sun in elliptical orbits with the Sun at one focus.*

Kepler's Second Law: *The line connecting a planet to the Sun sweeps out equal areas in equal amounts of time.*

Kepler's Third Law: *The time required for a planet to orbit the Sun, called its period, is proportional to the long axis of the ellipse raised to the 3/2 power. The constant of proportionality is the same for all the planets.*

However, Kepler could not explain physically what forces were responsible for these laws. This explanation was provided by Isaac Newton (1642–1727). He published his three laws of motion and the law of universal gravitation in the *Principia* in 1687, stating the following:

Newton's First Law of Motion: *A body stays at rest or moves with a constant velocity on a straight line unless acted upon by an external unbalanced force.*

Newton's Second Law of Motion: *An applied force is equal to the time rate of change of momentum.*

Newton's Third Law of Motion: *All forces occur in pairs, and these two forces are equal in magnitude and opposite in direction.*

Newton's Universal Law of Gravitation: *The force of gravity between two bodies is directly proportional to the product of their masses and inversely proportional to the square of the distance between them.*

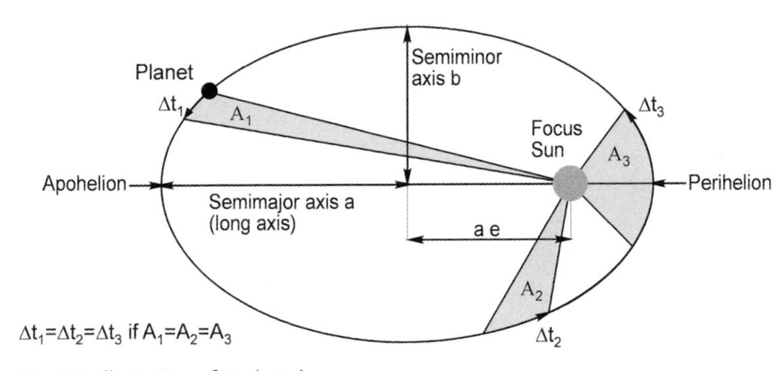

Fig. 7.4 Illustration of Kepler's laws.

$$F = \gamma \frac{mM}{r^2} \qquad (7.12),$$

where F = force of gravity; m and M = mass of the two bodies; r = distance between the two bodies; γ = gravitation constant.

$$\gamma = 6.670 \cdot 10^{-11} \frac{m^3}{kgs^2} \qquad (7.13).$$

These laws explain why a satellite, a moon, or a planet stays in orbit around the Earth, a planet, or the Sun, respectively. Very often it is argued that the gravitational force is counterbalanced by a centrifugal force. This is easy to understand but is physically incorrect, because according to Newton's first law of motion the body should then maintain a constant velocity on a straight line, which is obviously not the case.

The correct answer is that gravity forces the satellite to fall towards the Earth; however, because of its velocity it will never touch the Earth's surface, meaning that the satellite is in a continuous free fall "around" the Earth. Today, it is known, thanks to Albert Einstein (1879–1955), that even this answer is not the complete truth. According to Einstein's theory of relativity, the answer should be that gravity mass causes a curvature in space-time, in which the satellite is caused to move.

By combining Newton's second law of motion with his universal law of gravitation, an equation is obtained for the satellite's acceleration vector $\ddot{\vec{r}}$ if it is assumed that gravity of only one central point mass is the only force:

$$\ddot{\vec{r}} + \frac{\gamma M}{r^3} \vec{r} = \vec{0} \qquad (7.14).$$

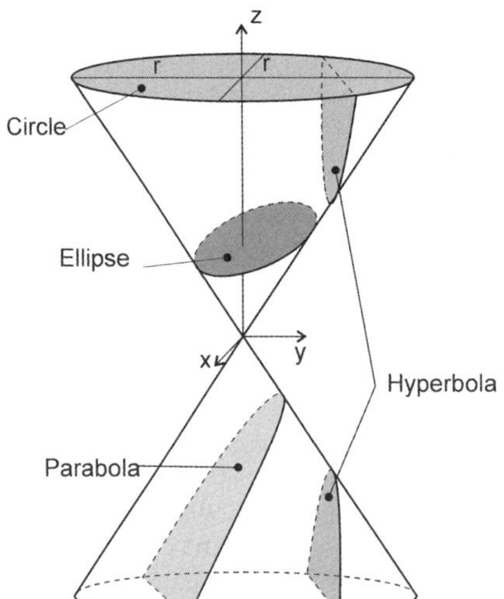

Fig. 7.5 Satellites can orbit a central mass in any of the four conic sections: circle, ellipse, parabola, or hyperbola.

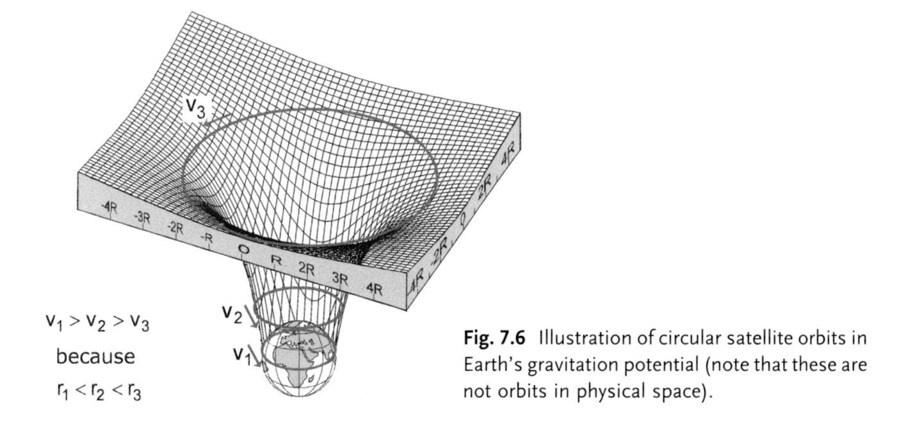

Fig. 7.6 Illustration of circular satellite orbits in Earth's gravitation potential (note that these are not orbits in physical space).

This equation is called the "two-body equation of motion," giving the motion of a satellite's position vector $\vec{r}$ in his orbit. A solution to this equation is the polar equation of a conic section:

$$r = \frac{a(1 - e^2)}{1 + e\cos\theta} \qquad (7.15),$$

where a is the semi-major axis, e is the eccentricity, and θ is the so-called true anomaly, an angle measured from the closest point of the orbit to the current position (see also Fig. 7.4). A conic section is a curve formed by cutting a plane through a right circular cone, as illustrated in Fig. 7.5. The angular orientation of the cutting plane determines whether the conic section is a circle, ellipse, parabola, or hyperbola.

7.3.2
The Energy Conservation Equation

In order to illustrate the physical meaning of the possible orbits, another equation that also can be derived from the two-body equation of motion can be used, the so-called "vis-viva" or "energy conservation equation":

$$\underbrace{\frac{1}{2}v^2}_{\substack{kinetic\\energy}} - \underbrace{\frac{\gamma M}{r}}_{\substack{potential\\energy}} = \text{const.} = \varepsilon = -\frac{\gamma M}{2a} = \frac{1}{2}v_\infty^2 \qquad (7.16),$$

where ε is the total mass-specific mechanical energy of the satellite and v is the magnitude of the velocity. The term for the potential energy defines the potential energy as zero at infinity and negative at any radius less than infinity (Fig. 7.6). Using this definition, the specific mechanical energy of elliptical and circular orbits is always negative, and a velocity at infinite distance to the central body mass cannot be defined. As the energy increases and approaches zero, the semi-major

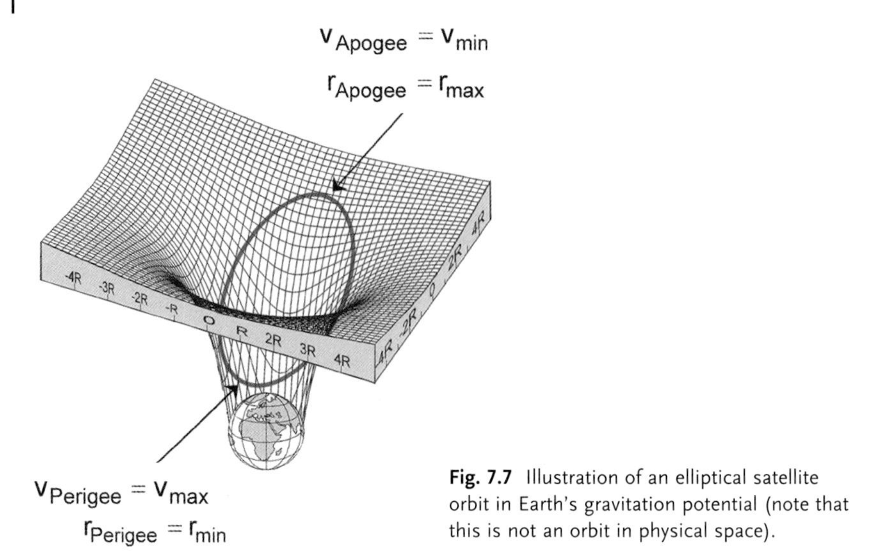

Fig. 7.7 Illustration of an elliptical satellite orbit in Earth's gravitation potential (note that this is not an orbit in physical space).

axis of the ellipse gets larger and larger, approaching a parabolic orbit. The velocity on a parabolic orbit is thus high enough to escape from the central mass, having at infinity a zero velocity relative to this mass. If the energy is further increased to positive values, the velocity is then high enough to escape from the central mass, having at infinity a finite velocity relative to this mass, leading to hyperbolic orbits.

The orbit period P for elliptical orbits also can be derived from the energy conservation equation as

$$P = 2\pi\sqrt{\frac{a^3}{\gamma M}} \tag{7.17}$$

For circular orbits, the semi-major axis a simply equals the orbit radius r. (Recall that this is just Kepler's third law including the constant of proportionality.)

7.3.3 Some Typical Velocities

The velocity of circular orbits is given by the energy conservation equation as

$$v_{circle} = \sqrt{\frac{\gamma M}{r}} \tag{7.18}$$

Therefore, the velocity needed to maintain a circular orbit is higher the closer the satellite is to the central body. This is illustrated in Fig. 7.6. Notice that the curves shown in Fig. 7.6 are not orbits in physical space but only in potential energy.

The velocity on elliptical, parabolic, or hyperbolic orbits at a certain distance from the central mass can be calculated by

$$v = \sqrt{\frac{2\gamma M}{r} - \frac{\gamma M}{a}} \qquad (7.19).$$

Thus, the velocity reaches a maximum value for the closest orbit point and a minimum for the most distant point of the orbit (Fig. 7.7).

Now, staying on Earth's surface, what kind of orbits can be achieved and what velocities are needed? Assuming a launch location on a big mountain and that the Earth is a perfect sphere without an atmosphere, the velocity to which a satellite must be accelerated perpendicular to Earth's gravitational field is simply

$$v_{LowEarthOrbit} = \sqrt{\frac{\gamma M_{Earth}}{R_{Earth}}} \approx 7.91 \frac{km}{s} \qquad (7.20).$$

This value is known as the first cosmic velocity. For slightly lower values, the satellite would fall in an elliptical trajectory back to Earth; for higher values it would reach an elliptical orbit around the Earth. However, if the velocity is further increased, reaching an infinite value for the semi-major axis *a*, then the satellite would be on a parabola. The velocity needed for this at Earth's surface is called the second cosmic velocity and has a value of

$$v_{Parabola} = \sqrt{\frac{2\gamma M_{Earth}}{R_{Earth}}} \approx 11.2 \frac{km}{s} \qquad (7.21).$$

Figure 7.8 depicts these findings. If losses resulting from atmospheric drag or from the fact that satellites cannot be launched directly perpendicular to Earth's gravitational field are taken into account, about 15–20% must be added in order to

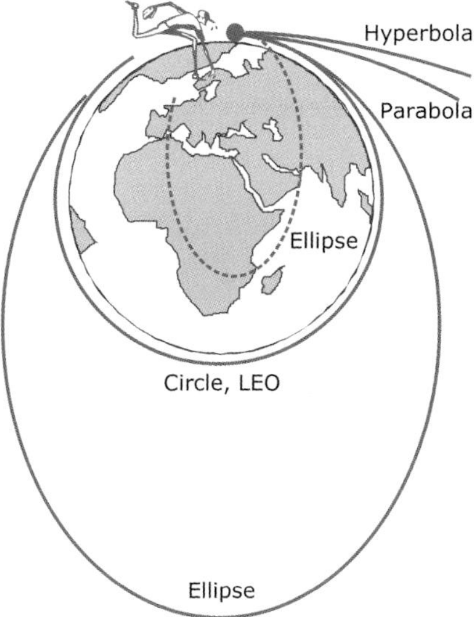

Fig. 7.8 Illustration of velocities needed for reaching a LEO and for escaping from the gravitational pull, assuming that Earth is a perfect sphere without atmosphere (and that the pitcher has enough thrust). See text for more details on the required velocities for the different orbits.

estimate the necessary Δv to get a satellite into orbit. Thus, for low Earth orbits, this Δv demand is about 9–10 km s^{-1}. Figure 7.8 also illustrates that for establishing elliptical orbits other than those going through the launching point, additional maneuvers or burns are necessary.

7.4 Orbital Maneuvers

7.4.1 High-thrust Maneuvers

To change a satellite's orbit, its velocity vector in magnitude and/or direction must be changed. Compared to the orbital period, most chemical propulsion systems operate for a short time only, so the maneuver can be treated as an impulsive change in velocity while the position is the same. Therefore, any maneuver must occur at a point where the old orbit intersects with the new. If there is no intersection, then an intermediate orbit must be used, which intersects both. Generally, the change in velocity needed to go from one orbit to another is

$$\Delta\vec{v} = \vec{v}_2 - \vec{v}_1 \qquad (7.22).$$

As a first example, an orbit change from an Earth equatorial orbit to a polar orbit at the same altitude is considered. The required velocity change would be

$$|\Delta\vec{v}| = |\vec{v}_2 - \vec{v}_1| = \sqrt{2}v_1 \qquad (7.23).$$

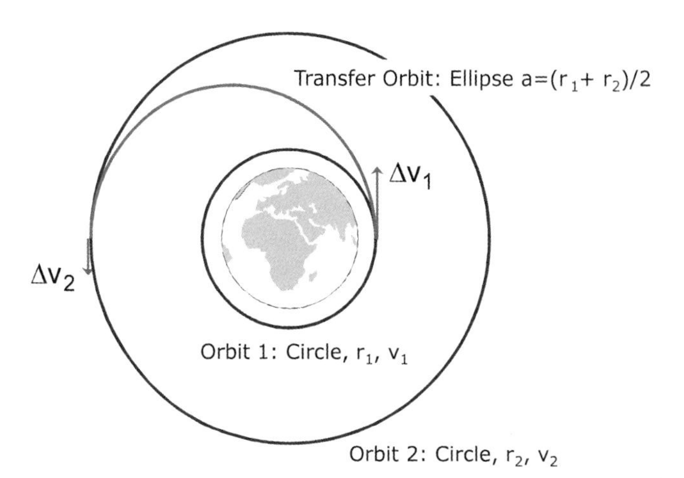

Fig. 7.9 The Hohmann transfer ellipse between two circular orbits. Note that the velocity on the final orbit 2 is lower than on the initial orbit 1, but the satellite must be accelerated twice for the transfer.

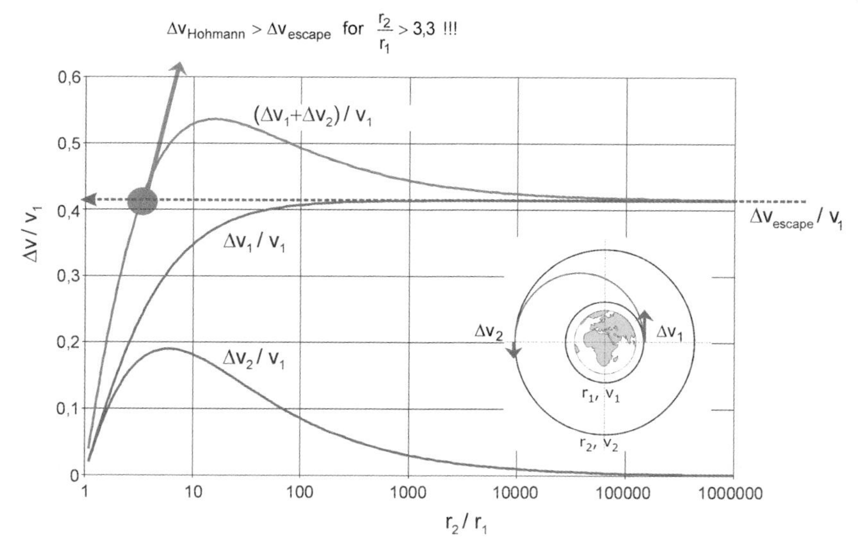

Fig. 7.10 Required Δv for a Hohmann transfer ellipse between two circular orbits.

Performing this maneuver in low Earth orbit would mean that the required Δv is $\sqrt{2}v_1 \cdot 7.91\,\mathrm{km\,s^{-1}} = 11.2\,\mathrm{km\,s^{-1}}$. Recall that this is just the value required to escape from Earth's gravitation when launched from the ground! Out-of-orbital-plane changes are therefore performed only when absolutely required.

In order to establish an orbit that is higher in altitude (e.g., a geosynchronous orbit [GEO]) than the initial one (e.g., a LEO), another typical maneuver is performed. The most efficient direct transfer orbit is an ellipse that tangentially intersects both orbits (Fig. 7.9). This transfer ellipse is called the Hohmann transfer orbit. The total change in velocity required for the transfer is the sum of the velocity changes at the perigee and apogee of the transfer ellipse. Because the velocity vectors are collinear, the velocity changes are simply the differences in magnitudes of the velocities in each orbit, namely, for circular initial and final orbits:

$$\Delta v_{\mathrm{Hohmann}} = \Delta v_1 + \Delta v_2 = \left(v_{\mathrm{ellipse},1} - v_{\mathrm{circle},1}\right) + \left(v_{\mathrm{circle},2} - v_{\mathrm{ellipse},2}\right) \qquad (7.24).$$

The result of Eq. (24) is depicted in Fig. 7.10 as a function of different ratios of the initial and final radii. An important and somewhat surprising result is that the required Δv for a transfer between two circular orbits around the same central mass reaches values above the required Δv for escaping from gravitational attraction. This is, for example, the case for a transfer between LEO ($r_{\mathrm{LEO}} \approx 6\,600$ km) and GEO ($r_{\mathrm{GEO}} \approx 42\,000$ km). The reason for this is that at the first burn the fuel mass needed for the second burn has to be accelerated and lifted in the gravitational field. Thus, from an energetic point of view, it is less expensive to launch a satellite from LEO into an interplanetary orbit than to GEO. Finally, in most cases it turns out that the required Δv for an orbit change is higher the more maneuvers (burns) are performed.

7.4.2 Low-thrust Maneuvers

It is often argued that thrusters such as electrical or ion thrusters can achieve a much higher exhaust velocity, and therefore a higher overall velocity change Δv, than chemical rockets. This is in many cases true, although the structure fraction would certainly be increased because of the needed additional power support. However, the total thrust achieved is usually not sufficient to allow for an application in the first stages of a rocket. Nevertheless, when using electrical thrusters, e.g. as a third stage, the total Δv can be increased as depicted in Fig. 7.11.

On the other hand, because of the low thrust, the operation time is also increased, leading to the situation of having an almost infinite amount of small pulses (as approximation for a continuous long-term burn). Considering the discussion of the previous chapter, this would lead to an increase of the required Δv to accomplish the same orbit transfer. As an example, Fig. 7.12 shows the result for a transfer between two circular Earth orbits in comparison to the Hohmann transfer. Thus, it is important to have not only a high Δv capability of a rocket but also a low Δv requirement according to the orbit maneuvers that are possible with the used propulsion systems. An overview of the required Δv to accomplish a certain mission is given in Table 7.3.

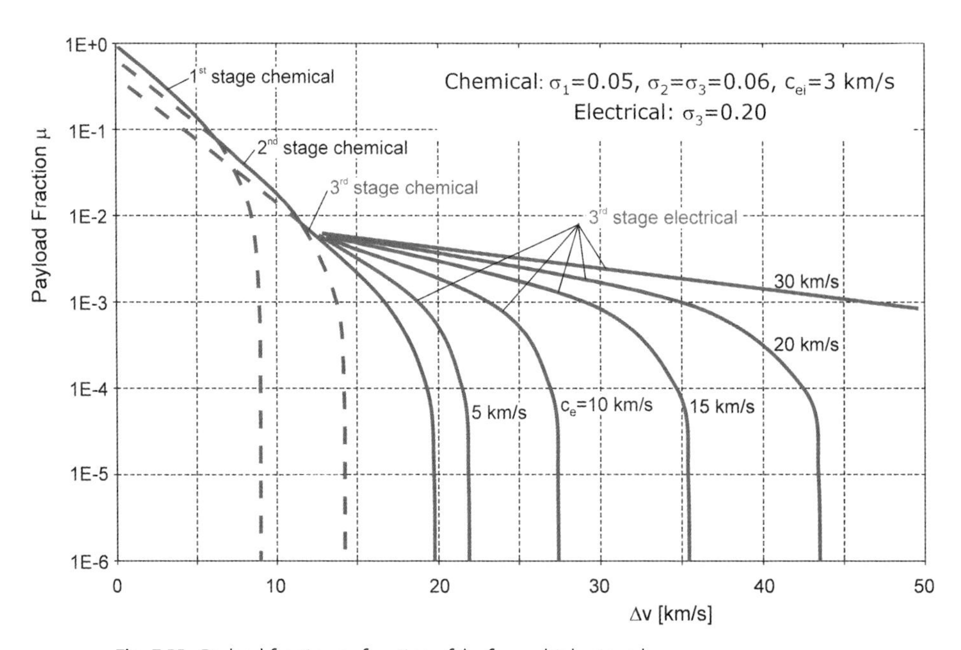

Fig. 7.11 Payload fraction as function of Δv for multiple staged rockets. Different combinations of chemical and electrical propulsion are compared. Note that electrical thrusters have higher exhaust velocity than chemical thrusters and that generally $\sigma_{chemical} < \sigma_{electrical}$ and $c_{e,chemical} < c_{e,electrical}$, but the thrust $F_{electrical} << F_{chemical}$.

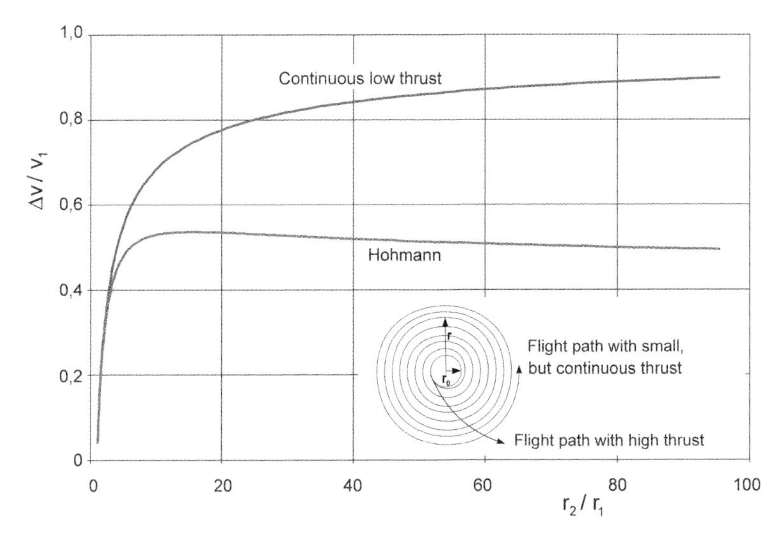

Fig. 7.12 Comparison of the Δv requirement for a Hohmann transfer and a transfer with continuous low thrust.

Table 7.3 Typical Δv requirement and resulting typical payload fraction μ for different mission types.

Mission type	Δv (km s^{-1})	μ
LEO	≈9–9.5	0.04
Escape from Earth's gravitation	≈12.5	0.015
LEO + Hohmann transfer to GEO	≈13	0.01
Earth–Moon–Earth	≈18	0.0025
Grande tour (Earth–Jupiter–Saturn–Uranus–Pluto)	≈23	0.00035
Earth–Mars–Earth	≈23	0.00025
Comet rendezvous	≈25	0.0002
Mission out of ecliptic plane (35 °)	≈26	0.00016
Earth–Mars–Earth with "soft" landing	≈34	0.00001
Crash into Sun	≈40	0.000005

LEO = Low Earth orbit
GEO = Geosynchronous Earth orbit

7.4.3
Gravity-assist Maneuvers

Table 7.3 illustrates that the Δv requirements for space missions are very high, and therefore – according to the Tsiolkovsky equation – the resulting payload fractions are very low. This is especially the case for interplanetary missions to the outer planets of our Solar System.

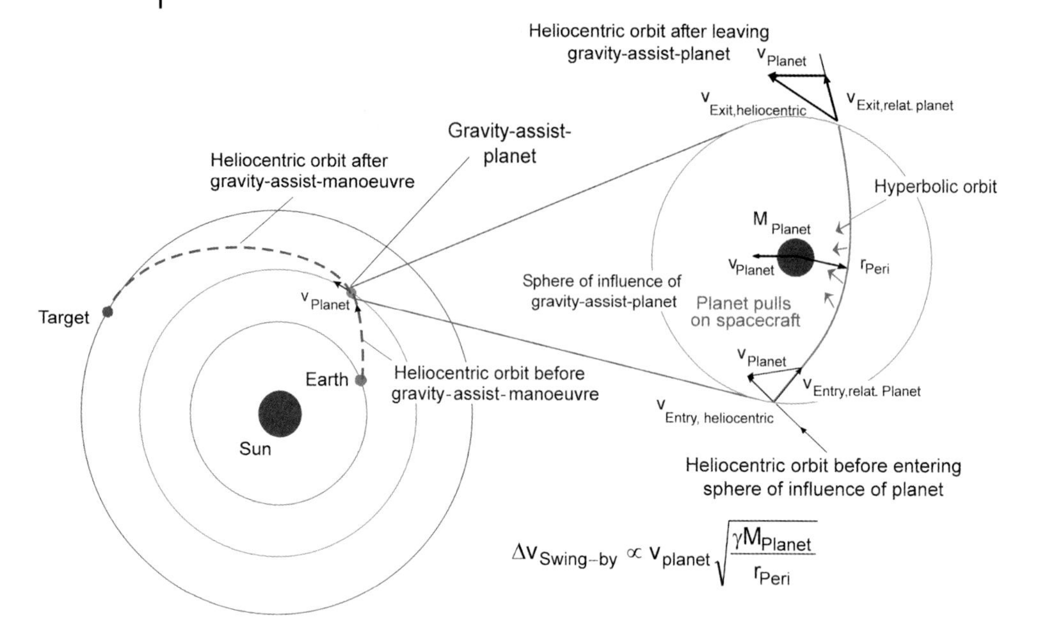

Fig. 7.13 Illustration of gravity-assist maneuvers. Left: in a heliocentric reference system, right: in a topocentric (planet) reference system.

Fortunately, there is an alternative that allows for increasing a vehicle's energy (and therefore velocity) relative to the Sun without requiring fuel: the "gravity-assist" or "swing-by" maneuver. This maneuver uses a planet's gravitational field and orbital velocity to "sling" the satellite, changing its velocity in magnitude and direction (Fig. 7.13). As the satellite enters the gravitational field on a hyperbolic orbit and passes behind the planet, the planet pulls it in the direction of the planet's motion. Thus, it gains energy with respect to the Sun, altering the satellite's original orbit around the Sun. If the satellite passes in front of a planet, it is pulled in the opposite direction, causing a slowdown in velocity relative to the Sun. Of course, because of Newton's laws (action – reaction) the same energy change occurs for the planet as well. However, because of the significantly smaller mass of the satellite, the effect on the satellite can be extreme, whereas for the planet it is negligible.

Gravity-assist trajectories often make the difference between possible and impossible missions. For example, the *Galileo* mission to Jupiter was launched towards the inner Solar System, performing first one swing-by maneuver at Venus followed by two subsequent swing-by maneuvers at Earth. Another example is the *Ulysses* probe, which performed a gravity-assist maneuver at Jupiter that sent it out of the ecliptic plane into a polar orbit around the Sun.

Of course, in order to establish such maneuvers, the constellation between the Earth, the gravity-assist planet, and the target planet must be in accordance. This often determines possible "launch windows," meaning the time frame in which the spacecraft has to be launched.

Finally, it is important to note that in both cases, either acceleration or deceleration of the satellite relative to the Sun, it leaves the planet's sphere of influence on a hyperbolic orbit (relative to the planet!). Thus, in order to establish an orbit around a planet, an impulse (thrust), and therefore fuel, is again required.

7.5 Example: Missions to Mars

In the following sections, some of the general findings from the previous chapters will be discussed regarding missions to our neighboring planet Mars. First, in order to establish an interplanetary trajectory to Mars, the satellite should be launched with a velocity higher than 11.2 km s^{-1} (the escape velocity from the Earth) relative to Earth. Further, it should be launched in the direction of the Earth's motion around the Sun; otherwise, the Δv requirement would be increased by the Earth's own velocity around the Sun of about 29.8 km s^{-1}. For a launch to Mars, often the Earth's own rotational velocity is also used, i.e., a launch towards the east and additionally during nighttime is performed (assuming a direct launch from Earth's surface and an injection into a Hohmann transfer). As already discussed, the most energy-efficient transfer trajectory is a Hohmann transfer joining Earth and Mars' orbit around the Sun. Of course, as soon as the satellite's orbit intersects with the Martian orbit, the planet should be at this position in order to set up a "rendezvous." A certain initial constellation between the two planets is therefore needed, as seen in Fig. 7.14.

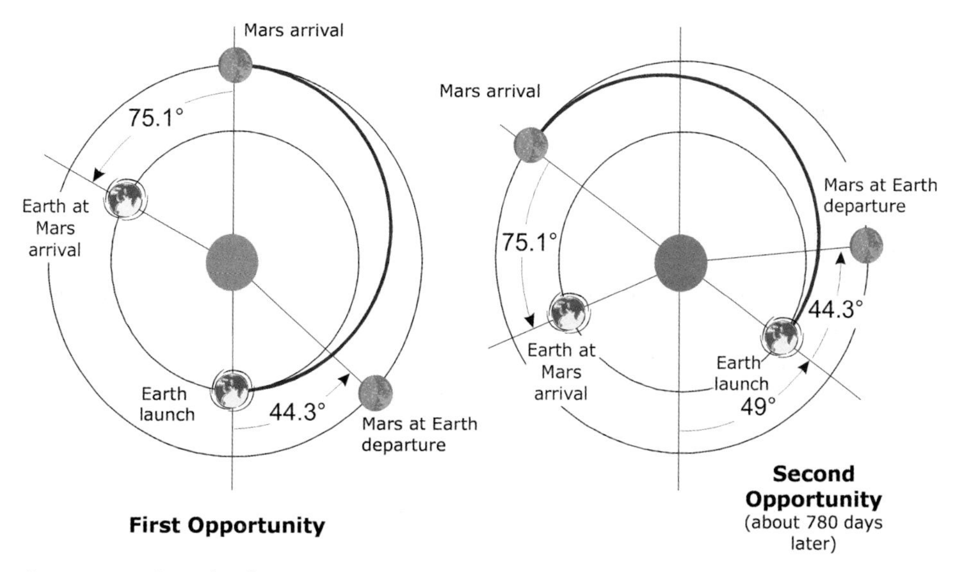

Fig. 7.14 Launch window for a mission to Mars.

All these issues define the launch window, which opens for just a few days or weeks and is additionally restricted in local time. If the window is missed, the next opportunity occurs after about 780 days.

Because of the limited payload fraction of the launch vehicle, the amount of fuel carried with the probe is limited, too. Thus, only a few minor interplanetary correction maneuvers (usually about four or five) are possible in order to adjust the orbit for a typical journey of about 500 000 000 km. The main fuel portion is then needed to slow down the spacecraft to a mostly high eccentric orbit around Mars, except, of course, if a direct landing without a prior velocity change is planned.

Many of the orbital missions to Mars performed in the last few years make use of so-called "aerobraking" maneuvers to establish a final orbit. Therefore, the closest point (periapsis) to the planet is chosen so that the spacecraft flies through the upper part of the atmosphere, which causes aerodynamic drag and further slows down the spacecraft, and subsequently its apoapsis (most distant orbital point). The effect on the apoapsis is higher than that on the periapsis, and after performing this aerodynamic break over several passages, an orbit with a lower eccentricity is achieved. Finally, a small burn at the apoapsis lifts the periapsis out of the atmosphere and the final orbit is established.

Again, the necessary fuel mass for the braking maneuvers together with the required satellite system mass, (e.g., energy, attitude, and orbit control; communication; thrusters; tanks), significantly limits scientific instrument mass. The situation is similar for entry and landing probes where some of the fuel necessary for orbit insertion can be saved, but additional system mass must be introduced, e.g., for the heat shield.

Because the scientific payload for interplanetary missions is so restricted in mass (and also in energy demand), an alternative to studying the planets or their surface material is to bring material from other celestial bodies, especially from Mars, to Earth for further in-depth investigations. Therefore, some basic boundary conditions and scenarios for a Mars sample return (MSR) mission will be discussed here (see Figs. 7.15 to 7.17). A more detailed discussion is provided by D. Boden and S. Hoffmann in *Human Spaceflight: Mission Analysis and Design*, edited by W. Larson and L. Pranke (1999).

The first part of an MSR mission, the way to Mars on a minimum energy transfer, has already been discussed. Assume that the spacecraft is already in a Mars orbit and is supplied with some material from the Mars surface that has been collected by use of a landing probe. Similar to the outbound part, it has to wait until the constellation between Mars and Earth allows for a return flight trajectory. For low-energy transfers, the Mars stay-time may reach values of more than one Earth year. Adding the transfer times, the total mission duration lasts for more than 2.5 years. The required Δv for the return injection is lower than for the trans-Mars injection; however, one has to recall that with current technologies the necessary fuel mass must be imported to Mars. As shown in Fig. 7.15, for the low-energy transfer from a LEO to a Mars orbit and a direct flight back to Earth, a Δv of about 8.2 km s^{-1} is needed. The Δv needed for a launch from Earth, the Δv for a launch from the Mars surface (about 4 km s^{-1}), and the Δv for any rendezvous or trajectory correction maneuvers

Mission parameter (example)	
Δv @ launch from Earth orbit	3.5 km/s
Outbound time	204 days
Δv for Mars orbit insertion	2.2 km/s
Stay time	553 days
Δv @ Mars launch	2.5 km/s
Return time	190 days
Total mission time	946 days
Total Δv	8.2 km/s

Assumptions:
- Trans-Mars Injection via launch from ISS-orbit
- Mars Orbit Insertion in 500 km circular orbit
- Direct Earth-entry upon return

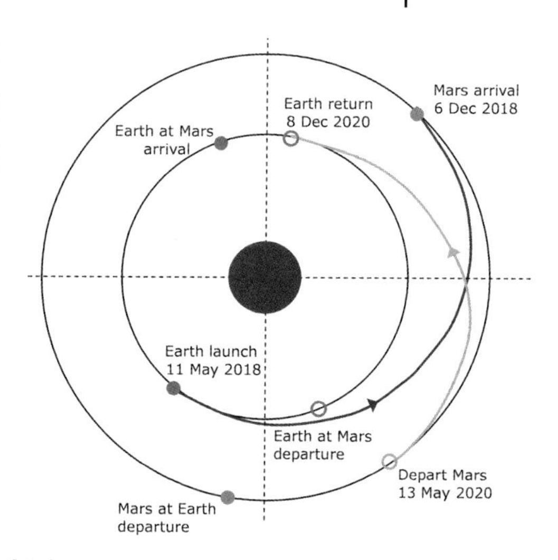

Fig. 7.15 Typical low-energy scenario (Hohmann transfer) for an MSR mission.

must be added to this value. Thus, even assuming a direct landing upon Earth return, which is extremely challenging because of the high return velocity, the overall Δv needed for such a mission is in the best case about 20 km s^{-1}. This high overall Δv value yields extremely low payload fractions, which could be returned to Earth (including the transfer vehicle with heat shield and other items).

Using fast, high-thrust transfers as depicted in Fig. 7.16 is not really an alternative for reducing the overall mission time, because the waiting time for the

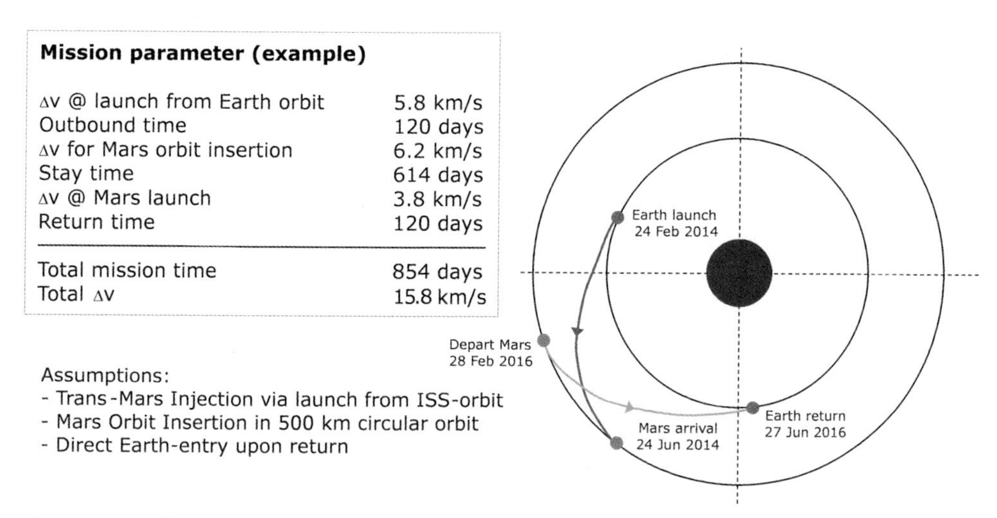

Mission parameter (example)	
Δv @ launch from Earth orbit	5.8 km/s
Outbound time	120 days
Δv for Mars orbit insertion	6.2 km/s
Stay time	614 days
Δv @ Mars launch	3.8 km/s
Return time	120 days
Total mission time	854 days
Total Δv	15.8 km/s

Fig. 7.16 Typical scenario for a short transit time for an MSR mission.

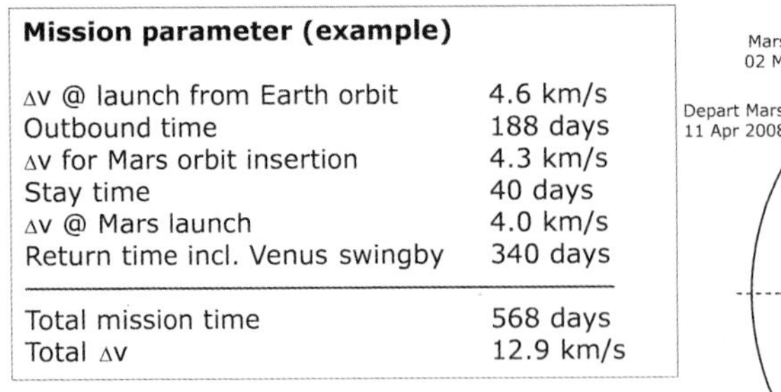

Mission parameter (example)	
Δv @ launch from Earth orbit	4.6 km/s
Outbound time	188 days
Δv for Mars orbit insertion	4.3 km/s
Stay time	40 days
Δv @ Mars launch	4.0 km/s
Return time incl. Venus swingby	340 days
Total mission time	568 days
Total Δv	12.9 km/s

Assumptions:
- Trans-Mars Injection via launch from ISS-orbit
- Mars Orbit Insertion in 500 km circular orbit
- Direct Earth-entry upon return

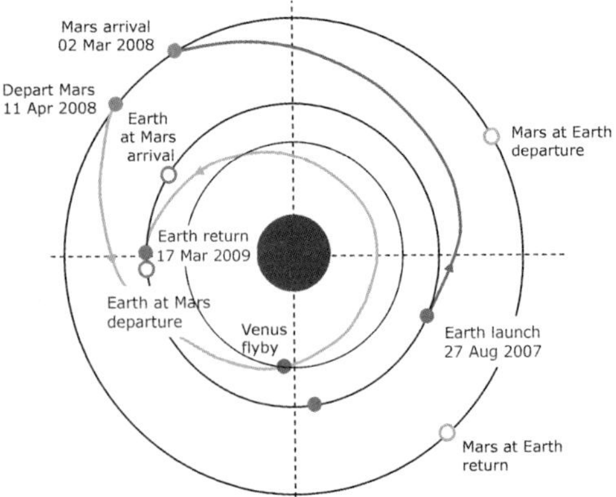

Fig. 7.17 Scenario for an MSR mission using a Venus gravity-assist maneuver for return.

return launch window is enlarged. Thus, only a small amount of the return flight time can be saved. However, the required Δv is dramatically increased, easily reaching values above 25 km s^{-1}.

What about using gravity-assist maneuvers? This possibility exists, as illustrated in Fig. 7.17. The alternative of performing a swing-by maneuver at Venus could shorten the overall mission time, while the total Δv requirement would be much lower than the high-thrust transfer, but nevertheless higher than a double low-energy Hohmann transfer (Fig. 7.15). However, the spacecraft would fly closer to the Sun than the orbit of Venus, which subsequently has an impact on the overall design, and especially on the thermal control system because of the higher solar radiation.

From this brief discussion of potential Mars sample return mission scenarios it becomes obvious that orbital mechanics provide a variety of possibilities. However, it is also obvious that generally only very small payload fractions can be realized and that each scenario has some specific benefits and drawbacks. In addition, technological issues not addressed here in detail must be taken into account, such as power supply, data processing, and communication requirements; automated guidance; navigation and control; system autonomy; entry, descent, and landing technologies; robotics and mechanisms; propulsion, etc. Thus, the payload mass used to calculate the payload mass ratio of a rocket (see Section 7.2) includes all spacecraft subsystems that are needed for survival and control of the science payload. Table 7.4 gives an overview of the subsystems of a spacecraft that are required to service the science payload.

Finally, for potential biological payloads, planetary protection requirements, as discussed in Chapter 13, must be added, making the situation even more complex and challenging, but not unsolvable.

Table 7.4 Overview of typical spacecraft subsystems for servicing scientific payloads (e.g., scientific instruments).

Spacecraft system	Subsystems of spacecraft system
Electrical power system (EPS)	Electrical power generation Conditioning, storage, and supply (e.g., solar generators, batteries, fuel cells, radio isotope generators, power control units) Electrical power regulation and distribution
Structure and mechanism system (SMS)	Carrying shock and vibration loads associated with acceleration during ground operation, launch, in-orbit operations, re-entry and landing Holding of payload, electronics, sensors, antennas, etc., in stable positions Protection against influences from space environment
Thermal control system (TCS)	Temperature control for all spacecraft components, e.g., batteries, electronics, propellant tanks, sensors, actuators, solar arrays, structures, etc. Passive thermal control systems via heat conduction and radiation, e.g., radiators, coatings, multilayer insulation (MLI), heat sinks, joint fillers, etc. Active thermal control systems, e.g., heat switches, active radiators, heat pipes, single- and two-phase fluid loops, electrical heaters, etc.
Communication system (CS)	Transmitters and receivers or transponders Antenna systems and antenna pointing Signal-to-carrier frequency modulation and multiplexing, signal demodulation, etc.
Command and data handling (C&DH)	Payload data processing and recording Housekeeping and science data encoding and telemetry Error handling Signal decoding and telecommand handling and distribution to the appropriate devices Software and computer systems
Propulsion system (PS)	Thrusters (chemical, electrical) for launch, in-orbit insertion, orbit transfer, orbit and attitude maintenance, reentry and landing maneuvering, attitude control, etc. Fuel and oxidizer storage and handling
Attitude and orbit control system (AOCS)	Orbit determination and control Attitude determination and control = spacecraft stabilization in desired directions despite the disturbance torques, uses sensors and actuators (e.g., star trackers, sun sensors, magnetic actuators, momentum wheels, fly wheels, thrusters, etc.)

7.6 Further Reading

Anderson, J. D. *Introduction to Flight*, McGraw-Hill Inc., 1989.

Buedeler, W. *Geschichte der Raumfahrt*, Sigloch Edition, 1982.

Brown, C. D., *Spacecraft Mission Design*, AIAA Education Series, Washington D. C. 1992.

Dadieu, A., Damm, R., Schmidt, E. W. *Raketentreibstoffe*, Springer Verlag, 1998.

Griffin, M. D., French, J. R. *Space Vehicle Design*, AIAA Education Series, Washington D. C. 1991.

Hammond, W.-E., *Space Transportation: A Systems Approach to Analysis and Design*, AIAA Education Series, Washington D. C. 1999.

Humble, R. W., Henry, G. N. and Larson, W. J. *Space Propulsion Analysis and Design*, Space Technologies Series, McGraw Hill, 1995.

Larsen, W. J., Pranke, L. K. (Eds.) *Human Spaceflight: Mission Analysis and Design*, Space Technology Series, McGraw Hill, 1999.

Messerschmid, E., Fasoulas, S. *Raumfahrtsysteme*, Springer, Berlin, 2004.

Oberth, H. *Die Rakete zu den Planetenräumen*, R. Oldenbourg, München, 1923, 5th edition UNI–Verlag Dr. Roth-Oberth, Feucht, 1984.

Sellers, J. J. et al. *Understanding Space – an Introduction to Astronautics*, McGraw-Hill Education, 2000.

Sutton, G. P. *Rocket Propulsion Elements – An Introduction to the Engineering of Rockets*, John Wiley and Sons, Inc., New York, 1992.

Vallado, D. A., McClain, W. D. *Fundamentals of Astrodynamics and Applications*, Kluwer Academic Publishers, 2001.

Wertz, J. R., Larson, W. J. (Eds.) *Space Mission Analysis and Design*, 2nd edition, Kluwer Academic Publishers, 1999.

7.7 Questions for Students

Question 7.1

Why is the Δv_{ch} capability of a single-staged rocket limited?

Question 7.2

What is the basic idea behind the “multi-stage” concept?

Question 7.3

Why is the Δv needed for a Hohmann orbit transfer from a low earth orbit to a geostationary orbit higher than the Δv needed to escape from Earth’s gravitational field from low earth orbit?

Question 7.4

Why is it practically impossible to use electrical thrusters in the first stage of a rocket?

Question 7.5

Planetary gravity-assist maneuvers are usually performed in order to increase a spacecraft’s velocity (energy) relative to the Sun. (a) Where does this energy come from? (b) Do you think that it is also possible to decrease the velocity relative to the Sun? If yes, how would you perform the maneuver?

Question 7.6

What fuel mass is needed in order to accelerate a mass of 200 kg (payload, structure, and motor mass) with a single-stage rocket (in a force-free environment) from zero to the triple exhaust velocity ($3\,c_e$)?

Question 7.7

Calculate the orbit velocities and orbit periods of circular Earth orbits with the following altitudes H above the Earth's surface:

$H = r - R_0 = 200$ km;, 500 km, 1 000 km, or 10 000 km

What altitude would be needed for a geostationary satellite (note that a sidereal day is slightly smaller than 24 h, namely 23 h, 56 min, 4 s)?

Note:

Earth mass $M_E = 5.974 \times 10^{24}$ kg

Gravitational parameter: $\gamma M_E = 3.986 \times 10^{14}\ m^3\ s^{-2}$

Earth radius $R_0 = 6.378 \times 10^6$ m

Question 7.8

An unmanned spacecraft has been placed in a 500-km circular orbit around the Earth. (a) Calculate the Δv budget needed for a Hohmann transfer from this orbit to a 600-km circular orbit in which a docking maneuver with a satellite could take place. (b) Calculate the necessary relative phase angle between the two spacecraft at the time of injection into the Hohmann transfer ellipse so that the docking maneuver can take place at the same time as the spacecraft reaches the final orbit. Notice that the transfer time of the Hohmann transfer is simply half of the orbital period for this orbit. (c) How much time will pass between two possible constellations?

Question 7.9

Discuss some of the requirements coming from a scientific payload that have influences on the design of the spacecraft subsystems listed in Table 7.4.

Question 7.10

Imagine that a potential scientific payload needs two times more electrical power than a competitive, but not so sophisticated, payload. What influences will occur to the design of the electrical power system (EPS), the thermal control system (TCS), the structure and mechanism system (SMS), and the attitude and orbit control system (AOCS)?

8
Astrobiology of the Terrestrial Planets, with Emphasis on Mars

Monica M. Grady

The planets of particular relevance to this chapter are the four rocky planets in the innermost region of the Solar System: Mercury, Venus, Earth, and Mars. Each of these planets has the potential, to a greater or lesser extent, to host life; in the case of Earth, this potential appears to have been fully realized. In this chapter these four rocky planets are examined and the relative likelihood of life evolving on Mercury, Venus, and Mars is assessed. Most emphasis will be placed on Mars because, after Earth, it seems to have the highest potential to host life. Besides observations by telescope and spacecraft, Martian meteorites are discussed as an additional way of exploring Mars.

8.1
The Solar System

Our Sun and its planetary system are located approximately two-thirds of the way out along one of the arms of the (approximately 10–12 billion years old) Milky Way spiral galaxy. The Sun is sometimes described as an undistinguished star in an undistinguished neighborhood – it is a G class star surrounded by its family of planets and their satellites, plus asteroids and comets. Advances in understanding the processes that led to the formation of the Sun and Solar System have resulted from astronomical observations of star formation regions in molecular clouds and the recognition and observation of protoplanetary disks and planetary systems around other stars (see Chapter 2). If we wish to understand the processes by which life arose and became established on Earth, and whether there are possible locations beyond Earth where life might exist, then it is also necessary to understand the starting materials and processes that led to the formation of the Solar System.

The first stage in the history of the Solar System was the collapse of an interstellar cloud: gravitational instability within the cloud resulted in collapse of a fragment of the molecular cloud to form a protoplanetary disk, or solar nebula.

Complete Course in Astrobiology. Edited by Gerda Horneck and Petra Rettberg

ISBN: 978-3-527-40660-9

The mechanism that triggered cloud collapse is not clear. Several possibilities have been suggested. Examples include a shock wave from a nearby supernova or the ejection of a planetary nebula from a giant star; but whatever the mechanism, the collapse of the cloud and subsequent accretion of material must have taken place over a relatively short time interval, around 1–3 million years, as deduced by measurements on meteorites. As the cloud fragment collapsed, dust grains fell towards the center of the nebula, clumping together to form increasingly large bodies. The growth of planet-sized bodies from micron-sized dust grains was controlled by several factors, such as the nature of the initial grains – whether fluffy or compact – and the degree of turbulence within the nebula. End-member models for planetesimal (small planet) formation include coagulation of material by gravitational instability in a quiescent nebula or coagulation during descent to the midplane of a turbulent nebula. The aggregation of interstellar dust ($\leq 0.1\ \mu m$ in diameter), eventually forming kilometer-sized planetesimals and culminating in the asteroids and planets, took less than 30 million years following formation of the first solid materials.

The inner, rocky terrestrial planets (i.e., Mercury, Venus, Earth, and Mars; Table 8.1), formed sufficiently close to the proto-Sun that volatile elements were lost through vaporization and water did not condense to ice. Marking the boundary between the rocky planets and the outer gas giants is the asteroid belt. The asteroids are a swarm of several thousand planetesimals, the largest of which, Ceres, is about 1 000 km across; most asteroids are on the order of a few kilometers across.

Table 8.1 Properties of the terrestrial planets and the Moon.

Property	Planet				
	Mercury	Venus	Earth	Moon	Mars
Mass (10^{24} kg)	0.330	4.87	5.97	0.073	0.642
Diameter (km)	4879	12 104	12 756	3475	6794
Density (kg m^{-3})	5427	5243	5515	3340	3933
Gravity (m s^{-2})	3.7	8.9	9.8	1.6	3.7
Escape velocity (km s^{-1})	4.3	10.4	11.2	2.4	5.0
Rotation period (h)	1407.6	−5832.5[a]	23.9	655.7	24.6
Length of day (h)	4222.6	2802.0	24.0	708.7	24.7
Distance from Sun (10^6 km)	57.9	108.2	149.6	0.384[b]	227.9
Perihelion (10^6 km)	46.0	107.5	147.1	0.363[b]	206.6
Aphelion (10^6 km)	69.8	108.9	152.1	0.406[b]	249.2
Orbital period (days)	88.0	224.7	365.2	27.3	687.0
Orbital velocity (km s^{-1})	47.9	35.0	29.8	1.0	24.1
Orbital inclination (degrees)	7.0	3.4	0.0	5.1	1.9
Orbital eccentricity	0.205	0.007	0.017	0.055	0.094

Property	Planet				
	Mercury	Venus	Earth	Moon	Mars
Axial tilt (degrees)	0.01	177.4	23.5	6.7	25.2
Mean temperature (°C)	167	464	15	–20	–65
Surface pressure (10^5 Pa)	0	92	1	0	0.01
Number of moons	0	0	1	0	2
Global magnetic field?	Yes	No	Yes	No	No

Source: NASA planetary fact sheet (http://nssdc.gsfc.nasa.gov/planetary/factsheet/index.html)

a Negative numbers indicate retrograde (backwards relative to the Earth) rotation.

b Values represent mean apogee and perigee for the lunar orbit. The orbit changes over the course of the year, so the distance from the Moon to Earth ranges roughly from 357 000 km to 407 000 km.

The sequence of events that produced the innermost rocky planets from the protoplanetary disk also produced planets of an entirely different nature: the gas giants Jupiter, Saturn, Uranus, and Neptune. These planets probably started out in the same way as the terrestrial planets, as an accumulation of rocky material. Temperatures at this distance from the proto-Sun were sufficiently low that water ice would also accrete along with the dust; volatile elements would also be retained and not vaporized. Runaway accretion resulted in central cores that were sufficiently large to attract and retain significant masses of hydrogen (H) and helium (He) from the remaining nebula gas. Uranus and Neptune, however, have lower H and He and higher carbon (C), nitrogen (N), and oxygen (O) abundances than Jupiter and Saturn. It is thought that this indicates that Uranus and Neptune formed some time after Jupiter and Saturn, after the nebula gas had dissipated.

Beyond the giant planets lies Pluto, the smallest planet in the Solar System, with a radius even smaller than that of the Moon. Beyond Pluto, at 30–50 astronomical units (AU), is the Kuiper Belt, an accumulation of rocky and icy bodies more akin to comets than asteroids. Pluto, indeed, may be a Kuiper Belt Object, rather than a planet. At the outermost reaches of the Solar System, in fact defining the boundary of the Solar System, is the Oort Cloud. While the Oort Cloud has never been directly observed, its existence is firmly supported by the orbital characteristics of long-period comets. Comets are icy dustballs, or dusty iceballs. They are compact aggregates of silicate dust and ices, within which organic compounds are trapped. Comets were formed at the outer reaches of the solar nebula, way beyond the "snowline" where water ice could condense.

8.2 Terrestrial Planets

8.2.1 Mercury

Little is known about the planet Mercury: there are very few images of it, and because those that are available were taken during a flyby mission, namely, by the *Mariner 10* spacecraft in 1974, we have images of only one side of this planet (Fig. 8.1). That situation will change in the coming decade, with both NASA's Mercury Surface, Space Environment, Geochemistry, and Ranging (*MESSENGER*) mission and ESA's *BepiColombo* mission orbiting the planet Mercury. (*BepiColombo* is named after Professor Giuseppe (Bepi) Colombo (1920–1984), a mathematician and engineer from the University of Padua, Italy. He was the first to see that an unsuspected resonance is responsible for Mercury's habit of rotating on its axis three times for every two revolutions it makes around the Sun. He also suggested to NASA how to use a gravity-assist swing-by of Venus to place the *Mariner 10* spacecraft in a solar orbit that would allow it to fly by Mercury three times in 1974–1975). Mercury appears to be a highly cratered, gray, and lifeless body. It is a planet of extremes, in temperature and radiation. The dayside of this heavily cratered planet is baked and cracked by the heat of the Sun, with temperatures up to 450 °C, while the night side is bitterly cold, with temperatures as low as −180 °C. Its proximity to the Sun causes it to receive an incredibly high dose of radiation, and its lack of a significant atmosphere ensures that the radiation penetrates right down to the surface.

The potential for Mercury to host life is very low but not completely negligible. Although the temperature range experienced at its surface is extreme, there are some regions where ice has been predicted, or observed, to occur. Radar images of the Polar Regions have been interpreted as showing the presence of water ice, and where there is ice there is the possibility of water and thus the possibility of life. In addition, the lack of tectonic activity on Mercury might allow organic molecules (delivered by comets and asteroids) to be preserved on the surface within the

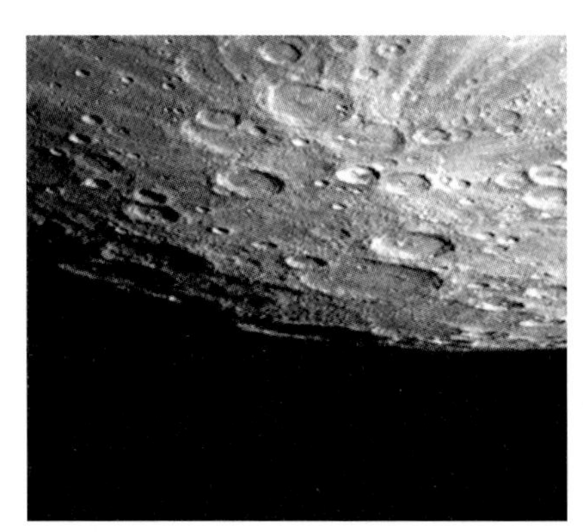

Fig. 8.1 South pole of Mercury, as photographed by *Mariner 10* in September 1974, showing bright craters that might contain water ice. The field of view is approximately 1 000 km across, and was taken from a distance of 85 800 km. (Image copyright: NASA/JPL/Northwestern University, Image No. PIA02941).

regolith. It is known from observations of extremophile microorganisms on Earth that there are some microbes that thrive in high-radiation environments (see Chapter 5). Indeed, there are some microorganisms that survive inside nuclear plants, and so the surface radiation of Mercury might hold no fears for such organisms.

8.2.2 Venus

Venus is Earth's sister planet, the closest planet to us, and the two planets have many similarities: size, composition, an atmosphere, and a similar gravity (Fig. 8.2). But in reality, Venus is very different from Earth. Venus can be summed up in three words: hot, smelly, and windy.

- Hot: the hottest surface in the Solar System, even hotter than Mercury, which lies much closer to the Sun. Temperatures at Venus' surface are about 460 °C.
- Smelly: although Venus' thick atmosphere, which is about 90 times thicker than the terrestrial atmosphere, is mainly carbon dioxide, it includes a very high percentage of sulfur dioxide.
- Windy: the thick atmosphere moves rapidly as the planet spins. The atmosphere is the key to why Venus is so hot – it traps the Sun's heat, so Venus is an extreme example of a runaway greenhouse.

Although the surface of Venus is unlikely to host life now, there has been speculation by Schulze-Makuch and Irwin that life might have existed in the past, when the Sun was dimmer and cooler. The same authors have also speculated about the type of organisms that might be able to exist in the lower cloud levels today (for a critical discussion, see Chapter 6). Results from ESA's Venus Express mission, which arrived at Venus in April 2006, will yield more information about the planet.

Fig. 8.2 False color global mosaic from the synthetic aperture radar instrument on the *Magellan* orbiter of the northern surface of Venus. The north pole is at the center of the image, with 0°, 90°, 180°, and 270° east longitudes at the 6, 3, 12, and 9 o'clock positions, respectively, of an imaginary clock face (Image No. PIA 00271, courtesy of NASA).

8.2.3
Earth

To date, Earth is the only planet within the Solar System that seems to have reached its potential for hosting life. There is evidence of life all over the planet, in the most extreme of environments. The time at which life arose on Earth is not known. At some point in its formation, between 4.57 and 3.5 billion years ago, Earth had stabilized to the extent that there were three dynamically interacting reservoirs: the atmosphere, the lithosphere, and the hydrosphere. In addition, plate tectonics was taking place, with a mechanism of plate movement close to that in operation today. The atmosphere was mildly reducing in composition and was composed of gases including CO_2 and CH_4 (see Chapter 3 for more details).

The presence of a biosphere at this early time is a matter of great debate: features within sediments that were interpreted as fossilized microorganisms have recently been reinterpreted as chemical traces, although there have been more recent reports of microbial biomarkers in 3.5-billion-year-old volcanic pillow laves (see also Chapter 1). Over the next billion years, the composition of Earth's atmosphere changed progressively. At around 2.3 billion years ago, the switch from a reducing to an oxidizing atmosphere became permanent, and the atmosphere had evolved to a composition closer to that observed today (see Chapter 6).

The unambiguous establishment of life can be tracked through the fossil record, with the first appearance of macrofossils about 600 million years ago. In the time intervening between the emergence of life and the present day, microbes colonized every environment in which it is possible for life to survive. The conditions prevailing in these "habitable niches" include extremes of temperature, salinity, and other extremes and are taken to define a "biological envelope" within which life can survive, grow, and evolve (see Chapter 5).

8.2.4
Mars

Mars is one of the key planets with a high potential for life. Like Earth, Mars is a rocky planet. It has a radius of approximately half and a mass around one-tenth that of the Earth; consequently, gravity on Mars is only about 40% of that on Earth. Mars' atmosphere is also different from that of the Earth: it is much thinner, around 600 Pa (6 mbar), compared with 10^5 Pa (1 000 mbar), and consists of predominantly carbon dioxide (about 95% CO_2) rather than nitrogen (N_2). The thin atmosphere provides the Martian surface with little protection from heat loss; thus, the average daily temperature is around –60 °C. Temperatures may reach +30 °C at the equator during daytime in summer and fall to –130 °C at the poles in winter.

8.2.4.1 Observing Mars

There have been many observations of Mars, by a variety of instruments and techniques, and we are currently experiencing an unparalleled period of Martian exploration. One of the earliest scientific observers of Mars was Giovanni Schiaparelli. In 1882 he described *canali*, an Italian word meaning *channels*, and went on to give detailed descriptions of linear features on Mars' surface. The most direct translation of *canali* in English is *canal*, implicating that the features were artificial, i.e., had been made by Martians. An observer in America, Percival Lowell, spent a great deal of his time at the beginning of the 20th century trying to show that there was a very large network of interconnecting canals on Mars. The most recent observations of Mars by telescope have been made by the *Hubble Space Telescope* (HST), which in 2001 produced detailed images showing the polar caps and the main features of the planet. When the HST image is matched to the maps produced by Schiaparelli in 1882 and Lowell in 1907, it becomes apparent that the earlier observers were, in fact, mapping areas of dust, not channels.

It was not until the 1960 s that spacecraft visited Mars, enabling further advances in understanding the features of the planet's surface. *Mariner 4*, in 1965, and *Mariner 6* and *Mariner 7*, in 1969, were flyby missions that recorded images of craters. Those missions also provided the first evidence that there were no artificial canals on Mars. The first clear pictures of the Martian surface came from *Mariner 9*, a spacecraft that orbited the planet for almost a year from 1971 to 1972. These first pictures of the surface of Mars from orbit showed in very great detail the channels and features that are now interpreted as river valleys. As of April 2006, there were four spacecraft in orbit around Mars: *Mars Global Surveyor* and *Mars Odyssey* (both NASA missions) and *Mars Express* (an ESA mission). *Mars Reconnaissance Orbiter*, launched by NASA in August 2005, arrived in the vicinity of Mars in March 2006 and will commence science operations in November 2006. Orbiting the planet yields information on a global scale; more detailed local information is gained from landing craft on the surface and making measurements *in situ*.

The first probes to land on Mars were *Viking 1* and *Viking 2* in 1976, followed by *Pathfinder* in 1997. There are two landers that, at the time of this writing (May 2006), are still in operation: NASA's Mars Exploration Rovers (MERs) *Spirit* and *Opportunity*. The two rovers have been in operation for one Martian year, which is equivalent to about two terrestrial years (1 Martian year lasts 687 days). They have returned detailed images and chemical analyses of the Martian surface. All of the landing sites are within about 15 °N and 15 °S of the Martian equator, but across a range of longitudes.

Much has been learned about Mars from telescope and spacecraft observations. But there is an additional way of exploring Mars that employs the highly precise instrumentation of the laboratory-based astronomer. There is a group of 37 meteorites that have come from Mars, and laboratory-based study by microscope and other instruments has yielded information on the thermal, fluvial, and atmospheric history of the planet. One of the most important pieces of information that came from these studies is an absolute chronology for Martian specimens (see Section 8.2.5).

Given that Mars formed from the same starting materials as Earth, and that it also suffered bombardment by comets delivering water and organic compounds throughout the Solar System, Mars is thought to have a high potential for supporting life. In the following sections, evidence will be considered that the planet has been molded by fluvial and thermal processes, such that water and heat are, or have been, important in Mars' history.

8.2.4.2 Evidence for Water

Since *Mariner 4* in 1965, there have been more than 30 missions to Mars, experiencing greater or lesser success (see Chapter 13, Table 13.5). This fascination with Mars comes from its potential for hosting life that is based on abundant evidence of water on Mars. Ever since Schiaparelli observed his *canali*, we have known about channels on Mars. Satellite images of the surface of Mars have shown many features that have been interpreted as having been produced by water: images from all of the orbiting spacecraft from *Mariner* 9 onwards show networks of channels cutting the Martian surface. Some of the channels appear to be narrow and deep, while others are broader and flatter (Fig. 8.3). They show all the features exhibited by rivers on Earth at different ages. Older rivers on Earth meander across flat plains, producing features such as cutoff lakes and terraces; these too can be seen on images from Mars. Large-scale evidence for a frozen sea has also been published, and it is thought that ice below the surface covered with dust created the features (Fig. 8.4). Further indication of subsurface ice has come from results from the MARSIS radar instrument on *Mars Express*.

The past five years have seen a dramatic increase in the number, resolution, and quality of images and compositional data from Mars' surface returned by both orbiting satellites and rovers. One result of these data has been a gradual change in our perception of the fluvial history of Mars. The paradigm for many years was the "warmer and wetter" scenario, in which water was stable over vast areas of Mars' surface, implying a thicker atmosphere and warmer temperatures during past epochs. It is now thought to be more likely that, apart from the very earliest period (the Noachian epoch, approximately the first 500 million years), Mars has been cold

Fig. 8.3 Martian river valley; part of the Reull Vallis, east of the Hellas Basin (41 °S, 101 °E) taken from a height of 273 km by the High-resolution Stereo Camera (HRSC) on *Mars Express*. The area is 100 km across; north is at the top (Image credit: ESA/DLR/FU Berlin, G. Neukum).

Fig. 8.4 Martian sea; part of the Elysium Planitia near the Martian equator (5°N; 150°E) taken by the HRSC on *Mars Express*. The area is a few tens of kilometers across (Image credit: ESA/DLR/FU Berlin, G. Neukum).

and dry for much of its history, with transient fluvial events, often short-lived catastrophic flooding, triggered by increases in magmatic activity.

Under a warmer-and-wetter regime, with pervasive and persistent bodies of standing water, there would be an extended hydrological cycle in operation. In contrast, a "cold and dry" climate for most of Mars' history would curtail any hydrological cycle. The difference between the two climate models is of great significance for the potential evolution of life on Mars.

Water alters the surface mineralogy of Mars: water-soluble salts are precipitated as evaporites (typically minerals such as anhydrite, gypsum, and carbonates) and primary silicate minerals are hydrated and altered to clay minerals and hydroxides. Such secondary alteration products have been identified by instruments onboard NASA's *Odyssey* orbiter (the thermal emission spectrometer [THEMIS] and the gamma-ray spectrometer [GRS]) and onboard ESA's *Mars Express* orbiter (the Observatoire pour la Mineralogie, l'Eau, les Glaces et l'Activité [OMEGA] and the *Mars Express* subsurface sounding radar altimeter [MARSIS]), as well as by direct analysis of Martian rocks and soil. Data from the GRS on *Odyssey* have shown that there are regions of Mars where H, assumed to be bound as water, is abundant. There is a high concentration of H at the poles from water ice in the polar caps. But there are also high concentrations of H around mid-latitudes and high northern latitudes. Those concentrations are thought not to be from water as ice but instead from water bound within minerals. This observation has been confirmed by data from the OMEGA instrument on the *Mars Express* orbiter, allowing the construction of a relative chronology for the formation of secondary minerals from different weathering regimes on Mars.

Additional evidence for alteration of the surface by water has come from the findings of NASA's MERs *Spirit* and *Opportunity*. The rovers have been exploring two different parts of the surface of Mars for a Martian year. *Spirit* landed in Gusev Crater, a shallow, flat-lying crater. As it explored the surface of the crater, it found sharp angular fragments of rock littering the surface. *Opportunity* landed in Meridiani Planum, some thousands of kilometers to the east of *Spirit*'s landing site. The panoramic camera aboard *Opportunity* acquired a panorama of the Payson outcrop on the western edge of Erebus Crater (Fig. 8.5). From the vicinity at the northern end of the outcrop, layered rocks are observed in the crater wall, which is

Fig. 8.5 Panorama of the "Payson" outcrop on the western edge of Erebus Crater acquired by the panoramic camera aboard NASA's Mars Exploration Rover (MER) *Opportunity* (Image caption and credit: NASA/JPL–Caltech/USGS/Cornell).

about 1 m thick. Some of those rocks had been disrupted by the crater-forming impact event and subjected to erosion over time. Closer examination of those rocks has shown them to contain small spherules. These spherules weather out of the rock and are present on the surface. It is thought that they were produced by the action of water. They are weathering and evaporation products. The Mössbauer spectrometer (see Chapter 12) on the rover *Opportunity* found the hydrated sulfate mineral jarosite present in the soil, showing that the rocks at Meridiani Planum had been altered by water.

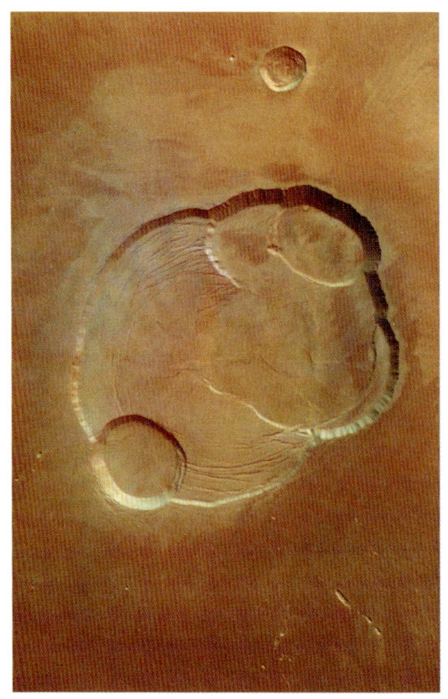

Fig. 8.6 The complex caldera at the summit of Olympus Mons (18.3 °N; 227 °E) with a depth of about 3 km. The image was taken from a height of 273 km by the HRSC on *Mars Express*. The image is about 100 km across; south is at the top (Image credit: ESA/DLR/FU Berlin, G. Neukum).

8.2.4.3 Evidence of Heat

As well as evidence of fluids on the surface of Mars, there is a great deal of evidence from satellite imagery that Mars has had a significant thermal history. Mars has a core-mantle structure similar to that of Earth, but it seems to have a rigid crust rather than the more flexible plate structure of the Earth, although recent results from the magnetometer on NASA's *Mars Global Surveyor* indicate evidence for limited tectonic spreading. This is shown by the presence, in the Tharsis region in the western hemisphere of Mars, of a cluster of enormous volcanoes, now presumably extinct. The largest of these volcanoes is Olympus Mons, which is the biggest volcano in the Solar System. It is approximately three times higher than Mount Everest and the area of the volcano would cover that of England plus Wales.

The presence of these volcanoes indicates that an enormous amount of molten rock or magma has erupted onto the Martian surface at times in the past. High-resolution images from the *Mars Express* camera show that Olympus Mons has a complex crater, indicating that magma was erupted in several different episodes. As Fig. 8.6 shows, there are complex fault structures on the crater floor, formed following eruption as the remaining rock in the underlying magma chamber cooled and solidified. There are four other smaller craters within the large crater area, indicating at least five periods of eruption.

8.2.5 Meteorites from Mars

As detailed in Section 8.2.4, much about Mars can be learned through studying the planet by telescope and spacecraft instrumentation. There is a third set of observations of Mars, complementary to those made by remote means: data acquired through laboratory analysis of Martian meteorites. Before detailing the understanding we have gained about Mars from Martian meteorites, it will be explained how we know that the meteorites do indeed come from Mars.

8.2.5.1 Why from Mars?

The meteorites in question were grouped together as the SNCs (after the type specimens of the three original groups, Shergotty, Nakhla, and Chassigny). In the following sections, these meteorites will be referred to as the SNCs until all the evidence surrounding their origins is detailed.

The first clue that the SNCs were not conventional asteroidal meteorites came from their age. The asteroids formed at the same time as the Solar System, some 4 570 million years ago. The SNCs have much younger crystallization ages (as young as 165 million years old), implying that they cooled and solidified after the bulk of the Solar System formed. In other words, they came from bodies that were still partially molten or still producing molten rock at least 165 million years ago. In order for molten rocks to exist on a body, the body must be very large, i.e., planet-sized. Thus, the SNCs come from a planet, not the asteroid belt. They do not originate from comets or Kuiper Belt Objects (including Pluto), which are assumed not to have melted.

Therefore, the SNC meteorites must have come from a planet, but which planet? Although the giant planets have rocky cores, Jupiter and Saturn are mainly gas and Uranus and Neptune are mixtures of gas and ice; the rocky cores are so deep below the thick atmospheres that it is unlikely that the meteorites came from the giant planets. It is also unlikely that the meteorites came from the satellites of those planets, because anything removed by impact would be captured by the giant planets themselves. On the basis of dynamical impossibility, the planets Jupiter, Saturn, Uranus, and Neptune, and their satellites, are rejected as potential source objects of these meteorites.

Moving to the innermost part of the Solar System, Mercury is a rocky planet very close to the Sun. It is very heavily cratered and thus it is clear that its surface is frequently struck by asteroids; it is equally clear that material would be able to escape from Mercury's small gravitational field. However, the proximity of Mercury to the Sun implies that much of the material leaving the Mercurian surface would be pulled straight into the Sun. There is, however, a small but finite probability that some material could leave Mercury and be ejected outwards from the Sun towards the Earth. There was a suggestion that the unusual basaltic meteorite NWA 011 might be from Mercury, because this meteorite has an oxygen isotopic composition different from any other meteorite group, but the suggestion has been neither verified nor taken seriously by meteoriticists.

In contrast to Mercury, Venus is a large planet (about the size of Earth) and has a very thick atmosphere (about 90 times thicker than Earth's). It is energetically very difficult to remove material from the Venusian surface by impact. For an asteroid to hit the surface of Venus, it must have sufficient energy to penetrate through the very thick atmosphere without being fully ablated in order to leave sufficient material available to create an impact crater on the surface. Following crater formation, impact ejecta would have to have sufficient energy to travel back through Venus' thick atmosphere and still maintain sufficient energy to escape from Venus' deep gravitational well. So, as in the case of Mercury, there is a small but finite chance of material being removed from Venus and impacting on the Earth. In fact, because Mercury is a small, atmosphere-less body, the probability of removing material from Mercury's surface is greater than acquiring material from Venus. Even so, it is not possible to discount Mercury and Venus as sources of the SNCs, although, as will be shown later, there is more compelling evidence that eliminates them as source objects.

This logically leaves only three bodies as potential parents of the SNC meteorites: the Earth, its Moon, and Mars. First, we will consider the Earth–Moon system. The Moon was long thought to be a possible source of meteorites. The Moon is cratered and, like Mercury, is a small and airless body. Objects that impact the Moon could produce craters from which material is readily ejected and, given the nearness of Earth, fall on the Earth. But the *Apollo* and *Luna* samples returned directly from the Moon are so different in age and composition from the SNCs that the latter cannot be from the Moon. However, lunar meteorites do exist, and the first one was recognized in 1983. The name of this meteorite is ALHA81005; it was collected in the Allan Hills region of Antarctica in 1981. Analysis of ALHA81005 showed, without dispute, that the object came from the Moon. Its mineralogy, mineral chemistry, and isotopic composition, and the gases trapped inside it, were identical to similar measurements made on the *Apollo* samples brought back from the Moon by astronauts in the 1970 s. There are now almost 50 lunar meteorites, and they are completely different from the SNCs. They are brecciated and rich in plagioclase, and all have ancient ages similar to the *Apollo* samples.

If not from the Moon, could the meteorites have come from Earth? We know the Earth is cratered, although many of the craters are hidden by vegetation or have been removed through plate tectonics and crustal recycling. It is easy to picture an asteroid impacting Earth's surface and ejecting material. Perhaps this material did not have quite sufficient energy to escape from Earth's gravitational well and fell back to the surface, acquiring a fusion crust on the way and appearing as a meteorite. However, origin as a terrestrial rock can be eliminated on the grounds of the oxygen isotopic composition of the meteorites. Oxygen has three isotopes: ^{16}O (the major isotope), ^{17}O, and ^{18}O. On a three-isotope plot, all terrestrial materials that contain oxygen have an isotopic signature that falls on a specific line, the terrestrial fractionation line (TFL). As Fig. 8.7 shows, the TFL is well defined, and the SNCs have oxygen signatures that do not fall on that line. This does not tell us where these meteorites came from, but it does tell us where they did not come from.

On the basis of oxygen isotopic composition, the SNCs did not come from Earth. The pattern of variation in the oxygen isotopic signature of the rocks, however, is similar to that of terrestrial rocks. In other words, they are spread out along a straight line that has a slope parallel to that of the TFL. This means that these rocks came from a single body, and that body is not Earth. It also is not the Moon, it is probably not Mercury or Venus, and it is certainly not the asteroids, nor Pluto, Jupiter, Saturn, Uranus, or Neptune. That leaves only one body: Mars.

On the basis of this negative deduction, the meteorites came from Mars.

The specific observations that linked SNC meteorites to Mars came with analysis of gases trapped within one of the samples, the Antarctic meteorite named Elephant Moraine (EET) A79001. This meteorite contains inclusions of black glass scattered throughout its mass. The glass was formed by shock melting of mineral

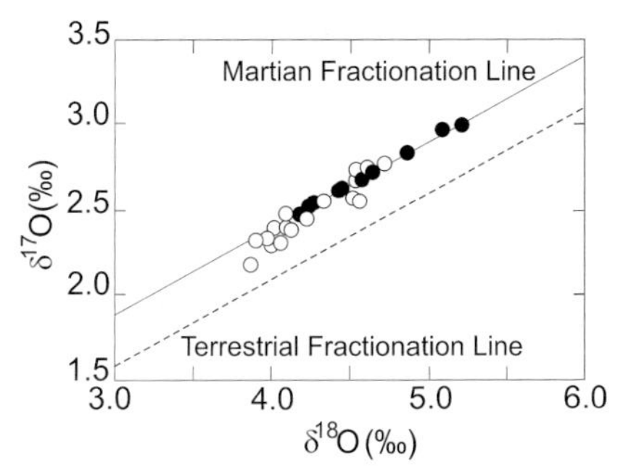

Fig. 8.7 Oxygen isotopic composition of SNC meteorites (filled symbols) compared with the TFL. The data appear to fall along a single fractionation line parallel to the TFL; this line is labeled as the Martian Fractionation Line (after Franchi et al. 1999).

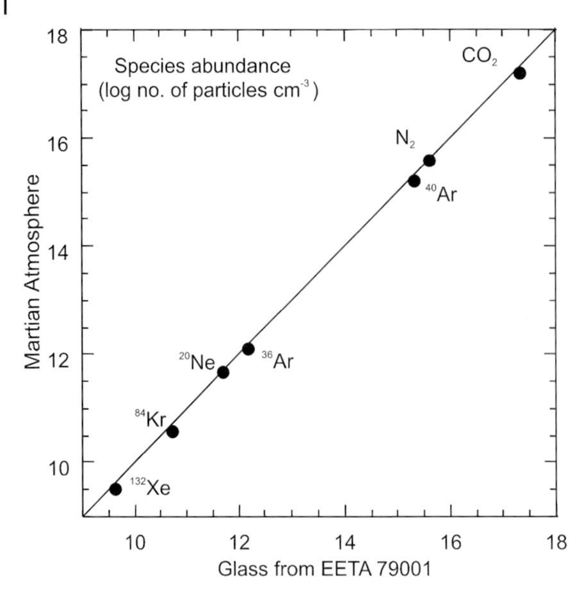

Fig. 8.8 Comparison of the composition of Mars' atmosphere with gas trapped in the shergottite EETA79001 (after Pepin 1985).

grains, presumably during the impact event that lofted the meteorite from the surface of its parent body. Analysis of gas trapped within the glass during the impact shock showed that it was identical in composition to that of the atmosphere on Mars, as measured by the *Viking* landers in 1976 (Fig. 8.8). The only way that this could happen is if EET A79001 came from Mars. Based on their oxygen isotopic composition, all the other SNC meteorites came from the same parent as EET A79001; therefore, they too must have come from Mars.

8.2.5.2 **What can we Learn About Mars from Martian Meteorites?**

The 37 different Martian meteorites are subdivided into four main groups. Collection of additional Martian meteorites from Antarctica and the Sahara Desert has extended the number of subgroups to five. The subgroups are rocks that formed in different locations at or below the Martian surface. The groups have different mineralogies and chemistries and cannot all have come from a single impact event. At least three craters, with minimum diameters of about 12 km, are required to produce the variety of Martian meteorite types. Measurements made on Martian meteorites complement data obtained from spacecraft exploration of Mars, and until we have a mission that returns material directly from Mars, these meteorites are the only objects that we have to help us understand Martian geology, in addition to *in situ* investigations by Martian rovers.

The largest of these groups of Martian meteorites is the shergottites, after the type specimen Shergotty, which is subdivided into three subgroups.

The *basaltic shergottites* are fine-grained cumulate rocks composed of almost equal amounts of clinopyroxene and plagioclase. The plagioclase has been converted, by shock, to maskelynite glass. Alignment of the minerals indicates that the rocks originated in a lava flow.

The *lherzolitic shergottites* are also cumulates but are more coarse-grained than the basaltic shergottites, indicating a slower cooling rate; they formed deeper below the Martian crust than the basaltic shergottites, and their terrestrial equivalent would be a peridotite. The main silicate is orthopyroxene, enclosing olivine grains, with minor plagioclase.

The *olivine-phyric shergottites* are composed of large olivine and orthopyroxene grains set in a finer-grained clinopyroxene matrix. They are thought to be from olivine-saturated magmas that were parental to those from which basaltic shergottites crystallized.

All three subgroups of shergottites have crystallization ages of 165–450 million years. However, it has been discussed that this age in fact could be an alteration age caused by reaction of surface fluids with the rocks. This is still a topic for discussion and research. At least two cratering events ejected shergottites from Mars into space; these events happened in much more recent times, 0.5–3 million years ago.

The second main group of Martian meteorites is that of the nakhlites, named for their type specimen Nakhla. Their dominant minerals are the silicates augite and olivine, which are altered to clay minerals along cracks and fractures. The nakhlites are almost unshocked rocks that formed at or near the Martian surface in a thick flow. They solidified from melts about 1.3 billion years ago and were ejected from the planet about 10–12 million years ago. The meteorites have been altered by weathering on Mars' surface, leading to the production of secondary minerals, e.g., clays, carbonates, and sulfates, which are associated with low concentrations of Martian organic material. It has thus been suggested that nakhlites might contain evidence for a Martian biosphere.

The recognition of complex assemblages of salt (halite) with carbonates and clay minerals in nakhlites has allowed interpretation of the scale and mode of fluid flow on the surface of Mars. Results from the nakhlites imply that when water was present on the surface, it was warm and briny and restricted in flow. In other words, it might have been locked in enclosed basins that occasionally overflowed in episodes of flash flooding.

The third group of Martian meteorites is the chassignites, named for the type specimen Chassigny. The mineralogy of the chassignites is different from that of the nakhlites and the shergottites. They are highly shocked olivine cumulates; if they were terrestrial rocks, they would be described as dunites. Like the nakhlites, the chassignites have a crystallization age of 1.3 billion years, implying some relationship between the two groups.

The final subset of Martian meteorites is composed of a single member, Allan Hills 84001. This unusual meteorite is predominately composed of orthopyroxene. It is riddled throughout with millimeter-sized patches of rimmed carbonates, showing that although this is an igneous rock, like the nakhlites, it too has suffered aqueous alteration. It is the oldest of all the Martian meteorites, having crystallized about 4.5 billion years ago, and has had a long and complex history of shock and thermal metamorphism. Because few hydrated minerals – such as those found in clays on Earth – have been identified among the alteration products in ALH84001, it has been proposed that the carbonates were produced at the surface of Mars in a region of restricted water flow, such as an evaporating pool of brine.

The seven groups of Martian meteorites are all igneous rocks, and from them we can learn about processes on Mars. There are three types of information we can gain from these rocks. The first is information about primary magmatism on Mars: the meteorites are igneous rocks, and they have crystallized from the flows of Martian volcanoes or within subsurface magma chambers. Therefore, studying the composition and mineralogy of these rocks helps us to understand the temperature and mode of volcanism on Mars.

The secondary alteration products in these meteorites help us to learn about fluid flow on the surface of Mars. Particularly in ALH84001 and in the nakhlites, the complex assemblages of secondary minerals shed light on the temperature and salinity of the water that flowed across Mars' surface. The zoned nature of some of the minerals tells us how fluid composition has changed, either in terms of temperature or in terms of the salts dissolved in the fluid. The restricted nature of some of these alteration assemblages also indicates the restricted nature of that fluid flow (perhaps fluid that was combined to an evaporating basin rather than to a river or a stream).

The third type of information that we can gain from Martian meteorites is associated with shock. The shergottites and the chassignites are shocked meteorites, and they contain patches of the glass maskelynite formed by the conversion of plagioclase during shock melting and quenching. Trapped within these melt pockets is Mars' atmosphere; as outlined above, this was the clue that finally confirmed the Martian origin of these rocks. By looking at Martian meteorites of different ages and at the gas trapped in the rocks of different ages, it is possible that we might be looking at samples of Mars' atmosphere trapped at different times. It is also possible that we might be able to trace the evolution of Mars' atmosphere by looking at these pockets of gas.

8.2.5.3 Microfossils in a Martian Meteorite?

The observation that at least one of the SNCs, EETA79001 (a specimen found in the Elephant Moraine region of Antarctica in 1979), contained indigenous, i.e., Martian, organic material associated with carbonate minerals sparked a debate on the possibility of Martian meteorites containing evidence for extraterrestrial life. In 1994 this work was followed by the discovery of similarly enhanced levels of organic carbon in close association with carbonates in a second Antarctic Martian meteorite, ALH84001.

ALH84001 became a further focus of interest in 1996 when a team of scientists led by David McKay of NASA's Johnson Space Center in Houston reported the description of nanometer-sized features within carbonate patches in ALH84001 and claimed to have found evidence of primitive "fossilized Martian biota." Identification of the features as nanofossils remains controversial because much of the evidence is circumstantial and relies on the coincidence between a number of otherwise unrelated characteristics of the meteorite, particularly the association of organic materials with carbonates. One specific aspect of the claim was that within the carbonates were tiny grains of magnetite, which seemed to form chains. Those chains of grains were interpreted as being reminiscent of traces from magnetotactic bacteria. However, other scientists have argued that these grains of magnetite are natural products within the meteorite and are not connected at all with any type of bacterium. Indeed, the formation conditions of both the carbonate and magnetite, the relevance of the organic compounds, and the interpretation of the morphology of the "nanofossils" have all been subject to detailed investigation by several groups of scientists.

More recently, the same team investigated carbonaceous features in Nakhla that they have posited might have a biological origin. Although many members of the scientific community are skeptical about interpretation of the structures in ALH84001 and Nakhla, discussion of the features has stimulated an enormous amount of interest in the possibilities of life on Mars.

8.2.6
Can We Detect Signatures of Life on Mars?

Whether or not the features in the Martian meteorites are remnant fossilized Martian bacteria, the debate has highlighted one of the difficulties of searching for fossil microorganisms on any planet, including Earth. The most ancient fossil life on Earth was, until recently, thought to be microfossils of the Apex chert in western Australia. However, recent work has reinterpreted these structures as non-biological. The point I am trying to make here is that on Earth one can return to the locality from which the microfossils were collected and obtain fresh material; one can study them *in situ* and in the context of the rocks and geology around them. One can use every technique available to analyze these features, but still one cannot say for certain whether they are biological or not. Thus, relying on the morphology, the appearance only, of microfossils, whether on Earth or on Mars, is not going to deliver the definitive answer of how to detect life.

So how should and how could one detect life on Mars? There was an instrument on the recent unfortunate *Beagle 2* lander (see Chapter 12) that would have performed the type of experiment that would help us understand whether life was present on Mars. The Gas Analyze Package (GAP) was a mass spectrometer on the *Beagle 2* lander; it was constructed to determine the isotopic composition of carbon in Mars's soil. The experiment would have taken a small chip of rock and heated it up slowly in the presence of oxygen (O); O would burn carbon (C). The temperature at which C burns gives information on the type of component present, whether it is a simple organic molecule or is more complex, whether it is a carbonate or is carbon trapped or dissolved in primary magmatic minerals. By looking at the speciation of C and its isotopic composition, GAP would have made the measurements that would allow us to build up a C cycle for Mars. We would have been able to put limits on the different reservoirs on Mars' surface and assess whether or not there was evidence for life on Mars.

One of the next chances for detecting signatures of life on Mars will be with the European mission *ExoMars*, to be launched in 2013 (see Chapter 12). The *ExoMars* rover will carry a comprehensive suite of analytical instruments, including the Pasteur payload, dedicated to searching for signs of past and present life on Mars, especially in subsurface samples, by a suite of coordinated investigations.

8.2.7 Conclusions: Life Beyond Earth?

To summarize, we have not yet found life (extant, dormant, or extinct) on Mars, or any evidence that it arose on Mars. There are several missions planned to Mars in the coming decade, but we do not know whether any one of them will be successful in finding life on Mars. Looking at the astrobiology of the terrestrial planets, it has been shown that Mars is indeed, at least within the rocky inner planets, the best place other than Earth to look for life.

Other targets to be considered in astrobiology are the non-terrestrial planets, or, more specifically, their satellites. Titan, the biggest moon of Saturn, has been described as a laboratory for prebiotic Earth, and results from the recent *Huygens* probe have confirmed that interesting chemistry is certainly occurring there (see Chapter 9). Results from the *Cassini* mission have indicated that Enceladus seems to have water, and Europa, one of Jupiter's Galilean satellites, was identified as probably possessing a subsurface ocean (see Chapter 10). The possibilities of life beyond Mars are discussed in other chapters within this book.

8.3 Further Reading

8.3.1 Concerning Planetary Formation and Chronology

Halliday, A. N., Kleine, T. Meteorites and the timing, mechanisms, and conditions of terrestrial planet accretion and early differentiation, in: D. Lauretta (Ed.) *Meteorites and the Early Solar System II*, Univ. Arizona Press, 775–801, 2006.

Lissauer, J. J. Planet formation, *Ann. Rev. Astron. Astrophys.* **1993**, *31*, 129–174.

Wadhwa, M., Russell, S. S. Timescales of accretion and differentiation in the early Solar System: the meteoritic evidence, in: A. Boss, V. Manning, S. Russell (Eds.) *Protostars and Planets IV*, University of Arizona Press, 995–1018, 2000.

8.3.2
Concerning Recent Results from Mars

Bibring, J.-P., Langevin, Y., Gendrin, A., Gondet, B., Poulet, F., Berthé, M., Soufflot, A., Arvidson, R., Mangold, N., Mustard, J., Drossart, P. Mars surface diversity as revealed by the OMEGA/Mars Express observations, *Science* **2005**, *307*, 1576–1581.

Christensen, P. R., Wyatt, M. B., Glotch, T. D., Rogers, A. D., Anwar, S., Arvidson, R. E., Bandfield, J. L., Blaney, D. L., Budney, C., Calvin, W. M., Fallacaro, A., Fergason, R. L., Gorelick, N., Graff, T. G., Hamilton, V. E., Hayes, A. G., Johnson, J. R., Knudson, A. T., McSween, H. Y., Mehall, G. L., Mehall, L. K., Moersch, J. E., Morris, R. V., Smith, M. D., Squyres, S. W., Ruff, S. W., Wolff, M. J. Mineralogy at Meridiani Planum from the Mini-TES Experiment on the Opportunity Rover, *Science* **2004**, *306*, 1733–1739.

Klingelhöfer, G., Morris, R. V., Bernhardt, B., Schröder, C., Rodionov, D. S., de Souza, P. A., Yen, A., Gellert, R., Evlanov, E. N., Zubkov, B., Foh, J., Bonnes, U., Kankeleit, E., Gütlich, P., Ming, D. W., Renz, F., Wdowiak, T., Squyres, S. W., Arvidson, R. E. Jarosite and Hematite at Meridiani Planum from Opportunity's Mössbauer Spectrometer, *Science* **2004**, *306*, 1740–1745.

Squyres, S. W., Grotzinger, J. P., Arvidson, R. E., Bell, J. F., Calvin, W., Christensen, P. R., Clark, B. C., Crisp, J. A., Farrand, W. H., Herkenhoff, K. E., Johnson, J. R., Klingelhöfer, G., Knoll, A. H., McLennan, S. M., McSween, H. Y., Morris, R. V., Rice, J. W., Rieder, R., Soderblom, L. A. *In situ* evidence for an ancient aqueous environment at Meridiani Planum, Mars, *Science* **2004**, *306*, 1709–1714.

Tokano, T. (Ed.) *Water on Mars and Life*, Springer Berlin Heidelberg New York, 2005.

8.3.3
Concerning Terrestrial and Martian Microfossils

Brasier, M. D., Green, O. R., Jephcoat, A. P., Kleppe, A. K., Van Kranendonk, M. J., Lindsay, J. F., Steele, A., Grassineau, N. V. Questioning the evidence for Earth's oldest fossils, *Nature* **2002**, *416, 76–81.*

McKay, D. S., Gibson, E. K., Thomas-Keprta, K. L., Vali, H., Romanek, C. S., Clemett, S. J., Chiller, X. D. F., Maechling, C. R., Zare, R. N. Search for past life on Mars: Possible relic biogenic activity in Martian meteorite ALH84001, *Science* **1996**, *273*, 924–930.

Schopf, J. W., Kudryavtsev, A. B., Agresti, D. G., Wdowiak, T. J., Czaja, A. D. Laser-Raman imagery of Earth's earliest fossils, *Nature* **2002**, *416*, 73–76.

8.3.4
Concerning Meteorites from Mars

Bogard, D. D., Johnson, P. Martian gases in an Antarctic meteorite? *Science* **1983**, *221*, 651–654.

Bridges, J. C., Catling, D. C., Saxton, J. M., Swindle, T. D., Lyon, I. C., Grady M. M. Alteration assemblages in Martian meteorites: implications for near-surface processes, *Space Sci. Rev.* **2001**, *96*, 365–392.

Franchi, I. A., Wright, I. P., Sexton, A. S., Pillinger, C. T. The oxygen-isotopic composition of Earth and Mars, *Meteorit. Planet. Sci.* **1999**, *34*, 657–661.

McSween, H. Y., Jr. What we have learned about Mars from SNC meteorites, *Meteoritics* **1994**, *29*, 757–779.

Nyquist, L. E., Bogard, D. D., Shih, C.-Y., Greshake, A., Stöffler, D., Eugster, O. Ages and geologic histories of Martian meteorites, *Space Sci. Rev.* **2001**, *96*, 105–164.

Pepin, R. O. Evidence of Martian origins, *Nature* **1985**, *317*, 473–475.

8.4
Questions for Students

Question 8.1

Outline the main reasons that the meteorite EET A79001 is believed to come from Mars.

Question 8.2

Why is Mars thought to be a possible place to search for life?

Question 8.3

What can we learn about Mars from Martian meteorites?

9
Astrobiology of Saturn's Moon Titan

François Raulin

The chapter presents the astrobiological case for Titan, the largest satellite of Saturn. Besides Jupiter's moon Europa, Titan appears to be the most interesting body for astrobiology in the outer Solar System because it has an active prebiotic-like chemistry, but liquid water is absent on its surface. This chapter starts with a description of the general properties of Titan and models of its formation and internal structure. Emphasis is placed on the astrobiological aspects, which are based on spacecraft observations, above all the recent Cassini–Huygens mission, on laboratory experiments simulating the atmospheric conditions of Titan, and on photochemical modeling. On the basis of those data, current astrobiological aspects of Titan are considered with regard to analogies with the early Earth, extraterrestrial organic (even prebiotic-like) chemistry, and possibilities for life.

9.1
Extraterrestrial Bodies of Astrobiological Interest

The following are different types of extraterrestrial bodies of prime interest for astrobiology:

- planetary bodies where life may be present, either extinct or extant,
- other bodies where a complex organic chemistry is taking place, and
- bodies that may show some similarities with our own planet Earth, particularly the early Earth, before the emergence of life.

The best examples of the first category are our neighbor planet Mars (see Chapter 8) and Europa, one of the Galilean satellites of Jupiter, because of its potential for an internal water ocean (see Chapter 10). In fact, with the exception of Io, the

Complete Course in Astrobiology. Edited by Gerda Horneck and Petra Rettberg

ISBN: 978-3-527-40660-9

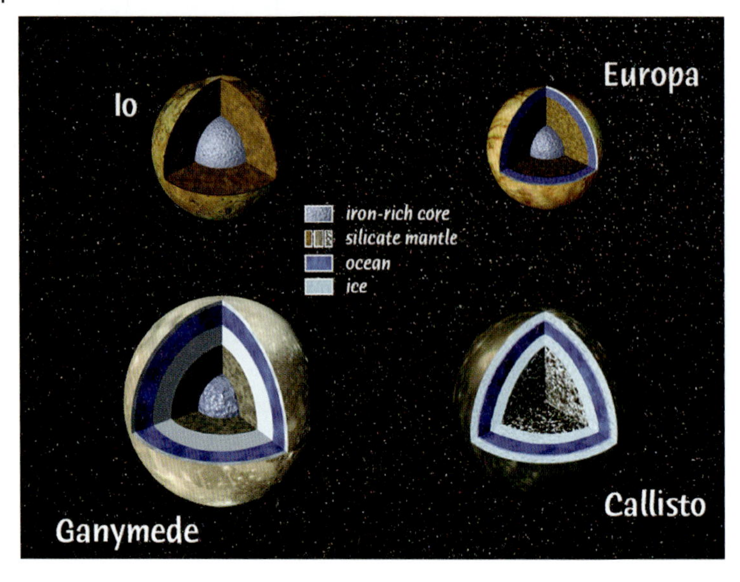

Fig. 9.1 Possible internal structure of the four Galilean satellites (adapted from NASA/JPL).

majority of the Galilean satellites seem to have a water ocean in their internal structure (Fig. 9.1) and thus are of interest for astrobiology. Recent data, provided by the infrared (IR) spectrometer of the *Cassini–Huygens* mission, identified Saturn's moon Enceladus as a body of high astrobiological interest. The temperature at its south pole is very high, comparable to that of the Earth, and may allow the presence of liquid water on its surface.

Comets and carbonaceous meteorites are the best example of the second category (see Chapter 3). Titan, the largest satellite of Saturn and the only satellite in the Solar System having a dense atmosphere (Fig. 9.2), combines all three aspects.

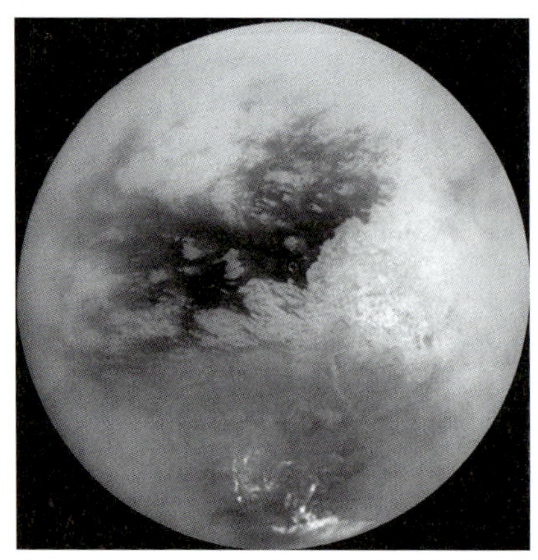

Fig. 9.2 Mosaic of Titan's surface from images taken by the narrow-angle camera of the *Cassini* spacecraft. The large bright-white area is called Xanadu. The resolution is about 1.3 km per pixel (image credit: NASA/JPL/Space Science Institute).

Therefore, Titan is considered one of the most interesting bodies of the Solar System with regard to astrobiology.

Many data have been obtained and are still expected from the *Cassini–Huygens* mission, a cooperative project of the U. S. National Aeronautics and Space Administration (NASA), the European Space Agency (ESA), and the Italian Space Agency (ASI). Future astrobiological exploration missions to Titan are under consideration. With an active organic chemistry, but in the absence of permanent liquid water on the surface, Titan already looks like another world: a natural laboratory for prebiotic-like chemistry.

9.2
Some Historical Milestones in the Exploration of Titan

On 25 March 1655, the Dutch astronomer Christiaan Huygens (1629–1695), with his newly constructed 2-inch diameter lens of 12-foot focal length (providing a 50 magnification), was observing Saturn and detected an enigmatic prominent object, three minutes of arc away from the giant planet. By continuing the observation over several nights, he noted that this object was following Saturn's trajectory. He thus deduced that it was a moon of Saturn. He was even able to calculate that this satellite orbits its planet within 16 terrestrial days. Much later, in 1847, this largest moon of Saturn was named "Titan" by John Herschel (1792–1871). According to the Roman mythology, Titan is one of the siblings of Saturn, the god of agriculture.

While observing Titan in 1908, the Spanish astronomer José Comas-Sola (1868–1937) reported seeing limb darkening around the disk of the satellite. This observation suggested the presence of an appreciable atmosphere around Titan. However, this report was received with skepticism within the scientific community, because Comas-Sola had reported similar observations for most of the Galilean satellites. Finally, his conclusion was taken seriously into consideration by the British astronomer James Jeans (1877–1946).

Jeans, author of the theory of thermal atmospheric escape, carried out a mathematical study of the possible escape of an atmosphere from planetary bodies. He determined that a body like Titan can keep a substantial part of its atmosphere, despite its relatively small size (with a diameter on the order of 5 000 km), if the temperatures of its environment are low enough (ranging from about 60 K to 100 K). More precisely, the escape rates for all species with a molecular mass exceeding about 16 Da will be very small at those low temperatures.

In the winter of 1943–1944, the American-Dutch astronomer Gerard Pieter Kuiper (1905–1973) observed Titan from the 82-inch McDonald Observatory (Mount Locke, Texas) using photographic emulsions sensitive up to the near infrared (0.7 μm). He obtained a spectrum showing the absorption bands of gaseous methane (CH_4). This was the confirmation that Titan indeed has an atmosphere, and it demonstrated that this atmosphere includes methane as one of its constituents. Kuiper was even able to calculate the abundance of methane and derived from his observations that the partial pressure of methane at Titan's surface should be about 0.1 bar (100 hPa).

The presence of methane in the gaseous phase was confirmed later on by many observations from ground-based telescopes, before the Titan flyby maneuvers of NASA's *Pioneer 11* spacecraft in 1979 and of *Voyager 1* in 1980, one of NASA's twin *Voyager* spacecraft.

9.3
General Properties, Formation and Internal Structure of Titan

9.3.1
Main Properties

In Table 9.1 the main general properties of Titan are listed. With a radius R of 2 575 km, Titan is the largest satellite of Saturn and is the second largest satellite of the Solar System, after Ganymede (R = 2 631 km), one of the Galilean Jovian moons. But if one takes into account its dense atmosphere and extends its radius up to the stratopause, Titan reaches a radius of 2 875 km, which makes it the largest moon in the Solar System. Like Saturn, Titan's mean distance from the Sun is about 9.5 astronomical units (AU) and its period of rotation around the Sun – as given by Kepler's third law (see Chapter 7) – is about 30 years. The solar flux at the location of Titan is about 1% of the flux at the level of Earth orbit (remember that the radiation decreases with the square of its distance from the source). The mean temperature of Titan, if considered as a black body, is about 82 K.

Table 9.1 Main characteristics of Titan (including HASI *Huygens* data).

Property		Value	
Surface radius		2 575 km	
Surface gravity		1.35 m s^{-2} (0.14 Earth's value)	
Mean volumic mass		1.88 kg dm^{-3} (0.34 Earth's value)	
Distance from Saturn		20 Saturn radii (~1.2×10^6 km)	
Orbit period around Saturn		~16 days	
Orbit period around Sun		~30 years	
		Atmospheric data	
	Altitude (km)	**Temperature (K)**	**Pressure (hPa)**
Surface	0	93.7	1 470
Tropopause	42	70.4	135
Stratopause	~250	~187	~1.5×10^{-1}
Mesopause	~490	~152	~2×10^{-3}

Titan is in fact the only satellite of the Solar System that has a noticeable atmosphere. Its atmosphere is made mainly of nitrogen, with a substantial percentage of methane. The surface pressure is 1.5 bar (1 500 hPa) and the surface temperature is about 94 K. This temperature is higher than the black body temperature because of a noticeable greenhouse effect of about 20 K caused by some of the atmospheric constituents (mainly methane and dihydrogen) and despite an anti-greenhouse effect of about 8 K produced by the haze particles. The consequence of these temperature and pressure (T, P) conditions is that Titan's atmosphere near the surface is about five times denser than the Earth's atmosphere.

The inclination of Titan on its orbit around Saturn is small (0.33 °). Because Titan's equatorial plane is practically in Saturn's ring plane, Titan has seasons like Saturn, which has an obliquity of 27 °. A complete rotation of Titan around Saturn takes 16 days, with synchronous rotation. But while the solid surface of Titan rotates slowly, the satellite exhibits a super-rotation of its atmosphere due to strong zonal winds (similar to the case of Venus, with its very dense atmosphere; see Chapter 8). Titan's orbit is slightly eccentric (e = 0.029): this provides some constraints on its internal structure and suggests the absence of a shallow ocean on its surface (which would tend to reduce the eccentricity value and change the orbit toward a circular one). Its distance from Saturn is about 20 Saturnian radii, far enough to avoid interactions with the Saturn rings and still small enough to allow Titan's atmosphere to interact regularly with the electrons of Saturn's magnetosphere.

Another detail of paramount importance to astrobiology is the mean density of the satellite, which is 1.88 g cm^{-3}. This value is equivalent to that of Ganymede (about 1.9 g cm^{-3}) and much smaller than that of Jupiter's satellite Europa (about 3 g cm^{-3}). The low density of Titan indicates an internal structure made of low-density materials such as water (in solid and possibly liquid phase) and, more generally, ices of a mean density of approximately 1 g cm^{-3}, mixed with high-density materials such as rocks and silicates with a mean density of about 3 g cm^{-3}. This information is important when modeling the formation of the satellite.

9.3.2 Models of Formation and Internal Structure

Like the other satellites of Saturn, Titan originated about 4.6 billion years ago from Saturn's subnebula. Titan grew up out of the accretion of small blocks and planetesimals present in the subnebula; during that process, the temperature increase generated mainly by gravity processes and radioactive decay was able to melt the icy components, and the rocks sunk toward the center of the planetary body. The resulting body was probably made of an undifferentiated inner silicate core, mixing rocks and ices, covered by a silicate outer core and a liquid mantle of a water–ammonia mixture, and dominated by a dense atmosphere. The origin and composition of this primordial atmosphere is still under debate. One classical scenario, developed by Prinn and Fegley in 1981, assumes that the early subnebula of Saturn was convective and stable and that carbon was mainly in the form of carbon monoxide (CO) and nitrogen in the form of N_2. At the location of Titan's orbit, which is 20 Saturn radii, the temperature and pressure conditions were high enough to allow the conversion of CO into CH_4 and N_2 into ammonia (NH_3). Then CH_4 and NH_3 would have been trapped in the form of hydrates in the planetesimals forming Titan. The specificity of Titan, compared to all of Saturn's other satellites, can consequently be explained by the particular position of its site of formation within Saturn's subnebula, as recently published by Sekine and coworkers (2005).

A different scenario, developed by Mousis et al. (2002), supposes that the subnebula of Saturn was turbulent and that CO and N_2 could not be efficiently converted into CH_4 and NH_3 under such conditions. The subnebula thus contained CO and some CH_4 and N_2 and some NH_3. Models of the evolution of such subnebula show that in the feeding zone of Saturn, were the planetesimals are formed, CH_4 and NH_3 were efficiently trapped as hydrates, contrary to CO and N_2. The relics of such CH_4/NH_3-rich planetesimals were thus used to form Titan. In this case, the composition of the atmosphere should reflect the volatile components of those planetesimals.

Both models assume that CH_4 and NH_3 were progressively released in the atmosphere and that NH_3 was converted into N_2 by photochemical processes in the upper regions of the atmosphere or even by impact shock–catalyzed dissociations. From such models of Titan's formation, it is possible to derive the resulting internal structure of Titan after 4.6 billion years of evolution.

After a relatively short period (about 100 million years) of a warm aqueous environment induced by efficient energy sources, those energies became less abundant and Titan's surface cooled down. Water started freezing out of the

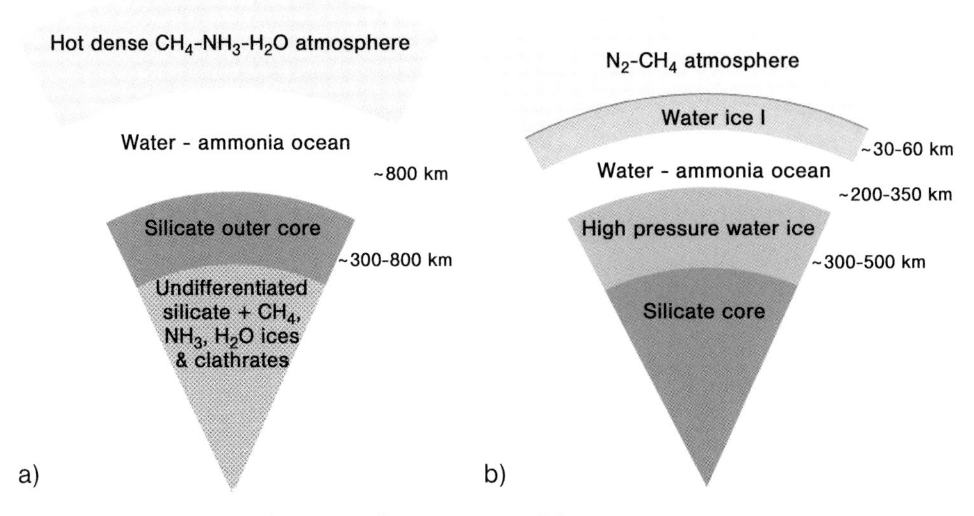

Fig. 9.3 Internal structure of Titan. (a) primordial Titan; (b) present Titan (adapted from Fortes 2000).

H_2O–NH_3 solution, thereby forming an ice I crust. (Ice I is the form of all natural snow and ice on Earth.) Its thickness rapidly increased, reaching 30 km in (only) 70 million years. The high pressure at the bottom of the ocean induced the crystallization of water ice VI and ammonia hydrates. (Ice VI is formed from liquid water at 1.1 GPa by lowering its temperature to 270 K.) This process should have enriched Titan's ocean in ammonia. The depth and composition of the resulting ocean depended on the heat flow from the core and its convective transport to the surface. Recent models by Tobie and coworkers (2005) indicate that the ocean might have contained up to 15 wt% of NH_3 over a depth of about 200 km (Fig. 9.3).

9.4 Atmosphere and Surface of Titan

As in many scientific fields, one can consider three main and complementary approaches to studying a planetary body.

- One approach is theoretical modeling, e.g., with regard to photochemistry, thermodynamics, and microphysics. In the case of planetary astrobiology, it allows one to follow the evolution of organics within time durations that are unreachable experimentally and to provide estimates of their concentrations in different space locations.
- The second approach is experimental, involving laboratory simulations with specific experiments. With these ad hoc tools, by developing experiments in the laboratory, one can obtain detailed information about the studied object (of astrobiological interest).
- The last approach is observation, which includes remote sensing from Earth and spacecraft and *in situ* measurements. This is an essential and perfect approach to constrain the models, both theoretical and experimental, and to provide, by coupling the observational data with the other information, an integrated comprehensive vision of the studied body.

9.4.1 Theoretical Modeling of Titan's Atmosphere

In a high-temperature system, one can derive the composition of the main and trace species by assuming thermodynamic equilibrium and calculating the distribution of species from their standard free energies of formation. The result of such calculation, based on relatively simple chemical equilibrium calculation, has the advantage of being independent of the reactions that have been considered. However, in the case of a low-temperature system, such as Titan's atmosphere, the equilibrium state is not reached. Therefore, this approach is more than questionable.

In the case of low-temperature systems, kinetic modeling is a safer method. It is possible to determine the steady-state composition that can be reached in any part of the system by the following approach. First, all chemical reactions involved in the system need to be identified, and their corresponding rate constants at the right temperatures must be obtained. Further, the spatial distribution of energy including the radiative transfer aspects must be known, as well as the diffusion constants for both molecular and eddy diffusions. From a set of chosen chemical reactions (which can be numerous!), a differential equation between concentration, time, and space coordinates (altitude z in a 1D model) can be written for each compound in a given location of the system. With adequate software these equations can simultaneously be solved. Thus, a steady state can be written for each constituent i (Eq. 1)), where n_i is the concentration of the constituent, P_i its production rate, L_i its loss rate, Φ_i its vertical flux, C_i its condensation factor, and z the altitude. The resolution of these equations gives the concentration of i as a function of the space coordinate (altitude z in a 1D model).

$$dn_i/dt = P_i - n_i L_i - d\phi_i/dz - C_i \tag{9.1}$$

Several kinetic models, mainly photochemical models, of Titan's atmosphere have already been published. Until now, the data (mainly the vertical concentration profiles) derived from these models were relatively poorly constrained by observational data, because only a few observations provided reliable concentration values at a given altitude. Furthermore, the models involved eddy diffusion coefficient, the profile of which is poorly known.

Nevertheless, these models provide a quite good understanding of the chemical processes involved in this complex atmospheric chemistry. The primary processes involve the dissociation of CH_4 and N_2 by the solar UV photons and the electrons

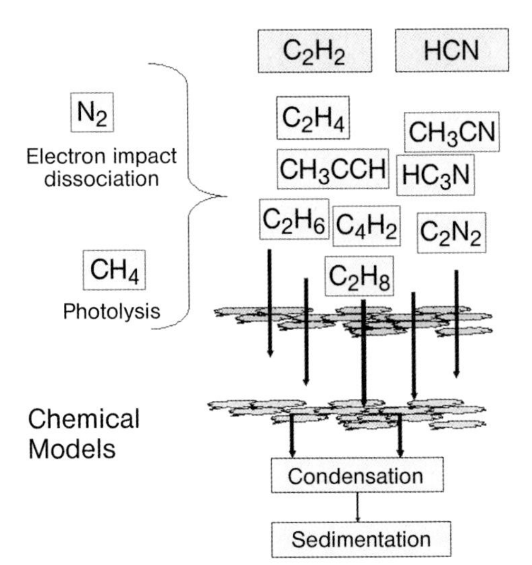

Fig. 9.4 Chemical and physical processes in Titan's atmosphere according to chemical modeling.

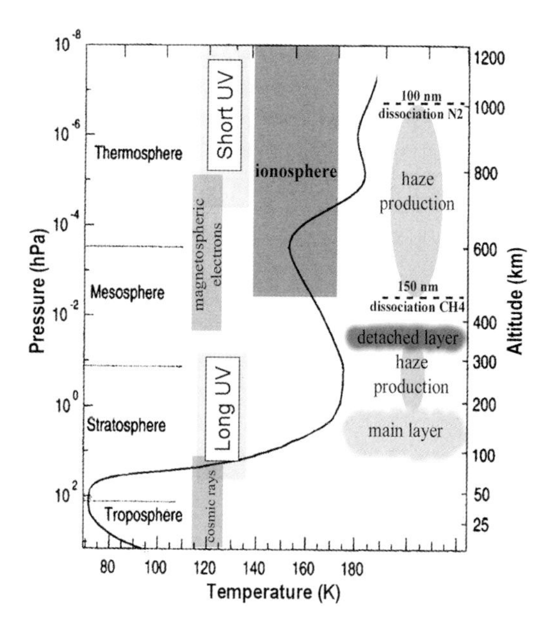

Fig. 9.5 Energy deposition in Titan's atmosphere.

from Saturn's magnetosphere. The resulting primary species formed in the high atmosphere yield to the production of simple hydrocarbons and N-compounds, especially C_2H_2 and HCN (Fig. 9.4). These two molecules play a key role in the general chemical scheme. Once they are formed, they diffuse down to the lower atmospheric regions, where the available UV photons are less energetic but are absorbed by the two compounds, which leads to their dissociation (Fig. 9.5). The resulting radicals allow the formation of more-complex hydrocarbons and nitriles. In the chemical processes, polyynes, such as C_4H_2 and C_6H_2, and cyanopolyynes, such as HC_5N, may play an important role in the formation of high-molecular-weight products. Additional CH_4 dissociation probably occurs in the low stratosphere through photocatalytic processes involving C_2H_2 and polyynes. The end product of this chemistry would be a macromolecular organic compound made mainly of C, H, and N.

If a species is omitted in a theoretical model, this will not change the results of thermodynamic equilibrium calculations. On the contrary, kinetic prediction may strongly depend on the nature of the assumed mechanism, and the omission of an important reaction may render the result invalid. Furthermore, contrary to thermodynamic data, many of the data feeding the photochemical models are not available. This is the case, for instance, with kinetic constants, especially at low temperatures and for complex species, as well as with UV and IR spectra and quantum yields and branching ratios of photodissociation reactions. The simple case of methane photodissociation with Lyman alpha photons (121.6 nm) clearly illustrates the problem. As shown in Table 9.2, the values of quantum yields and of branching ratios are quite variable from one author to another. This may induce strong uncertainties in the photochemical model, which then propagate.

Table 9.2 Quantum yield of the primary process of photodissociation of methane by Lyman alpha radiation (121.6 nm) obtained by different models.

Reaction	Quantum yield			
	Model A[a]	Model B[a]	Model C[b]	Model D[c]
$CH_4 + h\nu \rightarrow CH_3 + H$	0.51	0.49	0.41	0.291
$CH_4 + h\nu \rightarrow CH_2 + H + H$	0.25	0	0	0.055
$CH_4 + h\nu \rightarrow CH_2 + H_2$	0.24	0	0.53	0.584
$CH_4 + h\nu \rightarrow CH + H + H_2$	0	0.51	0.06	0.07

Source: Adapted from Hérbrard (2006).
a Model by Mordaunt and coworkers (1993).
b Model by Smith and Raulin (1999).
c Model by Wang and coworkers (2006).

A new model is currently being developed by this author's group to study the influence of these uncertainties in the model inputs on the output data of the model. This new 1-D model includes 121 neutral species, 20 photodissociations

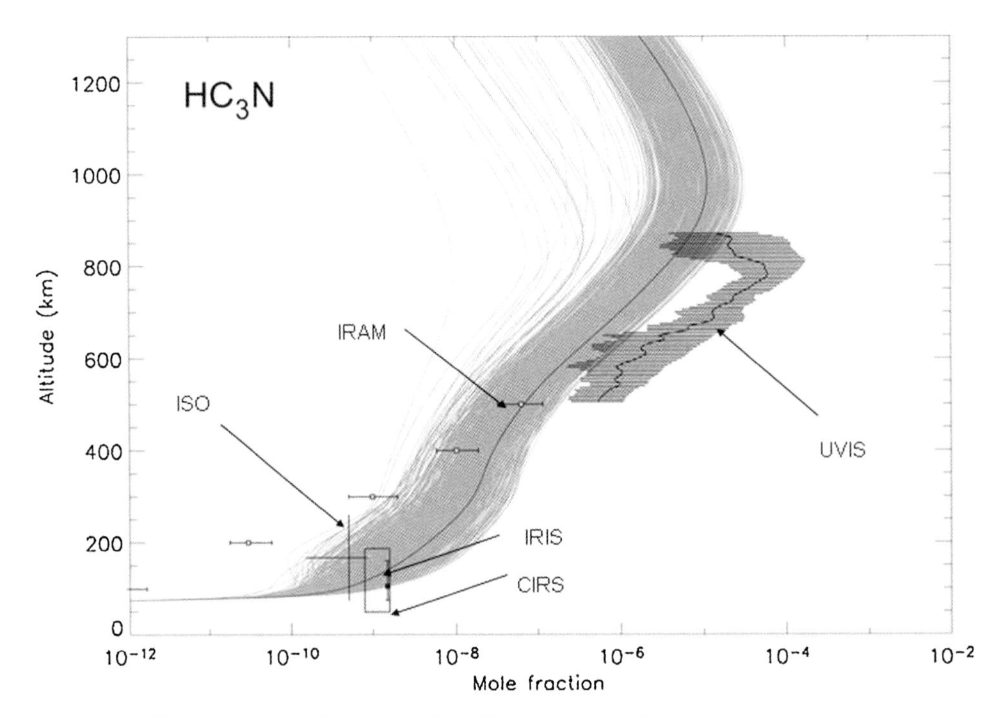

Fig. 9.6 1-D photochemical modeling of Titan: vertical distribution of HC_3N and associated uncertainties derived from the uncertainties in rate constants and photolysis rates. Several available observational data are plotted for comparison (from Hébrard 2006).

with 50 branches, and 664 chemical reactions with associated reaction rates. It uses 260 atmospheric layers and implies the resolution of 31 460 nonlinear, highly coupled equations. The propagation of chemical uncertainties in the model induces strong uncertainties in the vertical concentration profiles, as shown in Fig. 9.6 for HC_3N. These uncertainties have strong consequences on the conclusions that can be inferred from the models. With the resulting error bars, the modeled vertical profile fits with most of the observations, suggesting that it is difficult to constrain the chemical scheme of the model without having a better precision of the input parameters. In any case, the information provided by such work can be used for identifying the reactions for which kinetic data and other associated parameters are essential and need to be experimentally determined.

9.4.2
Experimental Approach

Different types of laboratory experiments related to the study of planetary environments can be considered. In particular, experiments are specifically designed for determining some of the parameters involved in the evolution of the planetary body, such as rate constant or thermodynamic equilibrium data or spectroscopic data (IR, UV, radio-wave, and so forth). Other experiments are developed to simulate partly or globally the evolution of the considered environment. Such so-called "simulation experiments" can integrate the many different physical and chemical processes involved. One of the first simulation experiments carried out in the laboratory is the classic Stanley Miller experiment, which is supposed to mimic the chemical evolution processes in the conditions of the primitive Earth environment (see Chapter 3). In the absence of any direct observational data of such an environment, this was indeed a unique tool. More recently, several groups have used a similar approach to study the evolution and organic chemistry of Titan's atmosphere.

This type of experiment requires a reactor with the following specifications:

- a judiciously chosen energy source that adequately mimics the sources really present in the atmosphere and that is able to break the chemical bonds of the molecules constituting the initial gas mixture representing the main composition of Titan's atmosphere;
- a good vacuum device to prepare and introduce this initial gas mixture into the reactor and to sample the resulting products; and
- efficient analytical tools for analyzing the products, in both the gas and condensed phases.

Through sampling and chemical analysis, the evolution of the system then can be followed and the results extrapolated to the real case. In fact, the design and development of the experiment is not so simple, and it requires careful preparation and great care. The right choice of the numerous parameters involved is essential, e.g., the abundance of the constituents in the starting gas mixture, the pressure, the energy flux, the temperature, etc.

Further, the design of the reactor, the nature of the energy source, and the analytical procedures are essential. The reaction vessel may induce some wall effect, which is of course absent in the atmosphere of Titan and which may bias the results. Simulating processes in the high atmosphere, where pressure is low and three-body collisions are not favored, requires the use of large reactors. Under such conditions, the products may be very diluted and difficult to analyze. On the other hand, when using small, closed reactors, the accumulation of products makes their analysis easier, but it may significantly alter the primary processes compared to the planetary conditions. Indeed, within such conditions, the composition of the gas mixture changes and no longer represents the main composition of the simulated atmosphere. This issue may be avoided by using a flow reactor, which is usually coupled to an electron impact energy source, such as a cold plasma discharge.

In Titan's atmosphere the main energy sources are solar UV radiation and mid-energy electrons coming from Saturn's magnetosphere (Fig. 9.5). Only UV photons of wavelengths shorter than about 150 nm can dissociate methane. The most abundant solar photons in this range of wavelengths are Lyman alpha (121.6 nm) photons. This radiation can easily be simulated in the laboratory by using monochromatic lamps. But the Lyman α line does not allow the photodissociation of N_2, which requires photons of a wavelength shorter than about 100 nm. UV radiation of such short wavelengths is difficult to obtain in the laboratory with irradiation systems compatible with closed reactors or even with flow reactors. For this reason, all laboratory experiments simulating the chemical evolution of Titan's atmosphere that have been carried out so far on N_2–CH_4 initial gas mixtures use an electron impact energy source that mimics Titan's atmospheric electrons impinging from Saturn magnetosphere. Another approach compatible with a short-wavelength UV source is the use of an initial gas mixture with another N-compound that can be photodissociated by UV photons of longer wavelengths and is also present in Titan's atmosphere – although at trace levels. Examples include the molecules HCN and HC_3N.

Since the flyby of Titan by the two *Voyager* spacecrafts in the early 1980 s (which provided a first determination of the main properties of its atmosphere, including its main composition and vertical temperature and pressure structure), many experiments have been carried out to simulate Titan's atmospheric chemistry in the laboratory.

In the gas phase, more than 150 different organic molecules have been detected in simulation experiments.

These global simulations of Titan's atmospheric chemistry use an open reactor with a continuous flow of a low-pressure N_2–CH_4 gas mixture (Fig. 9.7). The energy source is a cold plasma discharge producing mid-energy electrons (around 1–10 eV). The gas-phase end products (molecules) are analyzed by infrared Fourier transform spectroscopy (IRFTS) and gas chromatography and mass spectrometry (GC-MS) techniques. The transient species, i.e., radicals and ions, are determined by online ultraviolet-visible (UV–VIS) spectroscopy. The evolution of the system is theoretically described using coupled physical and chemical models.

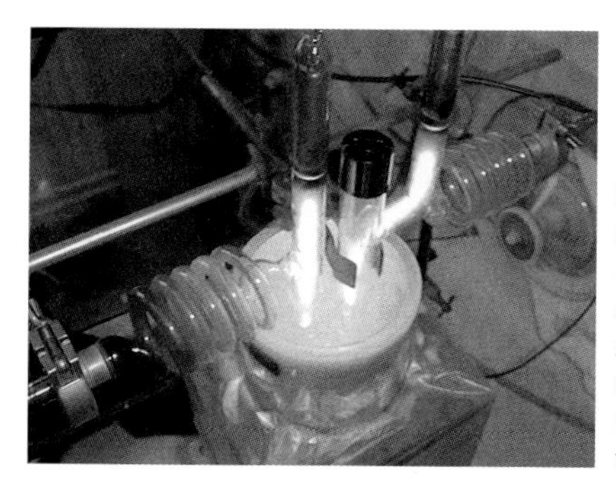

Fig. 9.7 Experimental setup for global simulation of Titan's atmospheric chemistry at LISA (Laboratoire Interuniversitaire des Systèmes Atmosphériques, Universités Paris 12 et Paris 7). A gas mixture of N_2 (98%) and CH_4 (2%) flows continuously through the open reactor and is maintained at low pressure (~1 hPa) and low temperature (~150 K) (image credit: P. Coll).

The identified organic products are mainly hydrocarbons and nitriles. The absence – at a detectable level – of molecules carrying amino groups, such as amines, with the exception of ammonia, must be highlighted. In these experiments all gaseous organic species observed in Titan's atmosphere were detected. Among the organics that were formed in simulation experiments, but that are not (yet) detected in Titan's atmosphere, one should note the presence of polyynes such as C_4H_2, C_6H_2, C_8H_2, and probably cyanopolyyne HC_4-CN. These compounds are also included in photochemical models of Titan's atmosphere, where they might play a key role in the chemical schemes allowing the transition from gas-phase products to aerosols. Of further astrobiological interest is the formation of organic compounds with asymmetric carbon.

Recent experiments by Bernard and coworkers on N_2–CH_4 mixtures including CO at the 100-ppm level show the incorporation of oxygen (O) atoms in the produced organics, with an increasing diversity of the products. More than 200 different compounds were identified. The main O-containing organic compound found is oxirane – also called ethylene oxide – $(CH_2)_2O$. Oxirane thus appears to be a good candidate to search for in Titan's atmosphere.

Further, simulation experiments produce solid organics, usually called tholins, a generic name invented by Carl Sagan in the late 1970 s. These "Titan tholins" are supposed to be laboratory analogues of Titan's aerosols. They have been extensively studied since the first work by Sagan and Khare more than 20 years ago. These laboratory analogues show very different properties depending on the experimental conditions. For instance, in the many published reports, the average C:N ratio of the product may vary between values of less than 1 and more than 11.

Recently, dedicated experimental protocols allowing a simulation closer to real conditions have been developed in this author's laboratory using low pressure and low temperature and recovering the laboratory tholins without oxygen contamination (from the air of the laboratory) in a glove box purged with pure N_2. Representative laboratory analogues of Titan's aerosols thus have been obtained and their complex refractive indices determined, for the first time with error bars. These data represent a new point of reference for modelers who are computing the properties of Titan's aerosols. Systematic studies have been carried out on the influence of the pressure of the starting gas mixture on the elemental composition of the tholins. They show that two different chemical-physical regimes are involved in the processes, depending on the pressure, with a transition pressure around 1 hPa.

The molecular composition of the Titan tholins is still poorly known. Several possibilities have been considered, including HCN polymers or oligomers, HCN-C_2H_2 co-oligomers, HC_3N polymers, and HC_3N-HCN co-oligomers. It is well established that they are made of macromolecules of largely irregular structure. Gel filtration chromatography of the water-soluble fraction of Titan tholins shows an average molecular mass of about 500–1 000 Da. Information on the chemical groups included in their structure has been obtained from their IR and UV spectra and from analysis by pyrolysis GC-MS techniques. The data show the presence of aliphatic and benzenic hydrocarbon groups, of CN, NH_2, and C=NH groups. Direct analysis by chemical derivatization techniques before and after hydrolysis allowed the identification of amino acids and their precursors. Their optical properties have been determined because of their importance for retrieving observational data from Titan. Finally, it obviously is of astrobiological interest to mention that the nutritious properties of Titan tholins for terrestrial microorganisms have also been studied, showing that some microorganisms can indeed use them as nutrients.

There is still a need for more-accurate experimental simulations, where the primary processes are well mimicked, including the dissociation of N_2 by electron impact with energies close to those of Titan's atmosphere and the dissociation of methane through photolysis processes. Such an experiment is currently under development at this author's laboratory, which is called SETUP (Simulation Expérimentale et Théorique Utile à la Planétologie).

9.4.3 Observational Approach

Of course, neither of the two previous approaches can be developed without any observational data. As mentioned above, much information on Titan has been obtained from ground-based observations, before and even after *Voyager*'s flyby of Titan in 1980 and 1981. Several instruments of the *Voyager* spacecraft provided further insight into Titan's atmosphere. For example, the determination of the atmospheric composition was obtained from observations by the infrared interferometer spectrometer and radiometer (IRIS) – in the IR range – and the ultraviolet spectrometer (UVS) – in the UV range. The vertical atmospheric structure from 200 km altitude to the surface was derived from the data of the radio-occultation experiment. Later observations from ground-based and Earth-orbiting space telescopes, such as NASA's *Hubble Space Telescope* (HST) and ESA's *Infrared Space Observatory* (ISO) provided additional data on Titan's atmospheric composition and even – for the first time – Titan's surface heterogeneities, by using dedicated spectral windows and adapted optics.

The NASA-ESA *Cassini–Huygens* mission is an excellent example of an observational approach to study Titan, because it combines both remote sensing and *in situ* observations. The *Cassini–Huygens* spacecraft was successfully launched from Cape Canaveral, Florida, USA, on 15 October 1997 by use of a Titan IVB/Centaur rocket.

The spacecraft – which consisted of a Saturn orbiter, the *Cassini* spacecraft, and a Titan atmospheric probe, the *Huygens* entry probe – was on a seven-year interplanetary trajectory toward Saturn that relied on four gravity-assist maneuvers at Venus (1998 and 1999), Earth (August 1999), and Jupiter (December 2000) to reach its final destination (Fig. 9.8; see Chapter 7 for more information on gravity-assist maneuvers). It reached Saturn in 2004, with a Saturn orbit insertion on 1 July 2004. Since that date, Saturn has a new – but artificial – satellite, the spacecraft *Cassini*. *Cassini* embarked on a four-year tour of the Saturnian system, which may last longer with an extended mission. At the end of 2004, after two initial Titan encounters, it released the *Huygens* probe on the third orbit around Saturn on 25 December 2004. *Huygens* penetrated Titan's atmosphere and parachuted down to the surface on 14 January 2005. The descent lasted about 2.5 hours, and *Huygens* continued functioning for three hours on the surface. *Cassini* recovered surface data for more than 1 hour and 10 minutes, before it went out of the field of view of the landed probe.

The *Cassini–Huygens* mission was designed to explore the Saturnian system in great detail, including the giant planet, its atmosphere, its magnetosphere and rings, and many of its moons, especially Titan, by using the 12 scientific instruments onboard the spacecraft (Table 9.3). Titan's exploration is one of the main objectives of the mission, which includes multiple opportunities for close remote sensing and observations of Titan by the *Cassini* orbiter. During the four years of

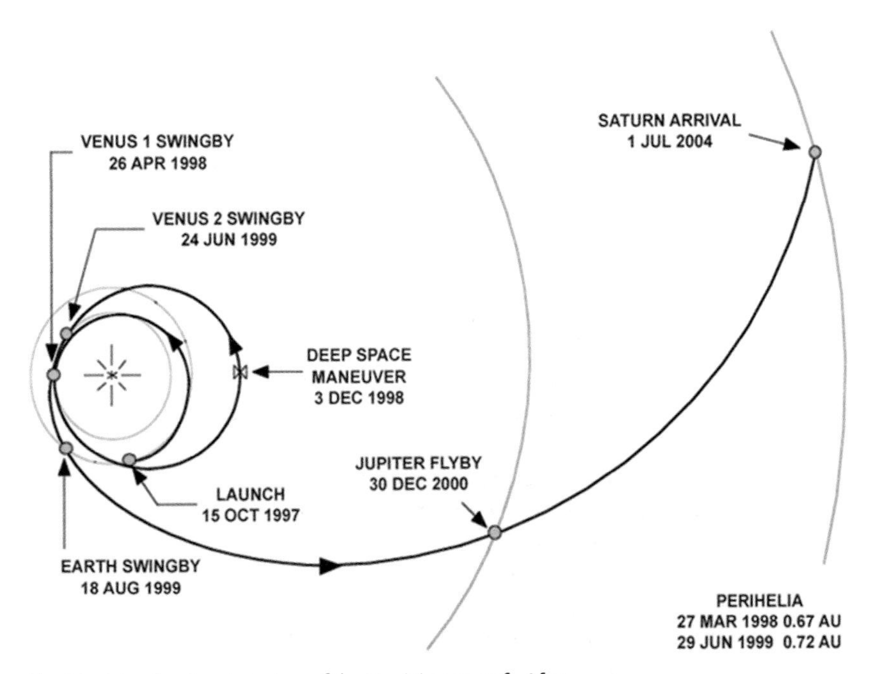

Fig. 9.8 Interplanetary trajectory of the *Cassini* spacecraft. After the insertion in the Saturn orbit on 1 July 2004, *Cassini* released the *Huygens* probe on 25 December 2004. The probe penetrated Titan's atmosphere on 14 January 2005 (image credit: NASA/JPL).

the nominal mission (mid-2004 to mid-2008), the planned 74 orbits around Saturn include 44 close Titan flybys.

Table 9.3 Instruments on the *Cassini* spacecraft, interdisciplinary programs (IDP), the leading scientists, and the potential for astrobiological return of their investigation.

Instrument or IDP	PI, team leader, or IDS[a]	Country	Astrobiological return
Optical remote sensing instruments			
Composite infrared spectrometer (CIRS)	V. Kunde, M. Flasar	USA	+++
Imaging science subsystem (ISS)	C. Porco	USA	+++
Ultraviolet imaging spectrograph (UVIS)	L. Esposito	USA	++
VIS/IR mapping spectrometer (VIMS)	R. Brown	USA	++
Fields, particles, and waves instruments			
Cassini plasma spectrometer	D. Young	USA	+
Cosmic dust analysis	E. Grün	Germany	+
Ion and neutral mass spectrometer	H. Waite	USA	+++
Magnetometer	D. Southwood, M. Dougherty	UK	Not applicable
Magnetospheric imaging instrument	S. Krimigis	USA	Not applicable
Radio and plasma wave spectrometer	D. Gurnett	USA	Not applicable
Microwave remote sensing			
Cassini radar	C. Elachi	USA	+++
Radio science subsystem	A. Kliore	USA	++
Interdisciplinary program			
Magnetosphere and plasma	M. Blanc	France	+
Rings and dust	J.N. Cuzzi	USA	+
Magnetosphere and plasma	T.I. Gombosi	USA	+
Atmospheres	T. Owen	USA	+++
Satellites and asteroids	L.A. Soderblom	USA	+
Aeronomy and solar wind interaction	D.F. Strobel	USA	++

a PI: principal investigator, IDS: interdisciplinary scientist.

Several instruments of the orbiter and most of the six instruments of the probe (Table 9.4) provide data of astrobiological interest (Fig. 9.9). The optical remote sensing instruments of the orbiter, especially the composite infrared spectrometer (CIRS), the ion and neutral mass spectrometer (INMS), and the ultraviolet imaging spectrograph (UVIS), determine the chemical composition of different zones of Titan's atmosphere. They are able, in particular, to detect many organics, including new species, and allow the determination of their vertical concentration profile. The cameras of the imaging science subsystem (ISS), the *Cassini* radar, and the visual and IR mapping spectrometer (VIMS) map Titan's surface through the haze layers and determine the morphology, the heterogeneity, the presence of craters, which may result from volcanic activity and from impactors, and information on the chemical composition of the surface.

Table 9.4 Instruments on the *Huygens* entry probe, interdisciplinary programs (IDP), the leading scientists, and the potential for astrobiological return of their investigation.

Instrument or IDP	PI or IDS[a]	Country	Astrobiological return
Gas chromatograph-mass spectrometer (GC-MS)	H. Niemann	USA	+++
Aerosol collector and pyrolyzer	G. Israël	France	+++
Huygens atmospheric structure instrument	M. Fulchignoni	Italy	++
Descent imager/spectral radiometer (DISR)	M. Tomasko	USA	+++
Doppler wind experiment	M. Bird	Germany	+
Surface science package (SSP)	J. Zarnecki	UK	+++
Interdisciplinary program			
Aeronomy	D. Gautier	France	++
Atmosphere/surface interactions	J.I. Lunine	USA	++
Chemistry and exobiology	F. Raulin	France	+++

a PI : principal investigator, IDS : interdisciplinary scientist

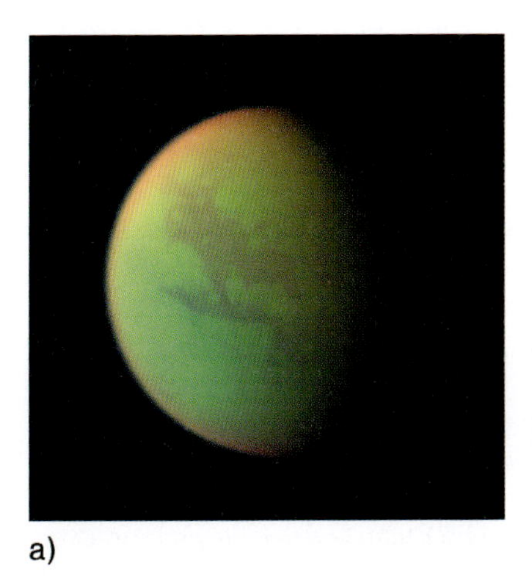

a)

b)

Fig. 9.9 Images of Titan taken from the *Cassini* spacecraft. (a) False-color composite image of Titan taken by the ISS wide-angle camera (combining near-IR and visible images) in April 2005 at about 170 000 km distance. Resolution in the images is approximately 10 km per pixel; (b) Ultraviolet image of Titan's night-side limb, taken by the narrow-angle camera. It shows many fine haze layers extending several hundred km above the surface (image credit: NASA/JPL/Space Science Institute).

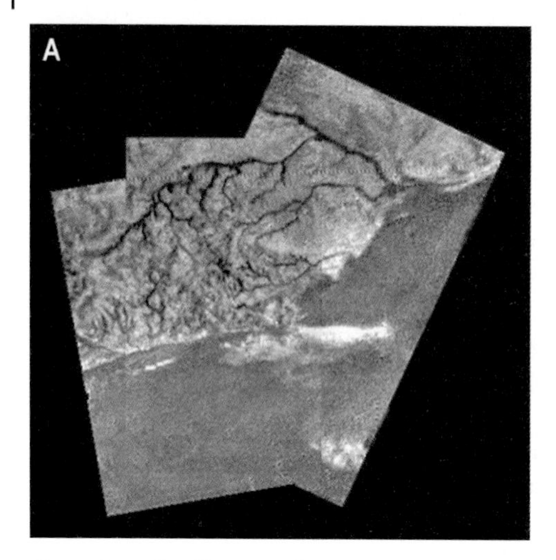

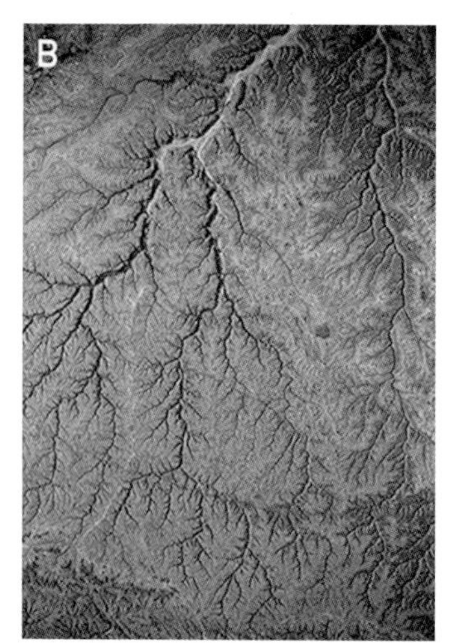

Fig. 9.10 Surface features of Titan and the Earth. (A) Channel networks, highlands, and dark-bright interface on the surface of Titan as seen by the DISR instrument on *Huygens* at 6.5-km altitude (image credit: ESA/NASA/JPL/University of Arizona); (B) A similar dentritic structure seen on the Earth in the Yemen desert (image credit: NASA).

The GC-MS instrument on the *Huygens* probe, a gas chromatograph (GC) with three GC capillary columns coupled to a quadrupole mass spectrometer (MS), performed a detailed chemical analysis of the atmosphere – including molecular and isotopic analysis – during the 2.5 hours of descent of the probe. After landing, it continued with measurements of the surface. The aerosol collector and pyrolyzer (ACP) experiment collected atmospheric aerosols and heated them up to high temperatures, which pyrolyzed the refractory part of the collected particles. Then the produced gases were transferred to the GC-MS instrument for molecular analysis. This was the first direct *in situ* molecular and elemental analysis of Titan's hazes. The *Huygens* atmospheric structure instrument (HASI) determined, in particular, the vertical profiles of pressure and temperature. The descent imager/spectral radiometer (DISR) measured the radiation budget of the atmosphere, investigated the cloud structure, and took images of the surface (Fig. 9.10A). The surface science package (SSP) provided information on the physical state and chemical composition of the surface.

A paramount scientific synergy is given by the complementarities of the *Cassini* observations, which on the one hand provide a global mapping of Titan by the *Cassini* spacecraft with observations of potential temporal and spatial variations and on the other hand provide very detailed information of one particular location of Titan from the data of the *Huygens* entry probe.

9.5
Astrobiological Aspects of Titan

9.5.1
Analogies with Planet Earth

Titan's dense atmosphere, which extents up to about 1 500 km, is composed mainly of N_2, like the atmosphere of the Earth. The other main constituents are CH_4, which amounts to about 1.6–2% in the stratosphere as measured by the CIRS on *Cassini* and the GC-MS on *Huygens*, and H_2, which amounts to about 0.1%. Despite the differences between Titan and the Earth, in particular in temperature ranges, there are several analogies between both bodies, as outlined below.

The first resemblances concern the vertical atmospheric structure (see Fig. 9.5). Although Titan is much colder, with a troposphere (~94 to ~70 K), a tropopause (70.4 K), and a stratosphere (~70–175 K), its atmosphere presents a similar complex structure and includes, like the Earth, as recently evidenced by *Cassini-Huygens*, a mesosphere and a thermosphere. With a much higher density in the case of Titan, the mesosphere extends to altitudes higher than 400 km (instead of only 100 km for the Earth), but the shape looks very much the same.

These analogies are linked to the presence of greenhouse gases and anti-greenhouse elements in both atmospheres. CH_4 has strong absorption bands in the medium- and far-IR regions, which corresponds to the maximum of the IR emission spectrum of Titan. CH_4 is transparent in the near-UV and visible (VIS) spectral regions. It thus can be a very efficient greenhouse gas in Titan's atmosphere.

H_2, which absorbs in the far IR (through bimolecular interaction of its dimers), plays a similar role. In the pressure–temperature conditions of Titan's atmosphere, CH_4 can condense but H_2 cannot. Thus, on Titan CH_4 and H_2 are equivalent to terrestrial condensable H_2O and non-condensable CO_2. In addition, the haze particles and clouds in Titan's atmosphere have an anti-greenhouse effect similar to that of the terrestrial atmospheric aerosols and clouds.

Methane on Titan seems to play the role of water on Earth, with a complex cycle that still is not well understood. Although the possibility that Titan is covered with hydrocarbons oceans has been ruled out, it is now likely that Titan's surface includes small lakes of methane and ethane, which seem to have been observed by the *Cassini* radar. The ISS camera on *Cassini* has also detected dark surface features near Titan's south pole that might be such liquid bodies (see Fig. 9.14). Moreover, the DISR instrument on *Huygens* has provided pictures of Titan's surface, which clearly show dentritic structures (Fig. 9.10A) that look like a terrestrial fluvial net (Fig. 9.10B). Because these features are free of crater impacts, it is suggested that they lie in a relatively young terrain and that in this area liquid was flowing recently on the surface of Titan. In addition, the *Huygens* GC-MS data show that the CH_4 mole fraction increases in the low troposphere (up to 5%) and reaches the saturation level at an altitude of approximately 8 km, allowing the possible formation of clouds and rain. Furthermore, GC-MS observed an increase of about 50 % in the CH_4 mole fraction at Titan's surface, suggesting the presence of condensed methane at the surface in the vicinity of the *Huygens* probe.

Other observations from the *Cassini* instruments clearly show the presence of various surface features of different origin that are indicative of volcanic, tectonic, sedimentological, and meteorological processes, similar to those that are found on Earth (Fig. 9.10). One can push the analogies between Titan and the Earth even further by comparing Titan's winter polar atmosphere and the terrestrial Antarctic ozone hole, although they imply different chemistries.

INMS on *Cassini* and GC-MS on *Huygens* have detected the presence of argon (Ar) in Titan's atmosphere. As in the Earth's atmosphere, the most abundant isotope is ^{40}Ar, which comes from the radioactive decay of ^{40}K. Its stratospheric mole fraction is about 3×10^{-5} as measured by GC-MS. The abundance of primordial argon (^{36}Ar) is about 200 times smaller. The other primordial noble gases have a mixing ratio of less than 10 ppb. This strongly suggests that Titan's atmosphere is a secondary atmosphere, produced by the degassing of trapped gases. Because N_2 cannot be efficiently trapped in the icy planetesimals that accreted and formed Titan – in contrast to NH_3 – this high abundance of N_2 indicates that Titan's primordial atmosphere was initially made of NH_3. Ammonia was then transformed into N_2 by photolysis and/or impact-driven chemical processes. The ^{14}N:^{15}N ratio measured in the atmosphere by INMS and GC-MS was 183 in the stratosphere, which is 1.5 times less than that of the primordial nitrogen. This low ratio indicates that the atmosphere was probably lost several times during the history of the satellite. Such evolution may also lead to methane transformation into organics (see Section 9.5.2), which may suggest large deposits of organics on Titan's surface.

Lastly, analogies can be drawn between the organic chemistry that is now very active on Titan and the prebiotic chemistry that proceeded in the environment of the primitive Earth (see Chapter 3). Despite the absence of permanent bodies of liquid water on Titan's surface, several of the organic processes that occur today on Titan imply the existence of some of the organic compounds considered to be key molecules in terrestrial prebiotic chemistry. Examples include hydrogen cyanide (HCN), cyanoacetylene (HC_3N), and cyanogen (C_2N_2). Indeed, a complex organic chemistry seems to be present in the three components of what one can call – by analogy with our planet – the "geofluids" of Titan: air (gas atmosphere), aerosols (solid atmosphere), and surface (oceans).

9.5.2
Organic Chemistry

Several organic compounds have already been detected in Titan's stratosphere (Table 9.5). The list includes hydrocarbons (with both saturated and unsaturated chains) and nitrogen-containing organic compounds, which are exclusively nitriles, as expected from laboratory simulation experiments. Most of these detections were performed by *Voyager* observations, with the exceptions of C_2 hydrocarbons, which were observed before; acetonitrile, which was detected by ground observation in the millimeter wavelength; and water and benzene, which were tentatively detected by the ISO spacecraft. Since *Cassini*'s arrival in the Saturn system, the presence of water and benzene has been unambiguously confirmed by the CIRS instrument. In addition, the direct analysis of the ionosphere by the INMS instrument during *Cassini*'s flybys of Titan shows the presence of many organic species at detectable levels (Fig. 9.11), despite the spacecraft's high altitude of 1 100–1 300 km.

Table 9.5 Main composition and detected trace components of Titan's stratosphere and comparison with products of laboratory simulation experiments.

Compound	Stratospheric mixing ratio	Location	Simulation experiments[a]
Main constituents			
Nitrogen N_2	0.98		
Methane CH_4	0.02		
Hydrogen H_2	0.0006–0.0014		
Hydrocarbons			
Ethane C_2H_6	1.3×10^{-5}	Equator	Major product
Acetylene C_2H_2	2.2×10^{-6}	Equator	Major product
Propane C_3H_8	7.0×10^{-7}	Equator	++
Ethylene C_2H_4	9.0×10^{-8}	Equator	++
Propyne C_3H_4	1.7×10^{-8}	North pole	+
Diacetylene C_4H_2	2.2×10^{-8}	North pole	+
Benzene C_6H_6	Few 10^{-9}		+
N-organics			
Hydrogen cyanide HCN	6.0×10^{-7}	North pole	Major product
Cyanoacetylene HC_3N	7.0×10^{-8}	North pole	++
Cyanogen C_2N_2	4.5×10^{-9}	North pole	+
Acetonitrile CH_3CN	Few 10^{-9}		++
Dicyanoacetylene C_4N_2	Solid phase	North pole	+
O-compounds			
Carbon monoxide CO	2.0×10^{-5}		
Carbon dioxide CO_2	1.4×10^{-8}	Equator	
Water H_2O	Few 10^{-9}		

a ++: abundance smaller by one order of magnitude than that of the major product; +: abundance smaller by two orders of magnitude than that of the major product.

Surprisingly, the GC-MS instrument onboard the *Huygens* probe has not detected a large variety of organic compounds in the low atmosphere. The mass spectra collected during the descent show that the medium and low stratosphere and the troposphere are poor in volatile organic species, with the exception of CH_4. This may be explained by the condensation of those high-molecular species on aerosol

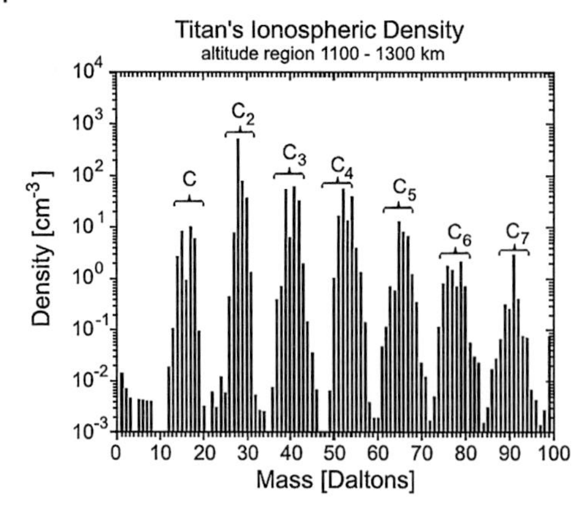

Fig. 9.11 Mass spectrum of Titan's ionosphere near 1200 km altitude. The spectrum shows signatures of organic compounds including those with up to 7 carbon atoms (image credit: NASA/JPL/University of Michigan).

particles. These particles, for which no direct data on the chemical composition were available before, have been analyzed by the ACP instrument. ACP was designed to collect aerosols during the descent of the *Huygens* probe on a filter in two different regions of the atmosphere. Then the filter was heated in a closed oven at different temperatures and the produced gases were analyzed by GC-MS. The results showed that the aerosol particles are made of refractory organics that release HCN and NH_3 during pyrolysis. This result strongly supports the tholins hypothesis (see Section 9.4.2). Based on these new and, for the first time, *in situ* data, it seems very likely that the aerosol particles are made of a refractory organic nucleus, covered with condensed volatile compounds (Fig. 9.12). The nature of the pyrolysates provides information on the molecular structure of the refractory complex organics. It indicates the potential presence of nitrile groups (-CN), amino groups ($-NH_2$, -NH- and -N<), and /or imino groups (-C=N-).

Furthermore, comparison of the data obtained from the first sampling, which consisted mainly of stratospheric particles, and the second one, which was taken in the mid-troposphere, indicates that the aerosol composition is homogeneous. This fits with some of the data obtained by DISR on aerosol particles, which indicate a relatively constant size distribution of the particles with altitude. Their mean diameter is on the order of 1 μm.

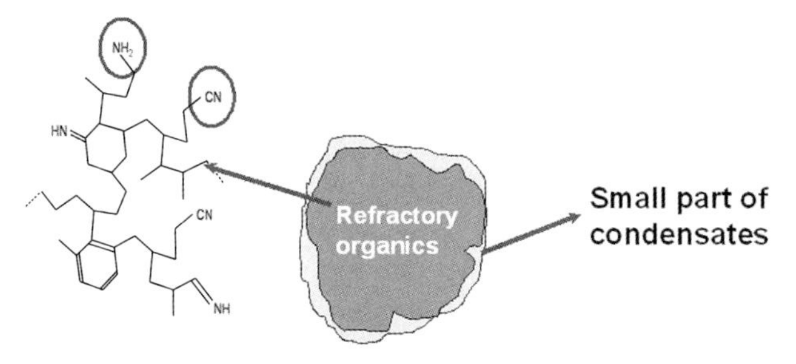

Fig. 9.12 Model of the chemical composition of Titan's aerosols derived from the *Huygens* ACP data.

Fig. 9.13 Surface of Titan as seen by the DISR camera on *Huygens* after it landed (image credit: ESA/NASA/JPL/University of Arizona).

These particles sediment to the surface, where they form a deposit of complex refractory organics and frozen volatiles. DISR collected the infrared reflectance spectra of the surface with the help of a lamp that illuminated the surface before the *Huygens* probe touched down. The retrieving of these infrared data showed the presence of water ice but no clear evidence – so far – of tholins. The presence of water ice is supported by the data from SSP. Its accelerometer measurements can be interpreted by the presence of small water ice pebbles on the surface where *Huygens* has landed, in agreement with the DISR surface pictures (Fig. 9.13). On the other hand, GC-MS was able to analyze the atmosphere near the surface for more than one hour after touchdown. The corresponding mass spectra showed the clear signature of many organics, including cyanogen, C3, and C4 hydrocarbons and benzene, indicating that the surface is much richer in volatile organics than the low stratosphere and the troposphere. These observations are in agreement with the hypothesis that in the low atmosphere of Titan most of the organic compounds are in the condensed phase.

Altogether, these new data of the *Cassini–Huygens* mission show the diversity of the locations where organic chemistry is taking place on Titan. Surprisingly, the high atmosphere looks very active, with neutral and ion organic processes; the high stratosphere, where many organic compounds had already been detected before *Cassini* arrived in the Saturn system, shows an active organic chemistry in the gas phase. In the lower atmosphere, this chemistry seems to be concentrated mainly in the condensed phase. Titan's surface is probably covered with frozen volatile organics together with refractory, tholin-like organic materials.

Cosmic rays reaching Titan's surface and hence these compounds may induce additional organic syntheses, especially if part of these materials is dissolved in some small liquid bodies made of low-molecular-weight hydrocarbons (mainly methane and ethane).

Despite the low surface temperatures, the presence of liquid water is not excluded. Cometary impacts on Titan may melt surface water ice, offering possible episodes of liquid water as long as about 1 000 years, as suggested by Artemieva and Lunine (2003). This scenario provides conditions for short periods of terrestrial-like prebiotic syntheses at relatively low temperatures. Low temperatures reduce the rate constants of prebiotic chemical reactions but may increase the concentration of reacting organics by an eutectic effect, which increases the rate of the reaction. In addition, the possible presence of a water–ammonia ocean in the depths of Titan, as expected from models of its internal structure, may provide an efficient way to convert simple organics into complex molecules and to reprocess chondritic organic matter into prebiotic compounds. These processes may have very efficiently occurred at the beginning of Titan's history – with even the possibility of the water–ammonia ocean exposed to the surface – allowing a CHON (carbon, hydrogen, oxygen, and nitrogen) prebiotic chemistry evolving to compounds that may be of astrobiological interest.

9.5.3 Life on Titan?

In the following sections, conditions are considered that may drive the chemistry in Titan's environment from organic chemistry to prebiotic chemistry or maybe even to biotic processes. The overarching question is whether Titan's conditions are compatible with those needed for sustaining life (see also Chapter 6). The surface of Titan is too cold and not energetic enough to provide the right conditions for habitability. However, the (hypothetical) subsurface oceans on Titan may be suitable for life, as suggested by Fortes (2000). The temperature of these hypothetical subsurface oceans is estimated as high as about 260 K. In addition, cryovolcanic hotspots are assumed with possible temperatures in the range of 300 K. These temperature conditions might allow the development of living systems. Even at a depth of 200 km, the expected pressure of about 500 MPa (5 kbar) is not incompatible with life, as shown by terrestrial examples. The expected pH of an aqueous medium made of 15% by weight of NH_3 is equivalent to a pH of 11.5. Some bacteria can grow on Earth at pH 12 (see Chapter 5). Even the limited energy resources do not exclude the sustaining of life.

Taking into account the potential radiogenic heat flow (about 5×10^{11} W) and assuming that 1% of this is used for volcanic activity and 10% of the latter is available for living system metabolism, Fortes (2000) has estimated an energy flux of about 5×10^8 W that may be available in Titan's subsurface oceans. Considering the terrestrial biosphere, such a flux corresponds to the production of about 4×10^{11} mol of ATP per year and about 2×10^{13} g of biomass per year. Assuming an average turnover for the living systems on the order of one year, the biomass density would be 1 g m^{-2}. This number is very small compared to the lower limit of the value of the biomass for the Earth, which amounts to about 1 000–10 000 g m^{-2}.

Nevertheless, these energetic considerations would allow a possible presence of limited, but not negligible, bioactivity on the satellite. Assuming that extraterrestrial living systems are similar to the ones we know on Earth, and based on the chemistry of carbon and the use of liquid water as solvent (see Chapter 6), the hypothetical biota on Titan, if any, might be localized in the subsurface deep ocean. Several possible metabolic processes have been postulated for the hypothetical subsurface oceanic life on Titan, such as nitrate/nitrite reduction or nitrate/dinitrogen reduction, sulfate reduction or methanogenesis (Simakov 2001), and catalytic hydrogenation of acetylene (Abbas and Schulze-Makuch 2002).

As expected, no sign of macroscopic life has been detected by the *Huygens* probe when approaching the surface or after landing. This can be concluded in particular from the many pictures taken by the DISR of the same location on Titan during more than one hour after landing. But this does not exclude the possibility of the presence of microscopic life. The metabolic activity of a corresponding biota, even if it is localized far from the surface in the deep internal structure of Titan, may produce chemical species that diffuse through the ice mantle, covering the hypothetical internal ocean, and feed the atmosphere. It has even been speculated in several publications that the methane seen in the atmosphere today is the product of biological activity. If this were the case, the atmospheric methane would be notably enriched in light carbon. On Earth, biological processes induce an isotopic fragmentation that produces enrichment in ^{12}C. The ratio ^{12}C:^{13}C increases from 89, which is the reference value taken from the Belemnite of the Pee Dee Formation, to about 91–94 depending on the biosynthesis processes. The ^{12}C:^{13}C ratio of atmospheric methane on Titan is 82, as determined by the GC-MS instrument on *Huygens*. Although we do not have a reference value for ^{12}C:^{13}C on Titan, this low value suggests that the origin of methane is likely to be abiotic.

9.6 Outlook: Astrobiology and Future Exploration of Titan

Although exotic life such as methanogenic life in liquid methane cannot be fully ruled out, the presence of extant or extinct life on Titan seems very unlikely. Nevertheless, with the new observational data provided by the *Cassini–Huygens* mission, the largest satellite of Saturn appears more than ever to be an interesting object for astrobiology. The several similarities of this exotic and cold planetary body with the Earth (see Section 9.5.1) and the complex organic chemical processes that are going on now on Titan (see Section 9.5.2) provide a fantastic means for a better understanding of the prebiotic processes, which are no longer reachable on the Earth at the scale and within the whole complexity of a planetary environment.

The origin and cycle of methane on Titan illustrate the whole complexity of the Titan system. Methane may be stored in large amounts in the interior of the satellite as clathrates (methane hydrates) trapped during the formation of the satellite from the Saturnian subnebula, where it was formed by Fischer-Tropsch processes (abiotic synthesis of hydrocarbons from CO and H_2 using metallic catalysts). It may also be produced through high-pressure processes, such as serpentinization, allowing the formation of H_2 by reaction of H_2O with ultramafic rocks, or by cometary impact. Interestingly, these processes have rarely been considered in the case of the primitive Earth, although they may have contributed to a possible reducing character of the primordial atmosphere of the Earth. That possibility was recently reexamined by Tian and coworkers (2005). This is an example of how the study of Titan is indeed bringing some new ideas to the understanding of chemical evolution processes on the early Earth.

In Titan's atmosphere, methane is photolyzed by solar UV radiation, whereby mainly ethane and tholin-like organic matter are produced. The resulting lifetime of methane in Titan's atmosphere is relatively short (about 10–30 million years). Thus, methane stored in Titan's interior may be continuously replenishing the atmosphere by degassing induced by cryovolcanism, which has been clearly evidenced by the first images of Titan's surface provided by the VIMS, ISS, and radar instruments on the *Cassini* spacecraft. It may episodically be released to the atmosphere, as recently suggested by Tobie and coworkers (2005). In any case, the methane cycle should indeed produce large amounts of complex organics accumulating on the surface as well as large amounts of ethane, which, when mixed with the dissolved atmospheric methane, should form liquid bodies on the surface or in the near subsurface of the satellite. Maybe the dark feature seen near Titan's south pole (Fig. 9.14) represents one of these expected liquid bodies.

From the current data, Titan looks like another world, with an active prebiotic-like chemistry but a lack of permanent liquid water on its surface. For this reason, Titan represents a natural planetary laboratory for prebiotic-like chemistry.

The *Cassini–Huygens* mission is still proceeding. It will continue its systematic exploration of the Saturnian system up to 2008, and probably to 2011 if the extended mission is accepted. Numerous data of paramount importance for astrobiology are still expected from several of its instruments (Table 9.3). The CIRS instrument should be able to detect new organic species in the atmosphere during future limb observations of Titan, especially at the pole. ISS and VIMS should provide more-detailed pictures of Titan's surface revealing the complexity and even

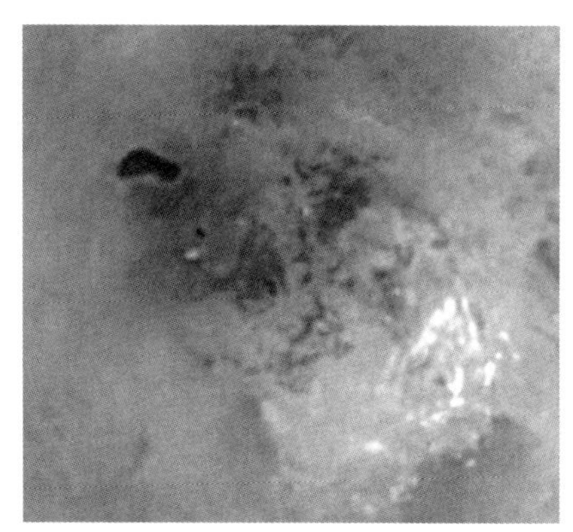

Fig. 9.14 Picture of Titan taken near the south pole (cross in the middle) by the ISS narrow-angle camera on *Cassini*. It shows, in addition to bright clouds, the presence of a dark feature on the upper left, which may be interpreted as a liquid hydrocarbon lake (image credit: NASA/JPL/Space Science Institute).

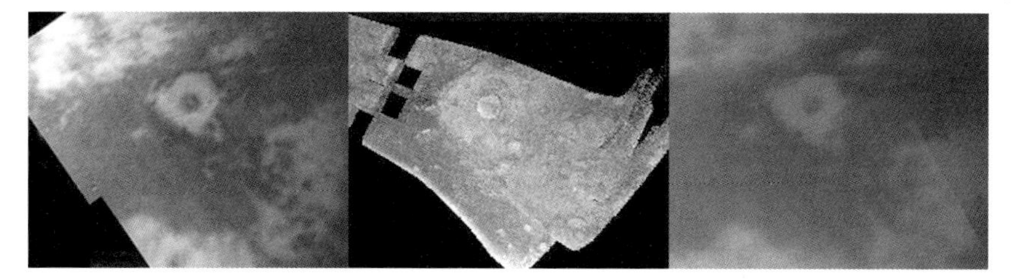

Fig. 9.15 One of the largest impact craters (about 80 km in diameter) observed on Titan's surface by two *Cassini* instruments: VIMS infrared (left) and VIMS false-color image (right) and radar image (center). The faint halo, slightly bluer and darker than the surrounding parts, is probably somewhat different in composition. Because it is made of material excavated when the crater was formed, this indicates that the composition of Titan's upper crust varies with depth (image credit: NASA/JPL/University of Arizona).

the physical and chemical nature of this surface and its diversity. Radar observation will continue the systematic coverage of Titan's surface, showing contrasted regions with smooth and rough parts that suggest shorelines. Observations of the same regions with these instruments coupled with those already performed with other instruments should bring a fruitful synergy (Fig. 9.15). In the longer term, future missions to Titan, including horizontal mobility to explore the surface in great detail and to analyze it at the molecular level and at many different locations, are already being considered. Such projects, if implemented, might bring essential information for our understanding of this astrobiological, exotic, and astonishing world.

9.7 Further Reading

9.7.1 Books and Articles in Books

Gargaud, M., Barbier, B., Martin, H., Reisse, J. (Eds.) *Lectures in Astrobiology*, Vol. 1, Springer, Heidelberg, 2005.

Harland, D. M. *Mission to Saturn, Cassini and the Huygens Probe*, Springer-Praxis Pub., Chichester, UK, 2002.

Lacoste H (Ed.), *Proceedings of the Second European Workshop in Exo-Astrobiology, Graz, Austria, 16–19 September 2002*, ESA SP 518, ESA-ESTEC, Noordwijk, The Netherlands, 2002, with the following articles:

- Abbas, O., Schulze-Makuch, D. Acetylene-based pathways for prebiotic evolution on Titan, *ESA SP-518*, 2002, pp. 345–348.
- Bernard, J.-M., Coll, P., Raulin, F. Variation of C/N and C/H ratios of Titan's aerosols analogues. *ESA SP-518*, 2002, pp. 623–625.

Lorenz, R., Mitton, J. *Lifting Titan's Veil- Exploring the Giant Moon of Saturn*, Cambridge University Press, Cambridge, UK, 2002.

Russell, C. T. *Cassini-Huygens Mission: Overview, Objectives and Huygens Instrumentarium*, Kluwer Academic Publishers, Dordrecht, The Netherlands, 2003.

Russell, C. T. *The Cassini-Huygens Mission: Orbiter Remote Sensing Investigations*, Kluwer Academic Publishers, Dordrecht, The Netherlands, 2005.

Schulze-Makuch, D., Irwin, L. N. *Life in the Universe. Expectations and constraints*, Springer, Berlin, Heidelberg, New York, 2004.

Simakov, M. B. The possible sites for exobiological activities on Titan, in: *Proceedings of the First European Workshop on Exo-/Astrobiology, Fracati, Italy, 21–23 May 2001*, Ehrenfreund, P., Angerer, O., Battrick, B. (Eds.) *ESA SP-496*, ESA-ESTEC, Noordwijk, 2001, pp. 211–214.

9.7.2
Articles in Journals

Artemieva, N., Lunine, J. Cratering on Titan: impact melt, ejecta, and the fate of surface organics, *Icarus* **2003**, *164*, 471–480.

Bernard, J.-M., Coll, P., Coustenis, A., Raulin, F. Experimental simulation of Titan's atmosphere detection of ammonia and ethylene oxide, *Planet. Space Sci.* **2003**, *51*, 1003–1011.

Coll, P., Coscia, D., Smith, N., Gazeau, M. C., Ramirez, S. I., Cernogora, G., Israel, G., Raulin, F. Experimental laboratory simulation of Titan's atmosphere: aerosols and gas phase, *Planet. Space Sci.* **1999**, *47* (10–11), 1331–1340.

Fortes, A. D. Exobiological implications of a possible ammonia-water ocean inside Titan. *Icarus* **2000**, *146*, 444–452.

Hébrard, E., Benilan, Y., Raulin, F. Sensitivity effects of photochemical parameters uncertainties on hydrocarbon production in the atmosphere of Titan, *Adv. Space Res.* **2005**, *36* (2), 268–273.

Imanaka, H., Khare, B. N., Elsila, J. E., Bakes, E. L. O., McKay, C. P., Cruikshank, D. P., Sugita, S., Matsui, T., Zare, R. N. Laboratory experiments of Titan tholin formed in cold plasma at various pressures: implications for nitrogen-containing polycyclic aromatic compounds in Titan haze, *Icarus* **2004**, *168*, 344–366.

Mousis, O., Gautier, D., Bockelée-Morvan, D. An evolutionary turbulent model of Saturn's subnebula: Implications for the origin of the atmosphere of Titan, *Icarus* **2002**, *156*, 162–175.

'The *Huygens* probe on Titan', News & Views, Articles and Letters, *Nature* **2005**, *438*, 756–802.

Prinn, R. G., Fegley, B. Kinetic inhibition of CO and N, reduction in circumplanetary nebulae: implications for satellite composition, *Astrophys, J.* **1981**, *249, 308–317.*

Raulin, F. Exo-astrobiological aspects of Europa and Titan: from observations to speculations, *Space Sci. Rev.* **2005**, *116* (1–2), 471–496.

Sekine, Y., Sugita, S., Shido, T., Yamamoto, T., Iwasawa, Y., Kadono, T., Matsui, T. The role of Fischer–Tropsch catalysis in the origin of methane-rich Titan, *Icarus* **2005**, *178*, 154–164.

Tian, F., Toon, O. B., Pavlov, A.A, De Sterck, H. A Hydrogen-rich early Earth atmosphere, *Science* **2005**, *308*, 1014–1017.

Tobie, G., Grasset, O., Lunine, J. I., Mocquet, A., Sotin, C. Titan's internal structure inferred from a coupled thermal-orbital model, *Icarus* **2005**, *175*, 496–502.

Wilson, E. H., Atreya, S. K. Current state of modeling the photochemistry of Titan's mutually dependent atmosphere and ionosphere, *J. Geophys. Res. – Planets* **2004**, *109 (E6)*, E06002.

9.7.3
Web Sites

NASA web site on *Cassini* mission: http://www.jpl.nasa.gov/cassini/

ESA web site on *Huygens* probe: http://sci.esa.int/huygens/).

9.8 Questions for Students

Question 9.1

Given the surface temperature (94 K) and pressure (1 500 hPa = 1.5 bar) of Titan, determine the relative density of its atmosphere near the surface compared to that of the Earth.

Question 9.2

(a) From the period of Saturn around the sun (29.4 years), deduce the mean distance of Titan from the Sun. (b) From the mean distance of Titan from the Sun determined in (a), determine the mean solar flux received at Titan compared to the Earth.

Question 9.3 (see Figs 9.16 and 9.17)

In Titan's atmosphere the main product of methane photochemistry is ethane, via the following reaction:

$$CH_3 + CH_3 + M \rightarrow C_2H_6 + M,$$

where M is a third body.

(a) This reaction does not occur in the high atmosphere of Titan. Why not? (b) In the low atmosphere, CH_3 radicals cannot be directly produced by methane photolysis, but C_2H and C_4H radicals are directly produced by acetylene and diacetylene photolysis. Why? (c) Give the mechanism of formation of CH_3 radicals in Titan's stratosphere photocatalyzed by acetylene. (d) Give the mechanism of formation of CH_3 radicals in Titan's stratosphere photocatalyzed by diacetylene.

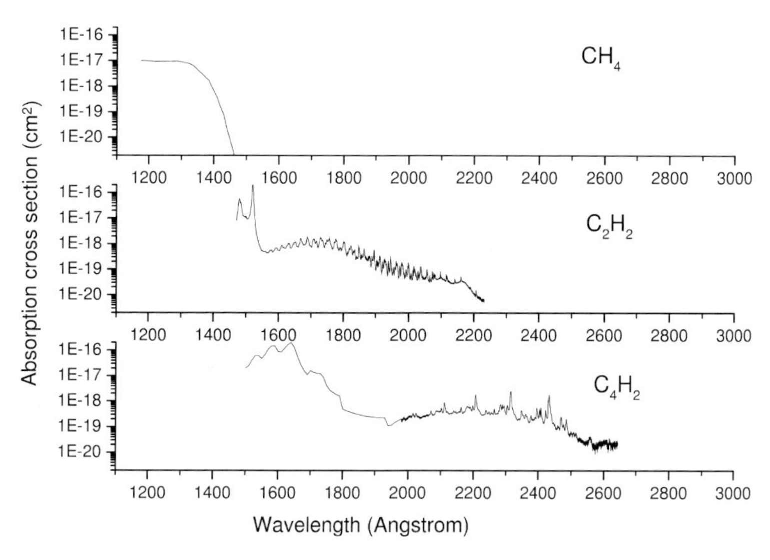

Fig. 9.16 UV spectra of methane, acetylene, and diacetylene.

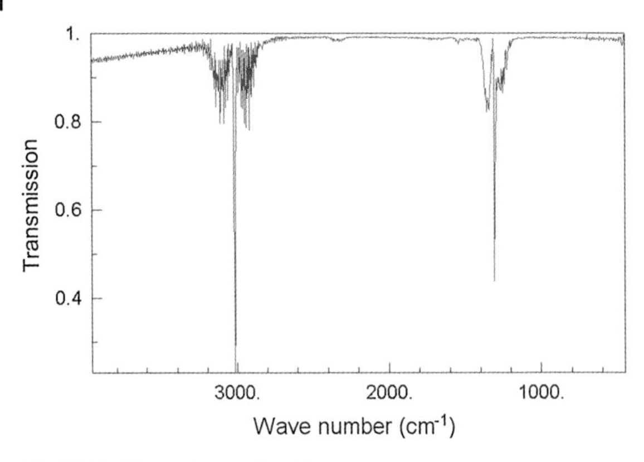

Fig. 9.17 IR spectrum of methane.

Question 9.4

Knowing

- the scale height of Titan's troposphere: Ha = 22 km,
- the surface pressure: Ps = 1.5×10^5 Pa,
- the mean mole fraction of methane in the troposphere: fm = 0.03,
- the global flux of destruction of methane in the atmosphere: $\Phi = 10^{10}$ molecules/cm^{-2} s^{-1}, and
- the value of the Boltzman constant: $k_b = 1.4 \times 10^{-23}$ J/K,

determine how long it will take for the whole atmospheric methane to disappear, in the absence of any source on Titan. What suggests this value?

Question 9.5

Methane does not absorb in the visible. Its UV and IR spectra are given in Figs 9.16 and 9.17. Explain why methane produces a greenhouse effect in Titan's atmosphere.

10
Jupiter's Moon Europa: Geology and Habitability

Christophe Sotin and Daniel Prieur

After reviewing the current knowledge of the geological processes and the surface composition of Jupiter's moon Europa, this chapter concentrates on the geophysical data that suggest below the ice crust the presence of a deep ocean that would be in contact with a silicate core. Models of Europa's thermal evolution are then described. Based on the ocean hypothesis of Europa, the chapter then considers the astrobiological aspects of Europa and the perspectives of potential life at the interface between the ocean and the silicate core. Reference is given to the adaptability of terrestrial life to extreme environments. As outlook, some experiments are described that could be carried out onboard future missions to the Jupiter system and its moon Europa in order to resolve some of the important issues related to putative habitability in the outer Solar System.

10.1
A Short Survey of the Past Exploration of Europa

The four largest moons of Jupiter – Io, Europa, Ganymede, and Callisto – were discovered in 1610 by Galileo Galilei, who used a self-made telescope. He recognized the importance of the Jupiter moons in supporting Kepler's conception of the Sun being the central body in our Solar System; therefore, in his honor, these four moons are now called the Galilean satellites. At nearly the same time (actually a few weeks earlier), the four moons were discovered by Simon Marius, a German astronomer, who gave them the names of the favorites of the ancient chief god Jupiter, except for Ganymede, who was the cupbearer of the gods.

Europa is the second Galilean satellite with regard to its distance to Jupiter. It orbits around Jupiter at a period of 3.55 days (Table 10.1). Its orbit is controlled by the so-called (1,2,4) Laplace resonance between Io, Europa, and Ganymede that maintains an eccentricity on the order of 1%.

Complete Course in Astrobiology. Edited by Gerda Horneck and Petra Rettberg

ISBN: 978-3-527-40660-9

Table 10.1 Characteristics of Jupiter's moon Europa.

Parameter	Quantity and dimension
Radius	1 560 ± 10 km
Density	3 018 ± 35 kg m^{-3}
GM[a]	3 202.86 ± 0.07 km^3/s^2
Gravity acceleration at the surface	1.31 m/s^2
J_2[b]	389 ± 39 10^{-6}
C/MR^2[c]	0.347
Period of rotation	3.55 days
Distance to Jupiter	0.6709 10^9 m
Mass of Jupiter	1.905 10^{27} kg
Eccentricity	0.96%

a G = gravitational constant, M = Europa's mass.
b J_2 = constant (see Eq. (10.1)).
c C = moment of inertia, R = Europa's radius (see Radau formula, Eq. (10.6)).

The first images of the moon Europa came from the *Voyager* missions, which reached the Jupiter system on 5 March 1979 at a distance of 349 000 km (*Voyager 1*) and on 9 July 1979 at a distance of 722 000 km (*Voyager 2*), before advancing to Saturn, Uranus, and Neptune, and further towards the outer edge of the Solar System. The *Voyager* missions were designed to take advantage of a rare geometric arrangement of the outer planets in the late 1970 s and the 1980 s that allowed for a four-planet tour at a minimum of propellant and trip time. The images of Europa obtained by *Voyager 2* showed a smoothness of most of the terrain and the near absence of impact craters. Only three craters larger than 5 km in diameter have been found.

But most of the information about Europa comes from the *Galileo* mission, which orbited around Jupiter from 8 December 1995 until 28 September 2003. The high-resolution images revealed intriguing tectonic features, such as domes, disrupted blocks, chaotic terrains, and fractures (Figure 10.1). The small number of impact craters suggests that Europa's surface has undergone some kind of resurfacing by tectonic and/or cryovolcanic processes.

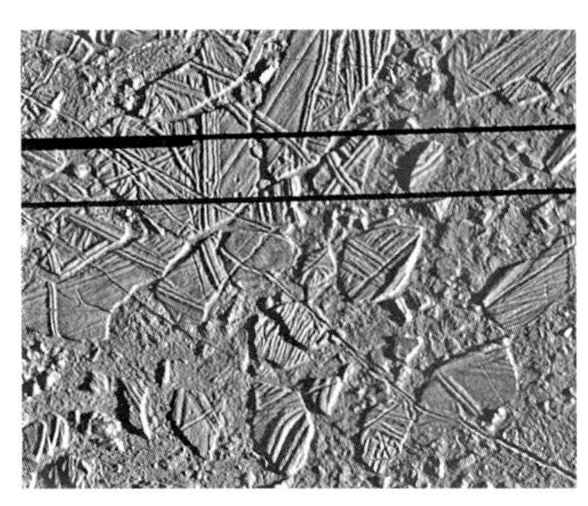

Figure 10.1 High-resolution image of the ice-rich crust of Jupiter's moon Europa, taken by the solid-state imaging system on board the *Galileo* spacecraft on 20 February 1997, from a distance of 5 340 km during the spacecraft's close flyby of Europa. The area shown is about 34 km by 42 km, centered at 9.4 degrees north latitude, 274 degrees west longitude, and the resolution is 54 m (Photo credit: NASA/JPL).

From an astrobiology point of view, the most interesting feature of Europa is the existence of a putative ocean of water. Life on Earth cannot exist without liquid water, which represents the major content of all known living organisms. Although certain organisms can form highly dehydrated resting stages–for example, spores, cysts, or seeds–that allow them to survive unfavorable environmental conditions (see Chapter 5), these organisms cannot carry out their whole life cycle in the absence of liquid water, which is needed at least for some time to allow metabolism and growth. Consequently, places in the Solar System where liquid water exists are considered potential targets for astrobiology investigations (see Chapter 6).

Among the four Galilean satellites, Europa is not the only body possibly harboring a significant volume of liquid water. But according to the models, Europa represents a single case in which liquid water is in contact with a silicate core (see also Chapter 9, Figure 9.1), whereas for Ganymede and Callisto the water layer is supposed to be encased between layers of ice. These unique conditions on Europa would allow chemical exchanges between the ocean and the rocks of the core, thereby providing to this environment a variety of chemicals that might play a role in sustaining putative life forms on the ocean floor. A similar situation exists on Earth in deep-sea hydrothermal vents, which are known for being inhabited by luxuriant communities of different types of organisms (see Chapter 5).

10.2 Geology of the Moon Europa

10.2.1 Surface Features

The images acquired by the *Galileo* mission show very intriguing tectonic features, including domes, disrupted blocks, and fractures. Crustal plates ranging up to 13 km across have been broken apart and "rafted" into new positions (Figure 10.1). They superficially resemble the disruption of pack ice on polar seas during spring thaws on Earth. The size and geometry of those features suggest that motion was enabled by ice-crusted water or soft ice close to the surface at the time of disruption.

These disrupted blocks, also called "rafts," have been interpreted by two different models. One model explains the formation and displacement of blocks as the result of a thin, icy crust (less than 2 km) above an ocean (Figure 10.2). The thin crust is periodically broken by faults that form in response to the tidal forces resulting from Europa's eccentric orbit around Jupiter. The blocks resulting from the faulting of the ice crust would then be entrained by the streams of the liquid ocean and move relative to each other. Because the surface temperature of about 100 K is much below the freezing point, the displacement of the blocks would be limited by the rapid freezing of any liquid that would be present between the blocks. Another reason for the crust to be disrupted is the possibility of localized sources of heat in Europa's silicate layer. It has been proposed that tidal forces would lead to partial melting in Europa's silicate layer and volcanism at the silicate–water interface. This scenario is inspired by the one that has been proposed in order to explain Io's very active volcanism. In this case, the water would be very hot during a volcanic eruption and would rise to the surface, melting the ice crust and inducing strong displacements. It must be noted that volcanism is not necessarily required for this process. Convection in Europa's silicate layer would create zones of higher temperature, and percolation of water within the silicates would create conditions favorable for hydrothermalism and subsequent displacement of hot water to the surface. Such a scenario is appealing to astrobiology, because the conditions at the silicate–ocean interface are very similar to those existing at several places on Earth's seafloors.

An alternative model assumes that the blocks and domes are related to subsolidus (solid-state) convection in the lower part of the crust (Figure 10.3). In this model, the domes are explained by the topography resulting from the rising of hot subsolidus icy plumes forming at the interface between the liquid layer and the icy core. It strongly suggests subsolidus convection in the icy crust. The topography of the domes is not well constrained and cannot be used to constrain the characteristics of the upwelling plumes. If this is the right explanation for the domes, it implies that the thickness of the ice crust is at least 20 km.

Europa's surface is covered by very long fracture zones, which were observed by the *Voyager* mission. Tidal forces driven by Jupiter have been supposed to be responsible for the formation of those cracks. Some cracks form chains of arcs that are connected at the cusps. Those cycloidal cracks might be formed in response to diurnal variations of the tidal forces.

The low number of impact craters suggests that Europa's surface is young compared to that of the Galilean satellites Ganymede and Callisto. However, it is difficult to obtain absolute ages of surfaces from the number of impact craters,

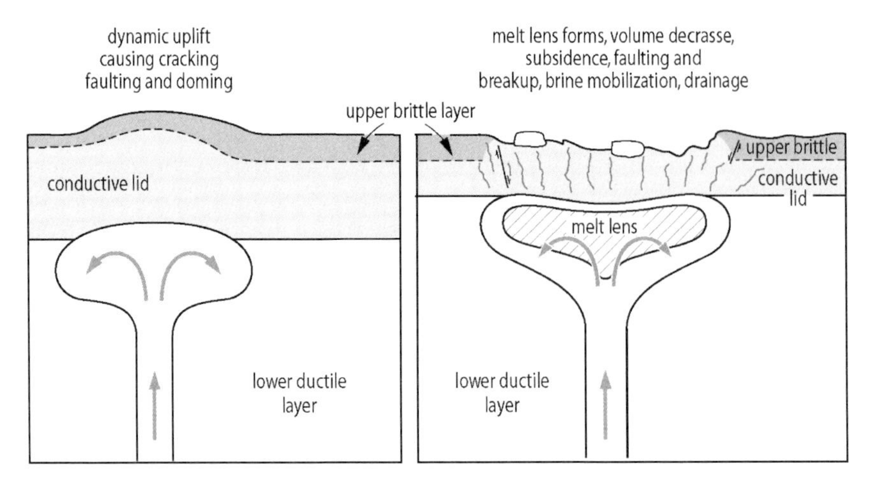

Figure 10.2 "Raft" model to explain the surface structures of Jupiter's moon Europa with hot water rising from the ocean-silicate interface (from Thomson and Delaney 2001).

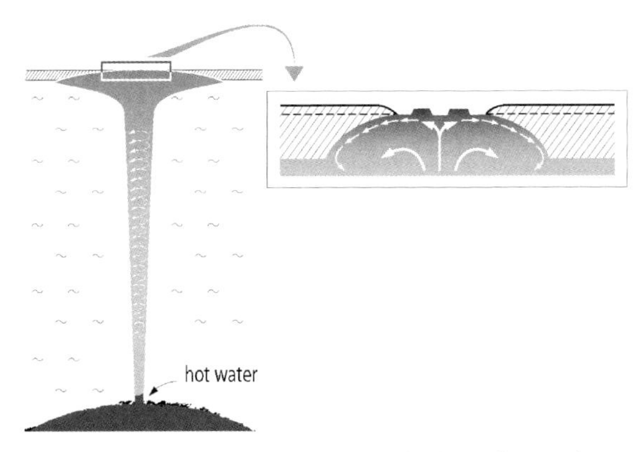

Figure 10.3 Subsolidus convection model for explaining the chaotic terrains and the displacement of blocks on Europa's surface (modified from Sotin et al. 2002).

because the cratering rate in the outer Solar System is not well known. Models suggest an age of between 10 million years and one billion years for the surface of Europa. Another possibility for assessing the age is the erosion rates by sputtering. However, these estimates are constrained by the measurements of the H_2O escape rate during the *Galileo* mission. If the sputtering has been operating for more than one billion years, it should have erased most of the topography; however, this is not the case for Europa. The remaining topography of Europa is consistent with an age of its surface in the range of some million years.

There are also a few multi-ringed structures on Europa, such as Tyre and Callanish, which are interpreted as impact features. Study of the topographic relaxation of those fractures suggests a low viscosity layer at about 5–10 km in depth. This layer may be either a liquid layer or a soft ice layer. This interpretation of the formation and relaxation of impact craters may be the strongest argument in favor of a deep liquid ocean below an icy crust.

The spacecraft *Galileo* finally plunged into Jupiter's crushing atmosphere on 21 September 2003. The spacecraft was deliberately destroyed for planetary protection reasons (see Chapter 13). The aim was to avoid a potential crash of the spacecraft with the moon Europa, in order to protect the possible ocean beneath the icy crust from terrestrial contamination.

10.2.2
Composition of the Surface

Data from the near-infrared mapping spectrometer (NIMS) onboard the *Galileo* spacecraft show that Europa's surface is made mostly of water ice. In order to explain the spectra obtained along the cracks, another component must exist in addition to water. It has been proposed that this additional component be hydrated magnesium sulfate. Such material might have been released by some kind of cryovolcanic activity related to the activity of the cracks.

10.3
Internal Structure of the Moon Europa

The density of Europa, at 3 018 kg m^{-3} (Table 10.1), lies between that of Io (3 530 kg m^{-3}) and that of Ganymede (1 936 kg m^{-3}) and Callisto (1 851 kg m^{-3}). Because its surface is made of water ice, Europa is likely composed of a mixture of water ice and silicates. Taking Io's density for the density of silicates and a value of 960 kg m^{-3} for H_2O (the latter value is assumed to lie between that of liquid water [1 000 kg m^{-3}] and that of iceI [920 kg m^{-3}]), the mass fraction of ice on Europa would be only 6%. Another way to interpret the density data is that Europa would be made of a rocky core of 1 450 km in radius overlain by a 110-km thick layer of H_2O (Figure 10.4).

Additional information about the internal structure of Europa comes from magnetic and geodetic measurements during the *Galileo* mission. Europa has a strong inductive magnetic field consistent with an electrically conducting interior with a radius equal to nearly the entire radius of the moon. However, it does not show any evidence of a permanent dipole. The induction field can be explained by a water ocean of similar conductivity to the Earth's oceans, provided this ocean has a thickness exceeding 10 km (Figure 10.4). One must note that another explanation is the existence of an iron core. At the present time, it is not possible to choose between these two options, because the depth of the induced magnetic field cannot be determined.

The gravity potential was measured during the 11 Europa flybys of the *Galileo* mission. The gravity potential can be expressed in spherical harmonics decomposition. For flybys such as those realized during the *Galileo* mission, only degrees 0 (GM/r) and 2 are kept. The relation is as follows:

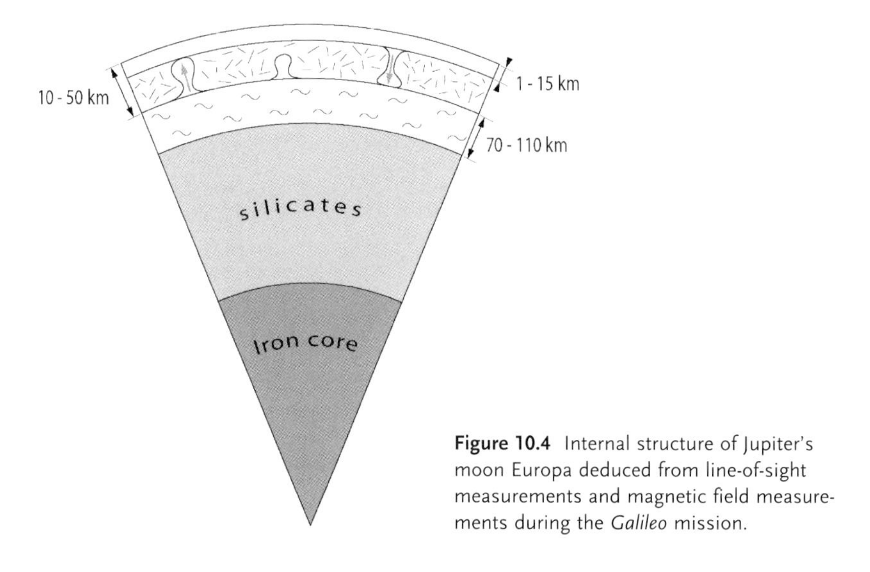

Figure 10.4 Internal structure of Jupiter's moon Europa deduced from line-of-sight measurements and magnetic field measurements during the *Galileo* mission.

$$U = \frac{GM}{r}\left[1 + \left(\frac{a_e}{r}\right)^2\left(-J_2 P_{2,0}(\sin(\lambda)) + J_{2,2}\cos(2\phi)P_{2,2}(\sin(\lambda))\right)\right] \qquad (10.1),$$

where U is the gravity potential, G is the gravitational constant, M is Europa's mass, r is the distance to the center of mass (CoM), ϕ is longitude, λ is latitude, a_e is the equatorial radius, J_2 and $J_{2,2}$ are constants to be determined, and $P_{2,0}$ and $P_{2,2}$ are associated Legendre polynomials. Assuming hydrostatic equilibrium, the J_2 value is related to the hydrostatic Love number k_S and the periodic Love number k_2 by the following equations:

$$J_2 = J_{2h} + \Delta J_2 = (1/2\,\alpha + 1/3\,q)\,k_S + 3/2\,\alpha\,k_2 e\cos(\nu) \qquad (10.2)$$

$$J_{2,2} = J_{2,2h} + \Delta J_{2,2} = 1/4\,\alpha\,k_S + 3/4\,\alpha\,k_2 e\cos(\nu) \qquad (10.3),$$

where J_{2h} and $J_{2,2h}$ are the hydrostatic parts of the degree 2 coefficients of Eq. (10.1). In Eqs. (10.2) and (10.3), e is the eccentricity and ν is the true anomaly. The coefficients α and q are defined by

$$\alpha = \frac{M^*}{M} \times \left(\frac{R}{D}\right)^3 \qquad (10.4)$$

and

$$q = \frac{\omega^2 R^3}{GM} \qquad (10.5),$$

where M^* is Jupiter's mass, D is the mean distance between Europa and Jupiter, R is Europa's radius, and ω is the rotation period.

Unfortunately, the flybys did not have a geometry that allowed $J_{2,2}$ to be determined. Assuming hydrostatic equilibrium, $J_{2,2}$ is equal to 0.3 J_2. With this assumption, it is not possible to determine the value of the periodic Love number k_2. Then, the hydrostatic Love number k_S is related to the moment of inertia C about the rotation axis by the Radau formula:

$$\frac{C}{MR^2} = \frac{2}{3}\left(1 - \frac{2}{5}\sqrt{\frac{4 - k_s}{1 + k_s}}\right) \qquad (10.6).$$

Assuming a one-dimensional model (density depends only on radius), the moment of inertia can be calculated for a given internal structure by the following equation:

$$C = \int_M x^2\,dm = \frac{8\pi}{3}\int_0^R \varrho(r) r^4\,dr \qquad (10.7),$$

where x is the distance of every elementary mass dm to the axis and ϱ is the density that depends on radius (r). In order to build an internal model of Europa, it has

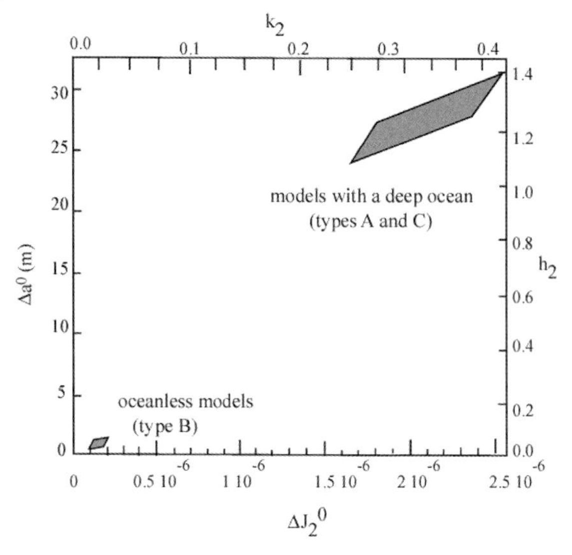

Figure 10.5 Variation of the periodic Love number k_2 for models with a deep ocean or without an ocean (from Castillo et al. 2000).

been assumed that the density depends only on the material that composes the layer. The variations due to temperature and pressure are negligible compared to those induced by the change of material.

Assuming that the mass M of a celestial body is known, the determination of the moment of inertia provides a relationship between the size of the core, its density, and the density of the H_2O mantle in a model where there are only two layers. However, the value found for Europa suggests that an iron core must be present. It is consistent with a fully differentiated moon that includes an inner iron-rich core (Figure 10.4).

It must be noted that the presence of an ocean strongly modifies the value of k_2 (Figure 10.5). Such a measurement of k_2 may become possible in a future mission to Europa, if the spacecraft orbits around Europa.

10.4
Models of Evolution of the Moon Europa

With the progress made in modeling thermal convection in planetary interiors, models describing Europa's thermal evolution have substantially improved in the last 20 years. Important improvements include the description of convection for complex viscosities by Deschamps and Sotin and information on the coupling between orbital characteristics and tidal heating by Tobie and coworkers. Heat sources include the heat of accretion, the heat produced by the decay of long-lived radiogenic elements, and tidal heating. The amount of tidal heating depends on the viscosity (Figure 10.6). The curve in Figure 10.6 has a bell-like shape with a

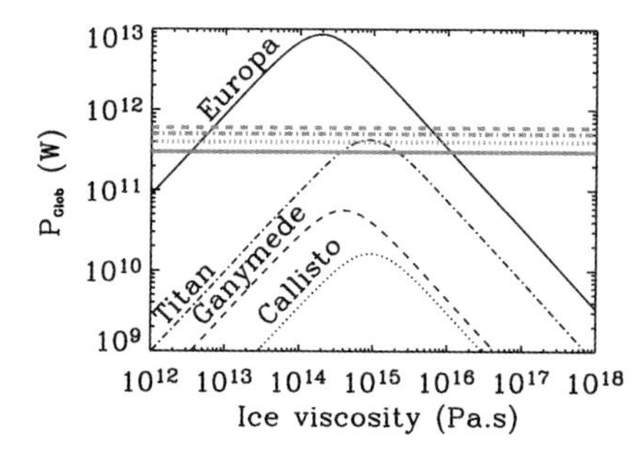

Figure 10.6 Amount of tidal heating versus viscosity for the three Galilean satellites Europa, Ganymede, and Callisto and for Saturn' moon Titan; the horizontal lines represent the amount of radiogenic power at present time in each satellite for comparison with tidal heating (from Sotin and Tobie 2004).

maximum viscosity around 10^{14} Pa s, which is the viscosity of ice close to its melting point. The amount of tidal heating is much larger for Europa than that for the two Galilean satellites Ganymede and Callisto or for Saturn's moon Titan. It may be more than an order of magnitude larger than radiogenic heating.

The first studies of the internal structure and thermal evolution of icy satellites were conducted prior to the *Voyager* mission in the late 1970 s by Consolmagno and Lewis. Subsolidus convection was not considered in those studies, and heat produced by the decay of radiogenic elements contained in a differentiated silicate core was supposed to be transferred by conduction. In a one-dimensional model, the surface heat flux (q_S) is proportional to the temperature gradient $(\delta T/\delta r)_{z=0}$ as follows:

$$q_S = \frac{Q}{4\pi R^2} = k\left(\frac{\partial T}{\partial r}\right)_{z=0} \qquad (10.8),$$

where Q is total power and k is the thermal conductivity that depends on temperature (Table 10.2). The thickness of the ice crust can be estimated very simply if one makes the following assumptions:

- there is no heat production in the ice crust,
- thermal conductivity does not depend on temperature,
- there is equilibrium between the radiogenic heating rate and the total heat power released, and
- the melting temperature is that of pure H_2O.

Table 10.2 Thermal properties of ice and silicates.

	Silicate	Ice	Mixture of ice and silicates[a]
Thermal conductivity (k in W/m/K)	4.2	$0.4685 + 488.12/T$	$fs \cdot k_S + (1 - fs) \cdot k_I$
Heat capacity (C in J/kg/K)	920	$185 + 7.037/T$	$xs \cdot C_S + (1 - xs) \cdot C_I$

a Index S stands for silicate, index I for ice.

These assumptions lead to a present heat flux around 8 mW m^{-2}, which provides a thickness of the ice crust of about 50 km. Subsequent studies by Deschamps and Sotin showed that the ice layer would be unstable to convection for such large values of the ice crust thickness.

Up to the mid-1990 s, thermal convection models were calculated using isoviscous scaling laws. Recent progress in the modeling of thermal convection processes for fluids having complex viscosity has changed the scaling laws that one must apply in order to investigate the cooling rate of a planet. If one ignores the possibility of tidal heating, the ice crust can be modeled as a viscous fluid bounded by two isothermal boundaries: the surface ($T = 100$ K) and the ocean ($T = T_m$). In the case of a fluid with only temperature-dependent viscosity, numerical simulations and laboratory experiments suggest that the amount of heat that can be transferred by subsolidus convection is driven by the instabilities, which form in the lower (hot) thermal boundary layer. Once stationary convection is reached, the thermal boundary layer is characterized by a constant value of the thermal boundary layer Rayleigh number (Ra_{TBL}) as follows:

$$Ra_{TBL} = \frac{\alpha \varrho g (T_m - T_{ice}) \delta^3}{\varkappa \mu} \quad (10.9),$$

where δ is the thickness of the thermal boundary layer, α is the coefficient of thermal expansion, μ is ice viscosity, $\varkappa$ is thermal diffusivity, and T_{ice} is the temperature of the convective ice layer, which is the layer located between the two thermal boundary layers. The temperature difference ($T_m - T_{ice}$) across the thermal boundary layer is proportional to a viscous temperature scale (ΔT_μ):

$$T_m - T_{ice} = 1.43 \Delta T_\mu = 1.43 \left(\frac{-1}{\left({}^{\partial Ln(\mu)}/_{\partial T} \right)_{T=Tice}} \right) \quad (10.10).$$

If one knows how ice viscosity depends on temperature, then one can calculate the temperature difference across the thermal boundary layer. Heat flux can be calculated using a law similar to Eq. (10.8):

$$q = \frac{Q}{4\pi R_m^2} = k \left(\frac{T_m - T_{ice}}{\delta} \right) \quad (10.11),$$

where δ is the TBL thickness determined by Eq. (10.9) and R_m is the radius of the liquid–ice interface. Viscosity of ice is the key parameter controlling the amount of heat that can be transferred by subsolidus convection. Taking values of 10^{14} Pa s for the viscosity of ice at its melting point and activation energies of 50 kJ per mole leads to a freezing of the ocean in less than one billion years for a simple model ignoring tidal dissipation.

If tidal dissipation is taken into account, it happens mostly in the convective layer. Because the total amount of heat is controlled by the instability in the top cold boundary layer, the heat flux at the base of the icy crust is limited to the difference between surface heat flux and tidal dissipation. This small heat flux prevents Europa from freezing completely. Tidal heating in a convective shell explains the presence of an ocean within Europa without invoking the presence of antifreeze material such as ammonia. The silicate core may eventually heat up if not enough heat can be removed through the ice layer. Another interesting feature related to tidal heating, as proposed by Sotin's group, is localized heating in upwelling plumes leading to partial melting of ice. Because liquid H_2O is denser than ice, it creates a depression that may explain the chaotic terrains (Figure 10.3). It also generates a negative buoyant force that prevents liquids from migrating to the surface.

The thickness of Europa's ice shell may be subject to changes with time, adjusting to the decrease in radiogenic heat and, more importantly, to variations in tidal dissipation that depend on orbital eccentricity. Although the amount of tidal heating depends on the thickness of the ice shell, creating a buffering effect on thickness variations, an important consequence of such a thermal-orbital coupling is that the thickness of Europa's ice shell may vary between a few kilometers and a few tens of kilometers on a time scale on the order of 100 million years. The observed diversity of geological surface features on Europa also may be due to these time-dependent variations of ice thickness.

10.5
Astrobiological Considerations about Possibilities for Life on the Moon Europa

Astrobiology aims at studying the origin, distribution, and evolution of life in the Universe. The major issue is that we, inhabitants of the Earth, only know life on our planet. For this reason, the search for life elsewhere in the Universe, and more simply in our Solar System, uses life on Earth as a reference (see Chapter 6). In this chapter, the knowledge about life on Earth is used for asking the following questions: Is or was there life on Europa and, if the answer is yes, why? In this study, only small-sized organisms, namely, microorganisms, will be considered, because early life forms on Earth also were admittedly represented by microorganisms (see Chapter 4). However, it is much more difficult to detect such microscopic life than large-sized organisms; their detection would probably not be a problem.

There are plenty of definitions for life on Earth, and this point could be the central theme of several books (see Chapter 1). More simply, all organisms living on Earth are organized on either a cellular basis–i.e., all animals, plants, and microorganisms–or a non-cellular basis–i.e., viruses and prions. All life forms are complex assemblages of atoms, particularly C, N, H, O, P, S, K, Mg, Na, Ca, and Fe, which constitute the macronutrients. These macronutrients form macromolecules (proteins, lipids, polysaccharides, and nucleic acids), which all contain carbon, are solved in water, and are present in all living forms. Water represents an ideal biological solvent, because of its polarity and cohesiveness; it is a necessary prerequisite for life as we know it. Briefly, life requires, at the very least, liquid water and carbon.

It is now generally accepted that liquid water may exist on Europa. Therefore, the following question now needs to be discussed: What else is required for life?

Still considering life on Earth as a reference, one can infer that all cellular and non-cellular organisms contain at least proteins, which are polymers of amino acids, each made of C, H, O, and N (see Chapter 3). Carbon is the major element present in large concentrations in all living forms, and one cannot raise the question of life on Europa without addressing the availability of carbon. Living forms on Earth use relatively simple carbon-containing molecules as a starting point to synthesize macromolecules. The carbon source can be carbon dioxide (inorganic) and other C_1 compounds (molecules containing only one atom of carbon, such as CO or CH_4) or organic carbon (consisting of one or more carbon atoms associated with H, O, and N, such as sugars, aldehydes, alcohols, fatty acids, amino acids, etc.).

Although the surface of the ice crust of Europa has been rearranged (and this is one of the arguments in favor of a liquid water ocean beneath the ice crust), a few impact craters (caused by meteorite impacts) are still visible. Taking into account the analysis carried out in many laboratories about the chemical composition of meteorites delivered to Earth (see Chapter 3), there is no doubt that organic carbon has been delivered to Europa, and probably in rather large amounts, especially if one considers the numerous impacts visible on Io, another Galilean satellite. Likewise, on early Earth, carbon and other atoms essential for life were delivered through meteorite bombardments.

At this step of considerations, all major ingredients for life are theoretically present on Europe. But is that enough? Considering again life on Earth, cellular organization prevails, and non-cellular organisms (viruses and prions) depend on a cellular host for their replication. A cell is an entity enclosed in a semipermeable membrane, which isolates the cell from its environment (including other cells) but allows exchange of material with this environment (in both directions, in and out). These materials include nutrients and particularly carbon sources but also energy sources (see Chapter 5).

What energy sources are used on Earth by living organisms? The two main energy sources are light energy from the Sun and chemical energy. Light emitted by the Sun may be converted into chemical energy: it is mostly used for the fixation of carbon dioxide and its conversion into organic carbon through photosynthesis. This process is used by plants, seaweeds, microalgae, and some bacteria.

The other energy source is of a chemical nature and is based on the transfer of electrons from a reduced compound (electron donor) to an oxidized compound (electron acceptor). Both the electron donor and the acceptor are taken by the cells from their surrounding environment. The electron transfer from the donor to the acceptor takes place within the cell membrane and is accompanied by an extrusion of protons. This leads to the formation of a gradient of protons across the membrane, with more protons outside than inside. This gradient or proton motive force is catalyzed by a specific enzyme, ATPase, which uptakes protons from outside to inside and forms ATP (adenosine triphosphate), a highly energetic molecule.

On Earth, there is a huge diversity of electron donors and acceptors that can be used by living organisms, provided they have the suitable set of enzymes, for example, electron carriers. Animals use organic molecules as electron donors and molecular oxygen as electron acceptor. But microorganisms are more versatile and may use hydrogen, sulfur, hydrogen sulfide, ammonium, nitrite, or ferrous iron as electron donor, while the electron acceptor may be carbon dioxide, sulfur, sulfate, nitrate, or ferric iron, depending on the reduction potential of the compound considered. It should be noted that molecular oxygen is not involved in these strictly anaerobic reactions, although molecular oxygen is an excellent electron acceptor that is used when it exists but is not a prerequisite for life. Indeed, life started on Earth well before photosynthetic organisms appeared that eventually released oxygen into the environment.

What kind of energy sources might be available on Europa? They might come from outside of the moon by meteorites or from its interior. The latter point will be considered in the following. As pointed out in Section 10.3, among the Galilean satellites that probably harbor an ocean of liquid water, Europa represents a single case where the ocean is in contact with the silicate core. If the temperature of the core is high and reaches up to its surface, then these conditions might foster the formation of hydrothermal circulations.

On Earth, such hydrothermal vents have been discovered since 1977 in many places of the oceans and particularly along oceanic ridges. In these tectonically active areas, seawater penetrates through cracks of the seafloor and approaches the magma chamber. At elevated temperatures, seawater interacts with the rocks. This leaching process enriches seawater with various inorganic compounds such as hydrogen, hydrogen sulfide, carbon monoxide, carbon dioxide, methane, and other small molecules. After a complex circulation, the hot fluid eventually vents out. This fluid is different from seawater because it contains various new chemicals, and it is lighter than seawater because density decreases with temperature. From a biological point of view, it is noteworthy that this fluid delivers into the ocean a plume of chemicals that may serve as electron donors and acceptors. A variety of microorganisms, Bacteria and Archaea (see Chapter 5), utilize these compounds for their energy metabolism and thrive optimally at those deep-sea hydrothermal vents. Furthermore, it has been proposed that some organic synthesis might occur under hydrothermal conditions through Fischer-Tropsch reactions, thereby providing an additional source of organic carbon (see Chapter 3).

At this point of the considerations, one may anticipate that the moon Europa might possess the major requirements for life, namely, liquid water, inorganic and organic carbon, and chemical energy sources. Similar ingredients were also available on the early Earth, when life started there (see Chapter 3). In the following, the putative environment of Europa is considered, and again life on Earth is taken as reference.

Organisms on Earth are dependent on the physicochemical conditions of their environment, in addition to the basic requirements of liquid water, carbon, and energy sources. Major parameters of potential influence are temperature, salinity, pH, and hydrostatic pressure. Those parameters also may be relevant in the context

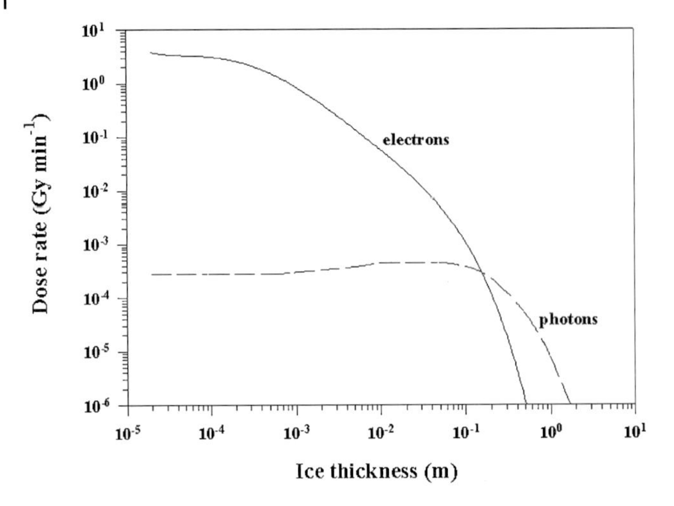

Figure 10.7 Rates of physically absorbed radiation doses, separate contributions of electrons and photons (bremsstrahlung), at the surface of Jupiter's moon Europa and below varying thicknesses of ice (from Baumstark-Khan and Facius 2002).

of Europa. High intensities of ionizing radiations, generated by the intense radiation belts of Jupiter, are also experienced at the surface of Europa, but because of the thickness of several kilometers of the ice crust, the ocean below is well protected from those radiations (Figure 10.7).

The temperature of the upper layers of the Europan ocean, that is, the boundary to the ice crust, should not be very different from 0 °C, depending on the chemical composition of the ocean; but it might be more elevated in the deepest parts, especially if the anticipated hydrothermal circulation exists. The salinity and pH of the ocean are also determined by its chemical composition. According to different models, the Europan ocean is probably salty, and the pH may be either acidic or alkaline. Finally, because of Europa's low gravity (Table 10.1), the depth of the ocean of about 100 km would not result in a very high hydrostatic pressure at its bottom. Pressures should not exceed 100 MPa.

On Earth, environmental parameters as described above for Europa's ocean are compatible with those supported or even required by a group of organisms called "extremophiles" (see Chapter 5). Concerning temperatures, it is well known that at very low temperatures macromolecules and even whole cells are preserved. For instance, gametes and embryos from mammals, seeds, or bacteria are commonly preserved in liquid nitrogen at −196 °C. A temperature of −12 °C has been identified as the lower limit for growth of certain microorganisms isolated from permafrost; but many microorganisms, bacteria as well as algae, are still able to metabolize and grow at temperatures close to 0 °C. At the top of the temperature scale, 113 °C is the maximum temperature for growth, as demonstrated by *Pyrolobus fumarii*, an Archaea isolated from a deep-sea hydrothermal vent. This organism grows optimally at 110 °C and is capable of surviving two hours in an autoclave under conditions that are commonly used for sterilization.

Elevated salt concentrations, up to 5.5 M (for instance, 32 % NaCl), still allow growth of Archaea, which are called "extreme halophiles." For optimal growth these organisms usually require concentrations of 1.5–4.0 M of salts, which corresponds to 9–23 % NaCl. In many cases, Na^+ ions constitutc the brines, but Mg^{2+} or Ca^{2+} ions also occur.

With regard to pH values, acidophilic microorganisms grow at low pH and even close to pH 0, for instance, the microorganism *Picrophilus oshimae*. Other bacteria and algae are adapted to grow at pH values around 11. These are called "alkaliphiles."

The hydrostatic pressure of about 100 MPa that is anticipated to prevail at the bottom of Europa's ocean corresponds to the pressure in the deepest trenches of Earth's oceans, for instance, 110 MPa at 11 000 m depth. Microorganisms have been isolated from such high-pressure environments, and some microorganisms appear to require pressures of least 40 MPa to start growth.

Considering the conditions that are expected for Europa's ocean (i.e., the availability of liquid water and carbon and energy sources and the physicochemical parameters), they are compatible with conditions that sustain life as we know it on Earth. These are arguments that justify proceeding with the search for life in that distant ocean. Of course, those conditions are still to be proven definitely, but the perspectives are a great challenge for future space missions to the Jovian system and especially to its moon Europa.

10.6 Summary and Conclusions

The discovery of an ocean on the Jovian moon Europa would be a major discovery of planetary exploration. It will be a major step toward the questioning of habitability of satellites and planets and a first step to the question of life wherever there is water. The results of the *Galileo* mission suggest that such an ocean may be present on Europa. However, each piece of evidence, taken separately, can be explained without the presence of an ocean. The ocean hypothesis is also strongly supported by models describing Europa's thermal evolution and its relationships with surface features. Europa is unique among the icy satellites because the liquid H_2O layer, if present, would be in contact with the silicate core. The conditions at the core–ocean interface would be similar to those existing at the terrestrial ocean seafloor, where living organisms develop without using any solar energy. Although it is very tempting to imagine that Europa's silicate core is volcanically active like that on Io, recent calculations on tidal dissipation suggest that there is not enough dissipation energy deposited in the silicate layer for silicates to reach their melting point. The potential presence of liquid water and volcanism puts Europa on the list of possible life-bearing bodies, though the probability is very low. On the one hand, the presence of an ocean is very likely, but on the other hand, siliceous volcanic activity is not predicted by current models. We lack measurements that would provide a definitive answer to these questions, and future missions will be developed to acquire such data.

10.7 Outlook and Plans for Future Missions

Jupiter's moon Europa belongs to the family of icy satellites. Two features make it special in this family. First, the small amount of H_2O makes possible an interface between liquid water and a silicate core. Second, the Laplace resonance maintains a high eccentricity that is responsible for the large amount of tidal heating. Calculations predict that tidal heating is mostly localized in the ice shell and that the amount is much larger than radiogenic heating. Tidal heating may prevent the total freezing of the satellite and would explain the presence of a liquid layer as suggested by the magnetic data. Europa may be the only place beyond Earth in the Solar System where a silicate–ocean interface exists. At this interface, conditions are close to those existing on Earth's seafloors. One important difference may be the lack of silicate volcanism at this interface. The only argument for the absence of volcanism comes from numerical modeling. It would be important to acquire data that can constrain such thermal evolution models.

Several arguments exist in favor of space missions dedicated to the search for life on Europa, and such a mission has been recommended to space agencies, particularly by the U. S. Academy of Sciences and in Europe by several assemblies, including the European Astrobiology Network Association (EANA). Several scenarios have been studied, from a simple orbiter to a lander including a cryobot, which is a vehicle able to reach the ocean and sample it after drilling 10 km through the ice crust. Novel data about Europa will obviously depend on the available technology and associated costs.

An orbiter mission to Europa should be able to map the surface with high resolution, to determine the thickness of the ice crust, and to definitively prove (or disprove) the existence of an ocean. Additional data on the contact region between the deep ocean and the core should be provided, as well as on the chemical composition of the surface ice with regard to the occurrence of inorganics, organics, and volatiles.

Finally, a landing mission would provide contact access to the surface and would allow ice sampling and analysis. However, because of the intense radiations at the surface of Europa (Figure 10.7), it would be necessary to have access to the ice subsurface of 1–2 m to reach areas that are protected from radiations.

Without a lander, it would still be possible to assess the presence of an ocean by measuring the gravity potential and the variations of Europa's equatorial radius as a spacecraft orbits around Jupiter (Figure 10.5). The thickness of the ice crust may be more difficult to measure. A laser altimeter would measure the topography of geological features, which would allow geologists to better assess the time of formation of the different geological structures that pave Europa's surface. Finally, infrared spectrometry and measurements of ions and neutrals along the orbit should allow determination of both the molecular and the elementary composition of the surface. Gravity potential measurements should allow determination of lateral mass variations in the silicate core if it has been subject to partial melting and volcanism. Such information would be important for assessing the possibility of life at the interface between the liquid layer and the silicate core.

10.8 Further Reading

10.8.1 Books and Articles in Books

Gargaud, M., Barbier, B., Martin, H., J. Reisse, J. (Eds.) *Lectures in Astrobiology*, Volume II, Springer, Berlin (in press), especially the following chapter therein:
- Prieur, D. An Extreme Environment on Earth: Deep-Sea Hydrothermal Vents. Lessons for Mars and Europa Exploration.

Greenberg, R. *Europa, the Ocean Moon*, Springer-Praxis, Chichester, UK, 2005.

Horneck. G., Baumstark-Khan, C. (Eds.) *Astrobiology – The Quest for the Conditions of Life*, Springer, Berlin, 2002; especially the following chapters therein:
- Greenberg, R., Tuffs, B. R., Geissler, P., Hoppa, G. V. Europa's Crust and Ocean: How Tides Create a Potentially Habitable Physical Setting, pp. 111–124.
- Stetter, K. O. Hyperthermophilic Microorganisms, pp. 169–186.
- Baumstark-Khan, C., Facius, R., Life under Conditions of Ionizing Radiation, pp. 261–284.

Jolivet, E., L'Haridon, S., Corre, E., Gérard, E., Myllykallio, H., Forterre, P., Prieur, D. The Resistance to Ionizing Radiation of Hyperthermophilic Archaea Isolated from Deep-Sea Hydrothermal Vents, in *1st European Workshop on Exo/Astrobiology*, 21–23 May 2001, Frascati, Italy. Ehrenfreund, P., Angerer, O., Battrick, B. (Eds.) ESA SP-496, ESA/ESTEC, Noordwijk, pp. 243–246, 2001.

Prieur, D. Deep-Sea Hydrothermal Vents: An Example of Extreme Environment on Earth, in *First Steps in the Origin of Life in the Universe*, Chela-Flores, J., Owen, T., Raulin, F. (Eds.) Kluwer Academic, Dordrecht, The Netherlands. pp. 187–194, 2001.

Prieur, D., Voytek, M., Reysenbach, A. L. Jeanthon, C. Deep-Sea Thermophilic Prokaryotes, in *Thermophiles: Biodiversity, Ecology and Evolution*. Reysenbach, A.L, Voytek, M., Mancinelli, R. (Eds.) Kluwer Academic/Plenum Publishers, New York. pp. 11–22, 2001.

Prieur, D. Microbiology of Deep-Sea Hydrothermal Vents: Lessons for Mars Exploration, in *Water on Mars and Life*, T. Tokano (Ed.), Springer, Berlin, pp. 299–324, 2005.

Reysenbach, A. L., Holm, N. G., Hershberger, K., Prieur, D., Jeanthon, C. In Search of a Subsurface Biosphere at a Slow-Spreading Ridge, in *Proceedings of the Ocean Drilling Program, Scientific Results, Vol. 158*, Herzig, P. M., Humpris, S. E., Miller, D. J., Zierenberg, R. A. (Eds.) pp. 355–360, 1998.

10.8.2
Articles in Journals

Brack, A., Ehrenfreud, P., von Kiedrowski, G., Lammer, H., Prieur, D., Szuszkiewicz, E., Westall, F. EANA trail guide in astrobiology: Search for a second genesis of life. *Internat. J. Astrobiology*, **2006**, *4*, 195–202.

Collins, G. C., Head, J. W., Pappalardo, R. T., Spaun, N. A. Evaluation of models for the formation of chaotic terrain on Europa, *J. Geohpys. Res.* **2000**, *105, 1709–1716.*

Consolmagno,G.J., Lewis, J. S. The evolution of icy satellite interiors and surfaces, *Icarus*, **1978**, *34, 280–293.*

Deschamps F., Sotin C. Thermal convection in the outer shell of large icy satellites, *J. Geophys. Res.* **2001**, *106*, 5107–5121.

Grasset, O., Parmentier, E. M. Thermal convection in a volumetrically heated, infinite Prandtl number fluid with strongly temperature-dependent viscosity: implications for planetary thermal evolution, J. *Geophys. Res.* **1998**, *103*, 171–181.

Hussmann H., Spohn T. Thermal-orbital evolution of Io and Europa. *Icarus*, **2004**, *171*, 391–410.

Khurana, K. K., Kivelson, M. G., Russell, C. T. Induced magnetic fields as evidence for subsurface ocean in Europa and Callisto. *Nature*, **1998**, *397*, 777–780.

López-Garcia, P., Moreira, D., Douzery, E., Forterre, P., Van Zuilen, M., Claeys, P., Prieur D. Ancient fossil record and early evolution (ca. 3.8 to 0.5 Ga), *Earth Moon Planets*, **2006**, *98*, 247–290. DOI 10.1007/s11038–006–9091–9.

McCord, T. B., Hansen, G. B., Clark, R. N. Martin, P. D., Hibbitts, C. A., Fanale, F. P., Granahan, J. C., Segura, M., Matson, D. L., Johnson, T. V., Carlson, R. W., Smythe, W. D., Danielson, G. E., The NIMS Team, Non-water-ice constituents in the surface material of the icy Galilean satellites from the *Galileo* near-infrared mapping spectrometer investigation, *J. Geophys. Res.* **1998**, *103*, 8603–8626.

McKinnon, W. B. Convective instability in Europa's floating ice shell, *Geophys. Res. Lett.*, **1999**, *26, 951–954.*

Pappalardo, R. T., Head, J. W., Greeley, R., Sullivan, R. J., Pilcher, C., Schubert, G., Moore, W. B., Carr, M. H., Moore, J. M., Belton, M. J. S., Goldsby, D. L. Geologic evidence for solid-state convection in Europa's ice shell, *Nature*, **1998**, *391*, 365–368.

Pappalardo, R. T., Belton, M. J. S., Breneman, H. H., Carr, M. H., Chapman, C. R., Collins, G.C, Denk, T., Fagents, S., Geissler, P. E., Giese, B., Greeley, R., Greenberg, R., Head, J. W., Helfenstein, P., Hoppa, G., Kadel, S. D., Klaasen, K. P., Klemaszewski, J. E., Magee, K., McEwen, A. S., Moore, J. M., Moore, W. B., Neukum, G., Phillips, C. B., Prockter, L. M., Schubert, G., Senske, D. A., Sullivan, R. J., Tufts, B. R., Turtle, E. P., Wagner, R., Williams, K. K. Does Europa have a subsurface ocean? Evaluation of the geological evidence, *J. Geophys. Res.*, **1999**, *104, 24015–24055.*

Pascal, R., Boiteau, L., Forterre, P., Gargaud, M., Lazcano, A., Lopez-Garcia, P., Maurel, M.-C., Moreira, D., Pereto, J., Prieur D., Reisse J. Prebiotic chemistry – biochemistry – emergence of life (4.4–2 Ga), *Earth Moon Planets*, **2006**, *98*, 153–203. DOI 10.1007/s11038–006–9089–3.

Smith, B. A. and the *Voyager* Imaging Team. The Galilean satellites of Jupiter, *Voyager 2* imaging science results, *Science*, **1979**, *206*, 951–972.

Sotin C. Tobie G. Internal structure and dynamics of the large icy satellites, *C. R. Acad. Sci. Physique*, **2004**, *5*, 769–780.

Sotin C., Head J. W., Tobie G. Europa: Tidal heating of upwelling thermal plumes and the origin of lenticulae and chaos melting, *Geophys. Res. Lett.* **2002**, *29, 8*, 74-1–74-4.

Thomson, R. E., Delaney, J. R. Evidence for a weakly stratified Europan ocean sustained by seafloor heat flux, *J. Geophys. Res.*, **2001**, *106, 12355–12365.*

Tobie G., Mocquet A., Sotin C. Tidal dissipation within large icy satellites: Europa and Titan, *Icarus*, **2005**, *177*, 534–549.

Zimmer, C., Khurana, K. K., Kivelson, M. G. Subsurface oceans on Europa and Callisto: Constraints from *Galileo* magnetometer observations, *Icarus*, **2000**, *147*, 329–347.

10.8.3 Web Sites

http://www2.jpl.nasa.gov/galileo/europa/

10.9 Questions for Students

Question 10.1

Knowing Europa's mass and assuming chondritic concentrations in radiogenic elements, determine the heat power provided by the decay of radiogenic species. If Europa is neither cooling nor heating up, determine the heat flux at the surface. Determine the thickness of the ice layer if the melting temperature is 270 K and the heat conductivity described by $k = 600/T$ where T is in K and k in W/m/K.

Question 10.2

Knowing the mass and the moment of inertia of Europa, draw the relationship between the density of the silicate mantle and the density of the H_2O layer. Comment on this curve.

Question 10.3

Determine the amount of heat that can be removed by convection if the ice layer can convect. Take a viscosity of ice at its melting point equal to 10^{14} Pa s and an activation energy of 60 kJ mol^{-1}. Different viscosities at the melting point can be taken in order to investigate the effect of this parameter.

11
Astrobiology Experiments in Low Earth Orbit: Facilities, Instrumentation, and Results

Pietro Baglioni, Massimo Sabbatini, and Gerda Horneck

This chapter introduces the space environment in Earth orbit with emphasis on those parameters of outer space that are of interest to astrobiology. Facilities and scientific instrumentation to tackle astrobiological questions in Earth orbit are described. They include exposed experiment facilities on the International Space Station and on the retrievable satellite Foton. These facilities are also used as examples to describe the constraints that instrument developers and experimenters have to face when developing hardware for space missions. The chapter concludes with results obtained from exposure experiments in Earth orbit.

11.1
Low Earth Orbit Environment, a Test Bed for Astrobiology

Our biosphere has evolved for more than 3 billion years under the protective blanket of the Earth's atmosphere, which shields terrestrial life against the hostile environment of interplanetary space. Space technology has now provided the tools for transporting terrestrial organic matter or organisms beyond this protective shield in order to study *in situ* their responses to selected conditions of space.

It was nearly 50 years ago that the first-ever manmade object reached outer space: the satellite *Sputnik* was launched by the USSR on 4 October 1957. *Sputnik's* radio signals, just simple "beeps," were rebroadcast by television and radio stations around the world and immediately captured the public's imagination. This was the initiation of the space age. One month later, on 3 November 1957, with *Sputnik II,* the dog Laika became the first animal sent to space. Less than four years later, on 12 April 1961, Yuri A. Gagarin became the first human to be launched into space, with the Soviet *Vostok* spacecraft, and made one elliptical orbit (181 km perigee, 327 km apogee) of the Earth in 89 min at an inclination of 65°. At that time, the space race between the U. S. and the USSR began, culminating in the U. S. landing of *Apollo 11* on the Moon in June 1969.

Complete Course in Astrobiology. Edited by Gerda Horneck and Petra Rettberg

ISBN: 978-3-527-40660-9

However, in the cases mentioned above, the biological systems, including humans, were protected from most of the hostile parameters of space, either by containment within a space capsule, which is a pressurized module with an efficient life support system (LSS), or at least by a space suit during extravehicular activity. In these cases, the only space parameters of interest, or concern, were microgravity and radiation.

Alternatively, if arriving in space without any protection, living beings are confronted with an extremely hostile environment, characterized by a high vacuum, an intense radiation field of solar and galactic origin, and extreme temperatures (Table 11.1). This environment, or selected parameters of it, is the test bed for astrobiological investigations, thereby exposing chemical or biological systems to selected parameters of outer space or defined combinations of them.

Table 11.1 Parameters of the environment of interplanetary space, of low Earth orbit, and of space simulation facilities that are relevant for astrobiological experiments.

Space parameter	Interplanetary space	Low Earth orbit	Simulation facility
Space vacuum			
Pressure (Pa)	10^{-14}	10^{-7}–10^{-4}	10^{-7}–10^{-4}
Residual gas (part cm^{-3})	1	10^4–10^5 H 10^4–10^6 He 10^3–10^6 N 10^3–10^7 O	Different values[a]
Solar electromagnetic radiation			
Irradiance ($W\ m^{-2}$)	Different values[b]	1360	Different values[c]
Spectral range (nm)	Continuum	Continuum	Different spectra[c]
Cosmic ionizing radiation			
Dose ($Gy\ year^{-1}$)	≤ 0.1[d]	400–10 000[e]	Wide range[c]
Temperature (K)	> 4[b]	Wide range[b]	Wide range[c]
Microgravity (g)	$< 10^{-6}$	10^{-3}–10^{-6}	0–1 000

a Depending on pumping system and requirements of the experimenter.
b Varying with orientation and distance to Sun.
c Depending on the radiation source and filter system.
d Varying with shielding, highest values at mass shielding of 0.15 g cm^{-2}.
e Varying with altitude and shielding, highest values at high altitudes and shield of 0.15 g cm^{-2}.

Most of these studies were done in low Earth orbit (LEO), which reaches an altitude of about 450 km. At this altitude, it takes about 90 min for a spacecraft to complete an orbit around the Earth. At an inclination of 52–62°, which is that of the orbit of the International Space Station (ISS) – two consecutive orbits are about 2 200 km apart (because of the progression of the Earth's rotation). Hence, for Earth observation purposes, these orbits cover most of the more inhabited areas of the globe.

In order to understand the physical parameters imposed on astrobiological test systems, it is first necessary to understand the space environment itself. The environment in LEO is characterized by an intense radiation climate, a high vacuum, and extreme temperatures.

11.1.1
Cosmic Radiation Field in LEO

Planets and moons of our Solar System are exposed to a complex radiation field of galactic and solar origin (Fig. 11.1). Galactic cosmic radiation (GCR) originates outside of our Solar System in previous cataclysmic events, such as supernovae explosions. When it enters our Solar System, its energies must be high enough to overcome deflection by the magnetic fields of the Solar System. Solar cosmic radiation (SCR) consists of two components, the low-energy solar wind particles that flow constantly from the Sun and the highly energetic solar particle events (SPE) that are emitted from magnetically disturbed regions of the Sun in sporadic bursts.

The surface of the Earth is largely spared from these cosmic radiations because of the deflecting effect of the Earth's magnetic field and the huge shield of 1 000 g m^{-2} provided by the atmosphere. The terrestrial average annual effective dose equivalent from cosmic rays amounts to 0.30 mSv, which is about 100 times lower than that experienced in LEO.

Sievert (Sv) is a measure of the dose equivalent, i.e., the biologically effective dose of radiation. It is the product of a quality factor specifying the biological effectiveness of a certain radiation quality and the absorbed dose. The absorbed dose is

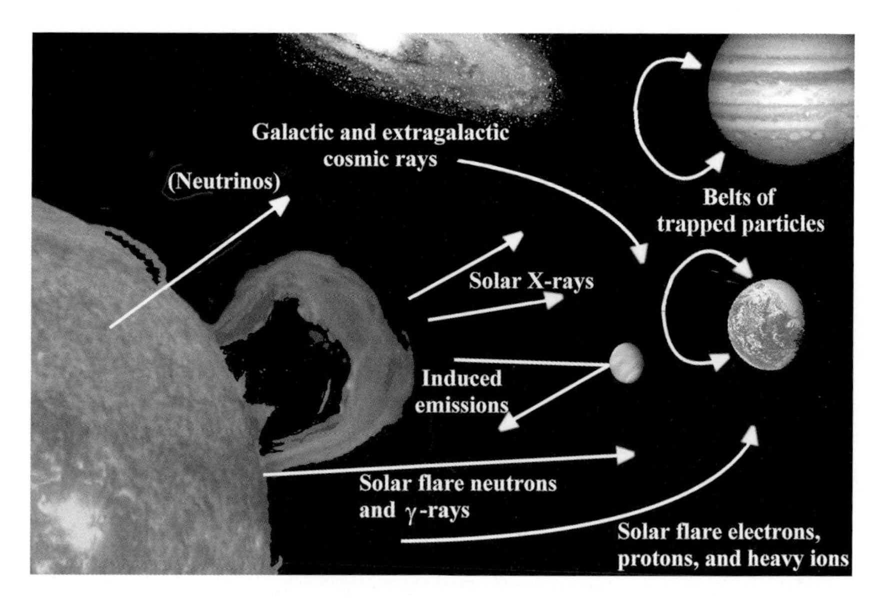

Fig. 11.1 Space radiation sources of our Solar System.

measured in Gray (Gy), with 1 Gy being equal to the net absorption of 1 J in 1 kg of material (water). In relation to the previously used unit rad, 1 Gy is equal to 100 rad.

11.1.1.1 Galactic Cosmic Radiation

Detected particles of GCR consist of 98% baryons and 2% electrons. Taking the baryonic component as 100%, then it is composed of 85% protons (hydrogen nuclei), with the remainder being alpha particles (helium nuclei) (14%) and heavier nuclei (about 1%). The latter component comprises the so-called HZE particles (particles of high charge Z and high energy), which are defined as cosmic ray nuclei of charges Z >2 and of energies high enough to penetrate at least 1 mm of spacecraft or of spacesuit shielding. Though they contribute to only roughly 1% of the flux of GCR, they are considered a potential major concern to living beings in space, especially for long-term missions at high altitudes or in high-inclination orbits or for missions beyond the Earth's magnetosphere. Reasons for this concern are based, on the one hand, on the inefficiency of adequate shielding and, on the other hand, on the special nature of HZE particle-produced lesions (explained in Section 11.4.3.3). The fluence of GCR is isotropic and energies up to 10^{20} eV may be present. When GCR enters our Solar System, it must overcome the magnetic fields carried along with the outward-flowing solar wind, the intensity of which varies according to the about 11-year cycle of solar activity (Fig. 11.2). With increasing solar activity, the interplanetary magnetic field increases, resulting in a decrease in the intensity of GCR of low energies. This modulation is effective for particles below some GeV per nucleon. Hence the GCR fluxes vary with the solar cycle and differ by a factor of approximately five between solar minimum and solar maximum, with a peak level during minimum solar activity and the lowest level during maximal solar activity.

The fluxes of GCR are further modified by the Earth's magnetic field. Only particles of very high energy have access to low-inclination orbits. At the poles,

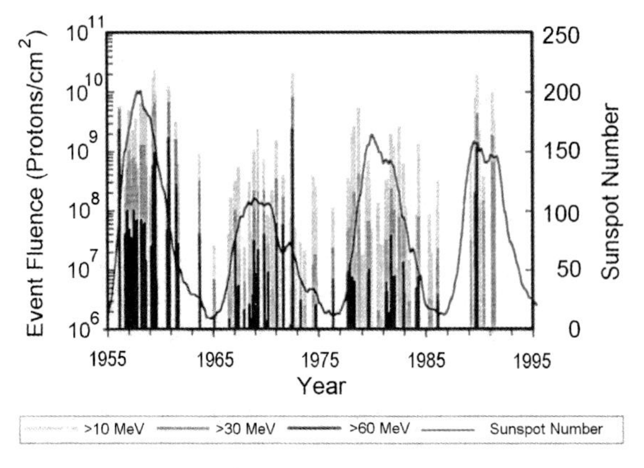

Fig. 11.2 Eleven-year cycle of solar activity; the solar proton fluxes are superimposed to the sunspot numbers.

particles of all energies can impinge in the direction of the magnetic field axes. Because of this inclination-dependent shielding, the number of particles increases in LEO from lower to higher inclinations.

11.1.1.2 Solar Cosmic Radiation

Our sun emits two types of radiation: (1) low-energy solar wind particles (mainly protons) that flow constantly from the sun and (2) the so-called solar particles events (SPEs) that originate from magnetically disturbed regions of the sun, which sporadically emit bursts of charged particles with high energies (Figs 11.1 and 11.2). These latter events are composed primarily of protons, with a minor component (5–10%) being helium nuclei (alpha particles) and an even smaller part (1%) heavy ions and electrons. SPEs develop rapidly and generally last for no more than a few hours; however, some proton events observed near Earth continued over several days. The emitted particles can reach very high energies, up to several GeV. In a worst-case scenario, doses as high as 10 Gy could be received within a short time, which is dangerous to both the spacecraft electronics and the astronauts. Such strong events are very rare, typically occurring about one time during the 11-year solar cycle. Although SPEs frequently occur during solar maximum (Fig. 11.2), they have rarely been observed at other intervals of the solar cycle. The next solar cycle (cycle 24) will peak around the year 2010. Spacecraft in LEO are largely protected from SPEs because the Earth's magnetic field provides a latitude-dependent shielding against SPEs. Only in high-inclination orbits do SPEs create a hazard to humans in space, especially during extravehicular activities. The mechanism is the same as for GCR: particles of all energies can impinge

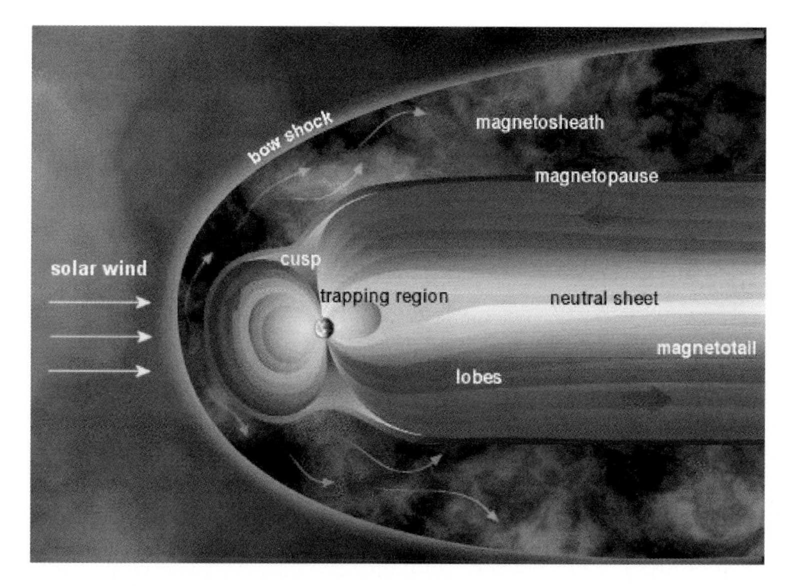

Fig. 11.3 Effect of solar wind on the Earth's magnetosphere.

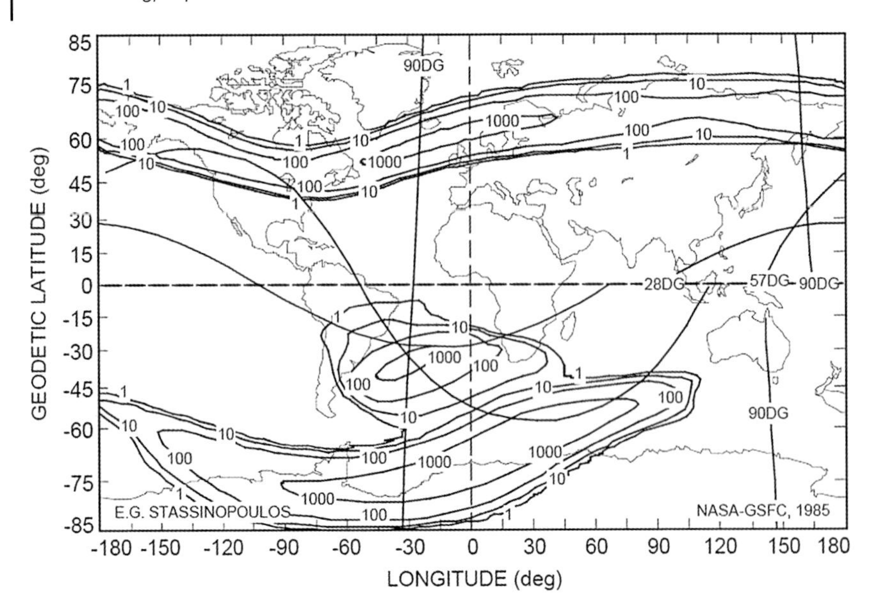

Fig. 11.4 Polar horns and South Atlantic Anomaly, regions of elevated radiation doses in low Earth orbit at 28°, 57° and 90° inclination; numbers give doses at an altitude of 500 km behind 0.2 g cm^{-2} aluminium shield (100 is equal to approximately 860 mGy per day) (source: http://parts.jpl.nasa.gov/mmic/10.pdf).

in the direction of the magnetic field axes. Hence, equatorial orbits provide the highest radiation protection.

11.1.1.3 Radiation Belts

In LEO, in addition to GCR and solar cosmic radiation (SCR), the radiation field comprises a third source of radiation: the van Allen Belts, which result from the interaction of GCR and SCR with the Earth's magnetic field and with the atmosphere. Above all, electrons and protons and some heavier ions are trapped by the geomagnetic field in closed orbits around the Earth (Fig. 11.3). These particles form two belts of radiation, an inner and an outer one, which differ in their type of formation. The main production process for the inner belt particles is the decay of neutrons produced in cosmic particle interactions with the atmosphere. The outer belt consists mainly of trapped solar particles. In each zone, the charged particles spiral around the geomagnetic field lines and are reflected back between the magnetic poles, acting as mirrors. Electrons reach energies of up to 7 MeV and protons up to about 200 MeV. The energy of trapped heavy ions is less than 50 MeV. The trapped radiation is also modulated by the solar cycle: proton intensity decreases with high solar activity, while electron intensity increases, and vice versa.

For space missions in LEO, depending on the orbit parameters and flight data, radiation doses in the range of 20 mSv per month may be received. Of special importance is the so-called South Atlantic Anomaly (SAA), where the fringes of the inner proton radiation belt reach down to altitudes of 400 km (Fig. 11.4). This behavior reflects the displacement of the axis of the geomagnetic (dipole) field by about 450 km with respect to the axis of the geoid (rotation axis), with a corresponding distortion of the magnetic field. This SAA region accounts for the dominant fraction – up to 90% – of the total radiation exposure of spacecraft in LEO.

11.1.2 Solar Extraterrestrial UV Radiation

The spectrum of solar electromagnetic radiation spans over several orders of magnitude, from short-wavelength X-rays to radio frequencies. At the distance of the Earth (1 astronomic unit = 1 AU), solar irradiance amounts to 1 360 W m^{-2}, the solar constant. Of this radiation, a large fraction is attributed to the infrared fraction (IR: >800 nm) and the visible fraction (VIS: 400–800 nm) and only about 8% to the ultraviolet range (UV: 100–400 nm).

On its way through the Earth's atmosphere, solar electromagnetic radiation is modified by scattering and absorption processes. Of special interest is solar UV radiation, because it has a high impact on the health of our biosphere. During the first 2.5 billion years of Earth's history, UV radiation of wavelengths >200 nm reached the surface of the Earth because of lack of an effective ozone shield. Following the rapid oxidation of the Earth's atmosphere about 2.1 billion years ago, photochemical processes in the upper atmosphere resulted in the buildup of a UV-absorbing ozone layer in the stratosphere. Today, the stratospheric ozone layer effectively absorbs UV radiation at wavelengths shorter than 290 nm (Fig. 11.5).

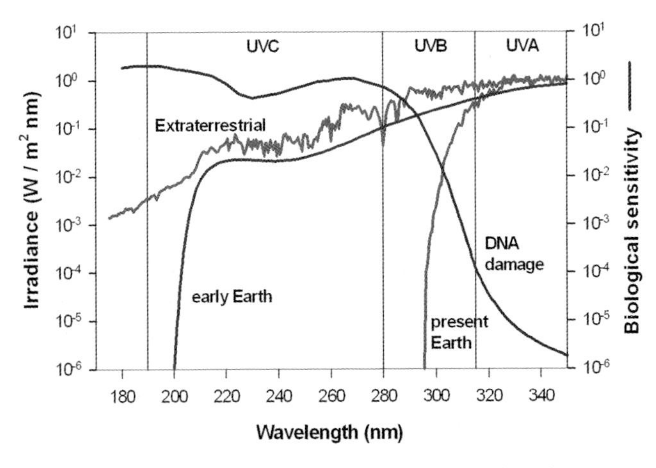

Fig. 11.5 Spectrum of solar UV radiation: extraterrestrial, at the surface of the early Earth (the first 2.5 billion years), at the surface of the present Earth, and spectral UV sensitivity of the genetic material DNA.

The spectrum of extraterrestrial solar UV radiation has been measured during several space missions, such as *Spacelab 1* and the *European Retrievable Carrier* (EURECA) mission. It can be divided into three spectral ranges: UV–C (200–280 nm), contributing 0.5%; UV-B (280–315 nm), contributing 1.5%; and UV-A (315–400 nm), contributing 6.3% to the whole solar electromagnetic spectrum. UV radiation of wavelengths shorter than 200 nm, which is absorbed by the molecules of the atmosphere, belongs to the vacuum UV. Although the UV–C and UV-B regions make up only 2% of the entire solar irradiance prior to attenuation by the atmosphere, they are mainly responsible for the high lethality of extraterrestrial solar radiation to living organisms (see Section 11.4). The reason for extraterrestrial solar UV radiation's high lethality – compared to conditions on Earth – lies in the absorption characteristics of DNA, which is the decisive target for inactivation and mutation induction by UV (Fig. 11.5). Its absorption maximum lies in the UV–C range. If photons of this UV range are absorbed, they induce a variety of photochemical reactions of the nucleic acid bases, such as dimerization of adjacent pyrimidine bases. If not properly repaired, these photoproducts in the DNA lead to mutation or even inactivation of the cells.

11.1.3 Space Vacuum

In free interplanetary space, pressures down to 10^{-14} Pa prevail. Within the vicinity of a body in space, the pressure may significantly increase due to outgassing. In LEO, pressure reaches 10^{-6} to 10^{-4} Pa. The major constituents of this environment are molecular oxygen and nitrogen as well as highly reactive oxygen and nitrogen atoms. In the vicinity of a spacecraft, the pressure further increases, depending on the degree of outgassing. For example, in the Shuttle cargo bay, a pressure of 3×10^{-5} Pa was measured.

If the pressure reaches values below the vapor pressure of a certain material, the material's surface atoms or molecules vaporize. Dehydration is the main process affecting biological samples in outer space. In addition to moisture, sealants, lubricants, and adhesives are the main substances that outgas from spacecraft. However, even metals and glasses can release gases from cracks and impurities. These gas products may condense again onto optical elements, thermal radiators, or solar cells, thereby obscuring them.

11.1.4 Temperature Extremes

In space, the temperature of a body, which is determined by the absorption and emission of energy, depends on its position with respect to the Sun and other orbiting bodies, as well as on its surface, size, mass, and albedo (reflectivity). In LEO, the energy sources include solar radiation (1 360 W m^{-2}), the Earth albedo (480 W m^{-2}), and terrestrial radiation (230 W m^{-2}). Periodically, an orbiting object can be shaded from the sun as it passes on the Earth's night side. Within an orbit lasting 90 min, the spacecraft is exposed to the sun for about 60 min and moves into the Earth's shadow for the residual 30 min. Therefore, in LEO, the temperature of a body can reach both extremely high and extremely low values within 90 min. For the International Space Station (ISS), tolerable temperature limits are defined as +120 °C being the highest value and –120 °C being the lowest.

There are at least five elements that influence the temperature a biological sample experiences in space:

- inclination of the orbit,
- experiment view angle relative to the Sun,
- absence of convection resulting from microgravity conditions,
- extent of conduction, which is influenced by the thermal link between experiment and spacecraft interface, and
- extent of radiation, which is influenced by the vicinity to other payloads on the spacecraft.

11.1.5 Microgravity

Because of the presence of masses in our Universe, true zero gravity does not exist, and even within the ISS at an altitude of 350 km, the level of gravity is only about 8% less than the gravitational field at the Earth's surface. In order to achieve a condition of reduced gravity, the spacecraft has to be accelerated so that the gravitational force and the centrifugal force are equal in amount but with opposite vectors. Actually, the spacecraft is in a state of freefall. The resulting acceleration condition is called "microgravity," because a residual acceleration in the range of 10^{-4} to 10^{-6} *g* cannot be excluded. Compared to the "gravity environment" on Earth, various physical phenomena behave differently in the theoretical state of zero acceleration. Sedimentation, hydrostatic pressure, and convection are linearly proportional to gravity and therefore become zero in weightlessness; however, diffusion still persists in microgravity. Several experimental and technical designs have been developed to study the effect of acceleration on the physiology of living systems and thus identify potential problems during long-term space flight.

11.2 Astrobiology Questions Tackled by Experiments in Earth Orbit

Space technology has opened a new opportunity for life sciences by providing the vehicle to transport terrestrial organic matter or organisms beyond the protective blanket of our atmosphere in order to study *in situ* their responses to selected conditions of space. Since the 1980 s, with the accessibility of *Spacelab*, the free-flying satellite *European Retrievable Carrier* (EURECA), and the Russian *Cosmos* and *Foton* satellites, the European Space Agency (ESA) has provided opportunities for experiments in Earth orbit. By utilizing outer space or special components of this environment as a tool, specific questions of astrobiology were tackled. These studies include but are not limited to:

- the understanding of the relevance of extraterrestrial organic molecules to the delivery of the building blocks of life on Earth or on any other planet;
- the understanding of the role of solar extraterrestrial UV radiation and cosmic radiation in the evolution of potential precursors of life in space;
- the reproduction of processes dealing with the prebiotic and biological evolution on Earth or any other celestial body of astrobiological interest;
- the likelihood of interplanetary transfer and the limiting factors of panspermia, the hypothesis of the distribution of life beyond its planet of origin;
- the simulation of the UV radiation climates of the early Earth or present Mars in order to assess their habitability;
- estimations on the upper boundary of the biosphere; and
- concepts of suprathermal reactions in space.

11.3
Exposure Facilities for Astrobiology Experiments

Following increasing requests from the scientific community to acquire data on the effects of the space environment on biological and material/component samples, the ESA promoted in the early 1990s the development of BIOPAN, a multi-user multi-mission facility conceived for space exposure recoverable experiments of short durations (two weeks). Provided with its own heat shield, a hinged motor-driven lid, and a package of scientific sensors, it has been flown six times already, experiencing only one mishap in 2002 with the *Foton M1* launch accident. It has a very good record of completed experiments, and a valuable amount of data has been collected so far. One more flight is currently planned in 2007 with *Foton M3*.

Another multi-user exposure facility, EXPOSE, is being developed at the time of this writing (2006). It is designed for the outer platforms of the International Space Station (ISS) and supports long-duration experiments (one to three years). Two attachment points are foreseen for EXPOSE:

- one of the external platforms of the European *Columbus* module as part of the European Technology Facility (EUTEF) and
- the URM-D platform, an external ISS facility at the Russian *Zvezda* module.

A more intensive utilization of these facilities is envisaged in the future in support of the new ESA Exploration Program.

11.3.1
BIOPAN

BIOPAN is based on the heritage of a low-tech Russian exposure container called KNA (Kontéjner Naúchnoj Apparatúry: Container of Scientific Equipment). KNA, used by the ESA for radiobiological experiments on three Russian BION missions in 1987, 1989, and 1992, was a passive experiment container with a hinged lid. The lid was open during launch – with the experiments protected only by the nose fairing of the launcher – and during orbital flight. Before landing, the spring-loaded lid was closed so that the experiments were safely stored during reentry.

The lessons the ESA learned from KNA directly resulted in the concept of BIOPAN. Although some basic characteristics of KNA were maintained (a pan-shaped structure, a movable lid, experiment positions in the bottom part and in the lid), BIOPAN is much more advanced in terms of versatility, controllability, size, experiment accommodation, and data acquisition. In BIOPAN, the lid is opened and closed by telecommand, housekeeping status data are down-linked during the flight, the space environment is monitored by a set of integrated sensors, and unwanted temperature excursions in orbit can be countered by active means (heaters), by passive means (insulation blankets), or by closing the lid.

BIOPAN hosts exobiology, radiobiology, and materials science experiments, aimed at evaluating the combined or individual effects of solar UV light, space radiation, vacuum, extreme temperatures, and microgravity on biological samples, materials specimens, and electronic components.

After a successful test flight in 1992, BIOPAN completed four operational missions in

- 1994 (*Foton 9*),
- 1997 (*Foton 11*),
- 1999 (*Foton 12*), and
- 2005 (*Foton M2*).

In 2002 a flight unit (flying as BIOPAN-4) was lost in the Soyuz launch failure of *Foton M1*. A seventh flight with an upgraded version of BIOPAN is scheduled for late 2007, onboard the Russian *Foton M3* satellite.

11.3.1.1 Technical Characteristics of BIOPAN

BIOPAN is a 27-kg experiment container, designed to fly with Russian spacecraft of the *Foton* class (Fig. 11.6). The first *Foton* was successfully flown in 1985. The satellite, typically launched with a Soyuz *U* vector, consists of three modules:

- the descent module,
- the battery pack, and
- the attitude control module.

The descent module is a pressurized spherical capsule where most of the scientific payloads are accommodated. *Foton* is capable of carrying up to 600 kg of scientific

Fig. 11.6 *Foton 12* spacecraft carried to the Soyuz launcher, in the MIK integration facility at Plesetsk Cosmodrome (Russia) in September 1999 (photo credit: ESA).

payload, with an average daily power budget available to the scientific payloads of 400 W (on a typical two-week mission). The residual acceleration levels onboard are between 10^{-4} g and 10^{-6} g. Protected by a heat shield, the pressurized spherical capsule is the only part of the satellite returning to Earth at the end of the mission. The payload is then retrieved and returned to the investigators within two days after landing.

BIOPAN is mounted onto the external surface of the descent module (Fig. 11.7), protruding from the thermal blanket, which is wrapped around the satellite shortly before launch. Power and data links are provided via cables whose connectors are strapped away at the separation of the satellite modules during atmospheric reentry. An internal battery provides energy to BIOPAN subsystems from the reentry phase onwards.

The upper shell of the container is a motor-driven lid, which, once in orbit (typically a few hours after liftoff, depending on the experiment requirements),

Fig. 11.7 Installation of BIOPAN onto the descent module of the satellite *Foton*, during the pre-launch operations (photo credit: ESA).

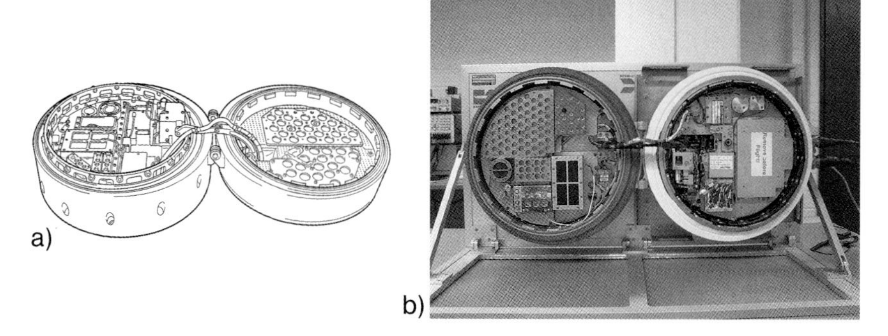

Fig. 11.8 Configuration of BIOPAN. Drawing (a) and photo (b) of BIOPAN fully open with the experiment hardware integrated, before flight (courtesy of ESA).

turns open at 180°, allowing the experiment samples be exposed to the space environment (Fig. 11.8).

At the end of the mission the lid is closed again by telecommand sent from ground, although a pre-programmed internal backup command ensures the proper execution in case of communication problems. In nominal operations, the lid remains open throughout the entire mission and is closed just before reentry of the capsule. To withstand the extreme heat during reentry, the entire BIOPAN structure is protected by an ablative heat shield of the same kind used for the capsule itself. For the safety of the facility and its closing device, the closure of the lid also may occur in orbit in case of temperatures exceeding the nominal range.

BIOPAN is provided with a variety of sensors to monitor and record the environmental history of the mission. Included are a UV sensor, a radiometer, an active radiation dosimeter, and a set of eight thermistors to measure the experiment temperatures at different locations. The sensor data are stored onboard during flight. Electrical heaters and experiment-customized thermal blankets (multilayer insulators) are used to keep the biological specimen within pre-determined temperature ranges and to provide minimum shielding against soft cosmic radiation.

11.3.1.2 Experiment Hardware Accommodated within BIOPAN

The experiment packages – up to a total mass of 4 kg, a maximum height of 50 mm, and a total available surface area of 1 100 cm^2 – can be individually designed by the investigators. They are installed on two mounting plates, one in the bottom part and one in the lid (Fig. 11.8). The research covers different scientific and technical fields, from astrobiology (see Section 11.4) and materials science to components tests and validation. In Tables 11.2 to 11.4, experiments are listed that so far have been flown on BIOPAN/*Foton* missions.

Table 11.2 *Foton*/BIOPAN missions since 1992.

Mission	Year	Payload
Foton-8/ BIOPAN-0	1992	Test flight with non-ESA-sponsored experiments
Foton-9/ BIOPAN-1	1994	6 experiments: Base, Shrimp, Mapping, Survival, Vitamin, Dust
Foton-10	1995	
Foton-11/ BIOPAN-2	1997	6 experiments: Base, Shrimp, Mapping, Survival, Vitamin, Dust
Foton-12/ BIOPAN-3	1999	4 experiments: Survival, Yeast, Dosimap, Vitamin
Foton-M1/ BIOPAN-4	2002	9 experiments – lost (launch failure)
Foton-M2/ BIOPAN-5	2005	9 experiments: Marstox, Letvar, Photo I, Rado, Lichens, Organics, Yeast II, R3D-B, Permafrost
In preparation:		
Foton-M3/ BIOPAN-6	2007	10 experiments: R3D-B2, UV-olution, Highrad, Yeast III, Life, LMC, Tardi A & B, Lithopanspermia, Marstox II, Rado II

Table 11.3 Experiments flown in BIOPAN in the period 1992–1999.

Research topic	Principle investigator
Radiation dosimetry	
Radiation measurements behind defined shieldings	G. Reitz, DLR, Cologne
Depth dosimetry	L. Adams, ESTEC, Noordwijk
Depth dosimetry	Yu. Akatov, RCSRS, Moscow
Neutron dosimetry	V. Dudkin, RCSRS, Moscow
UV dosimetry	V. Tsetlin, IBMP, Moscow
HZE and LET measurements	A. Vikhrov, RCSRS, Moscow
HZE and LET measurements	E. Benton, USA
Radiation biology	
Radiation damage in plant seed DNA bases *in situ*	J. Cadet, CEA, Grenoble
Radiation damage in embryos of the brine shrimp	A. Hernandorena, CERS, Biarritz
Effects of soft radiation in yeast cells	J. Kiefer, University of Giessen
Efficiency of radioprotecting drugs	J-P. Moatti and N. Dousset, Un. P. Sabatier
Astrobiology	
Effects of solar UV on peptides in artificial dust grains	A. Brack, CNRS, Orléans
Survival rate of biological organisms	G. Horneck, DLR, Cologne
Materials and components validation	
Bit flips in EEPROM cards exposed to space radiation	R. Harboe Sørensen, ESTEC
Stability of coating materials outside exposed in space	M. van Eesbeek, ESTEC
Stability of thermal-control coating materials exposed in space	E. Overbosch, Fokker Space, Leiden

Table 11.4 Experiments flown in BIOPAN 4 and 5 (2002 and 2005).

Experiment	Principal investigator
Marstox (D)	P. Rettberg and G. Horneck, DLR
Yeast II (D)	J. Kiefer and M. Loebrich, Saarland University
Photo I (I)	M.T. Giardi, CNR Rome
Organics (NL)	P. Ehrenfreund, Leiden University
R3D-B (D)	D. Haeder, Erlangen University
Lichens (E)	L. Sancho, Madrid University
Letvar (D, A)	N. Vana, Wien University
Permafrost (Ru)	D. Gilichinsky, RAS
Rado (Ru, Hu)	V. Shurshakov and J. Palfalvi, IBMP

The experiment temperature profile in orbit is selectable: either a non-controlled mode with temperatures freely oscillating between $\leq -35\,°C$ and $\geq +10\,°C$, in synchrony with the alternating periods of solar illumination and shadowing; or a controlled, stable temperature with a fixed set point in the range of 10–25 °C using heaters and dedicated thermal insulators.

The experiments flown in the past five BIOPAN missions have been accommodated in sample holders specifically designed by the scientific teams under ESA supervision. In addition to complying with the carrier interfaces, they are required to pass a number of qualification tests, which include a vibration test according to the *Foton* specification levels and a thermocycling test performed in a thermovacuum chamber.

The typical experiment environment offered by BIOPAN includes:

- a microgravity level of $<10^{-5}\,g$;
- a radiation level up to 6 Gy (600 rad) per day and up to 80 Gy (8 000 rad) per 15-day mission (depending on solar activity and shielding; see Section 11.1);
- solar electromagnetic radiation on the order of 40 solar constant hours per mission, and
- the possibility of controlled temperature profiles at those experiment packages that are thermally decoupled from each other.

At flight completion, when BIOPAN is carried back from the landing site to the ESA ESTEC clean room and opened in a controlled environment, these packages are dismounted, inspected, and returned to the investigators, who examine their contents in their own laboratories around Europe. Reference samples are stored in BIOPAN under identical conditions but shielded against UV. Additional control samples are maintained on the ground under controlled terrestrial conditions.

11.3.1.3 Operational Aspects of BIOPAN

The typical operational cycle of BIOPAN includes satellite interface tests three months prior to launch (L), experiment integration at $L-1$ week, two weeks of orbital flight, and return of the experiments to the investigators four days after landing. The main figures of a typical *Foton*-BIOPAN mission are shown in Table 11.5. Parameters such as flight duration, altitude, and energy available for the payload can be modified depending on the chosen mission profile and on the upgrading of the satellite (e.g., for the *Foton M* version).

Table 11.5 Main figures of a typical BIOPAN mission onboard a *Foton* satellite.

Flight parameter	Flight data
Flight frequency	Once every two years
BIOPAN's models	1–3
Flight duration	14–16 days
Altitude	220–390 km
Inclination	62.8°
Energy available (15 days)	3 kWh (164 Ah)

Integration of BIOPAN and interface testing with *Foton* take place approximately three months before the launch at the *Foton* construction site in Russia. At this stage, the manufacturing of the satellite and of its subsystems is completed, and all the necessary electrical and mechanical interface tests between carrier and payloads need to be carried out. The BIOPAN flight unit performances and its interfaces with *Foton* are verified through a sequence of mechanical and electrical tests. Dedicated electronic ground support equipment is used to support these activities.

During these tests, the satellite undergoes a simulated mission profile in which all the various commands are issued and each subsystem is checked. These tests are meant to qualify the satellite for the mission, to control the readiness of each payload, and to verify possible mutual interference between the carrier and scientific instruments. After successful completion of these tests, BIOPAN is dismounted, packed, and transported back to ESTEC. This operation is necessary to allow the integration of the biological samples at a very late stage at ESTEC, as demanded by the nature of these experiments. In contrast, most of the other scientific payloads remain inside *Foton* and are transported with the satellite to the launch site (Plesetsk Cosmodrome or Baikonour Cosmodrome).

The sensitivity of the biological samples – often including yeast cells, spores, bacteria, and plants seeds – to storage effects such as temperature variation, humidity, and/or aging requires their preparation in well-controlled environmental conditions shortly prior to launch. They are prepared in the investigators' laboratories up to 10 days in advance and delivered to ESA-ESTEC in thermally

Fig. 11.9 Final assembly of *Foton* with the Soyuz nosecone. BIOPAN is visible on the top of the descent module (photo credit: ESA).

controlled containers. They are integrated into BIOPAN one week before launch. This solution is preferred to an integration "onsite" at Plesetsk or Baikonour because of the lack of dedicated equipment and suitable laboratories there.

At $L - 4$ days, an ESA team carries the BIOPAN flight unit with the samples from ESA-ESTEC to the launch site, using a special Russian charter plane. Contained in a special battery-powered thermal container until the last minute, BIOPAN is finally installed onto the capsule at $L - 3$ days, a few hours before the final closing of the *Foton* hatches.

The spacecraft is then ready to be integrated with the launcher. Charged with its complete payload complement, and a mass of 6.5 tons, *Foton* is lifted from its supporting stand, lowered into the fairing coupling ring, and closed within the two half shells of the Soyuz nosecone (Fig. 11.9). In this configuration, it is then tilted horizontally and coupled to the third stage of the rocket, before being lifted up and launched.

Once in orbit, the telecommand to open the lid of BIOPAN is issued from the ground at a time that can be decided based upon the experiment needs (typically between 10 min and 20 h after liftoff).

In orbit, the position and the status of the *Foton* satellite and of BIOPAN are monitored via telemetry. The telemetry data include the main parameters, representative of the BIOPAN status (position of the lid, power and energy budgets), while the other housekeeping and scientific data from the sensors are stored onboard and analyzed after the mission (Fig. 11.10). The lid closure is again commanded from the ground approximately one day before landing.

After completion of the mission, the *Foton* satellite is de-orbited when its ground track is favorable for a landing within the Russian territory or in Kazakhstan.

The capsule reentry is assisted with a parachute system and a retro-rocket that decrease the landing shock down to values on the order of 35–40 g. This system is activated just a couple of meters before the final touchdown, via a gamma-ray source installed in the heat shield. Every payload, including BIOPAN, has to be

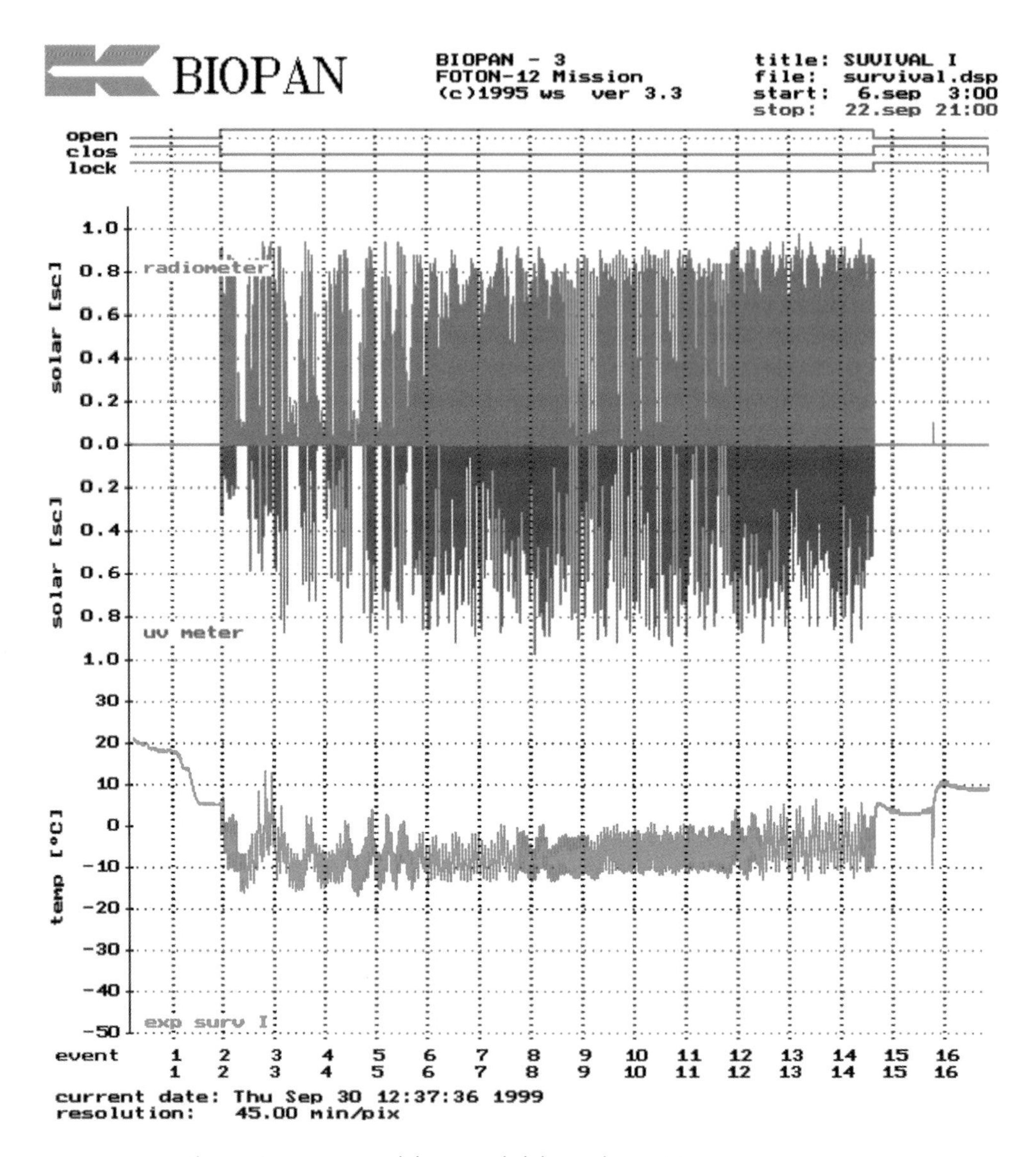

Fig. 11.10 Environmental data recorded during the BIOPAN-3/*Foton 12* mission. Top: Time profile of the radiation exposure measured by radiometer and UV meter. Bottom: Temperature data recorded in the lid of BIOPAN for the experiments without temperature control (courtesy of ESA).

Fig. 11.11 *Foton* descent module with BIOPAN at the landing site (photo credit: ESA).

designed to withstand a possible landing shock of 90 g (duration: 40–50 ms), which might occur in the case of retro-rocket malfunctioning.

After separation of the descent module from the battery pack (some 30 min before final landing), the main power is cut off from the instruments, which means that thermal control is no longer assured. Rescue teams reach the landing site with helicopters from the closest military base.

A prompt finding of the satellite in the Russian-Kazakh steppes (not always easy, especially with unfavorable weather conditions) and an early recovery of the experiments is fundamental for the mission success of BIOPAN (Fig. 11.11). BIOPAN is then dismounted from the burnt satellite heat shield and, still closed and locked, placed in its cooled transport container and carried back to Samara. Here, the mission data are downloaded and the payload is then transported back to ESA-ESTEC the day after. The opening of BIOPAN and experiment de-integration and delivery to the investigators are performed in an ESA-ESTEC clean room within four days after landing.

11.3.1.4 Orbital Characteristics of a BIOPAN Mission

The orbital parameters selected for *Foton*, the carrier spacecraft of BIOPAN, have not noticeably changed since its first flight in 1992 (Table 11.6). With the introduction of *Foton-M* in 2002, the orbital trajectory became slightly more circular in order to optimize the microgravity levels. From *Foton-M2* onwards, a different launch site is used (Baikonur rather than Plesetsk), but this has almost no effect on the orbital inclination. The consistency of the orbital parameters enables the BIOPAN investigators to repeat their experiments if necessary under conditions that – apart from the sun cycle events – are remarkably similar. It also means that the experiment environment for upcoming flights may largely be predicted from the recordings made during previous missions.

Table 11.6 Orbital parameters, flight duration, and launch site of the BIOPAN missions.

Mission	Spacecraft	Apogee	Perigee	Inclination	Duration	Launch site
BIOPAN-0	*Foton-8*	383	228	62.8°	15.6 d	Plesetsk
BIOPAN-1	*Foton-9*	385	227	62.8°	17.6 d	Plesetsk
BIOPAN-2	*Foton-11*	385	227	62.8°	13.6 d	Plesetsk
BIOPAN-3	*Foton-12*	405	225	62.8°	14.6 d	Plesetsk
BIOPAN-4[a]	*Foton-M1*	304	262	62.8°	15.6 d	Plesetsk
BIOPAN-5	*Foton-M2*	304	262	63°	16 d	Baikonur
BIOPAN-6	*Foton-M3*	304	262	63°	13 d	Baikonur

a Orbital parameters and flight duration of BIOPAN-4 as announced before the crash.

11.3.1.5 Environment of BIOPAN Experiments

Temperature, pressure, and solar irradiation at the experiment site are monitored by an integrated set of BIOPAN sensors. The data are stored onboard during the mission and can be accessed after the flight. Irradiation by cosmic rays is monitored by investigator-provided detectors. The microgravity environment is not measured inside BIOPAN itself but inside the *Foton* capsule by ESA-provided sensors.

The temperature history of the experiment samples in orbit depends on the amount of thermal insulation (single foils or multilayer blankets), the thermal coupling to the BIOPAN carrier plates, and the use of heaters. It was observed that a non-insulated experiment decoupled from an unheated carrier plate may experience extreme temperature fluctuations from –40 °C to +60 °C. In contrast, an insulated experiment that is tightly coupled to a heated carrier plate can be maintained at a stable temperature, selectable between +15 °C and +25 °C (Fig. 11.10).

The possibility of overheating during atmospheric reentry at the end of the mission was recognized early during the development of BIOPAN. Therefore, a quite massive heat shield was designed for BIOPAN. While the total weight of BIOPAN is close to 27 kg, including the experiments, the heat shield is responsible for 12 kg of that figure. As a result, the peak temperatures during reentry and after landing recorded inside BIOPAN never exceeded +30 °C.

BIOPAN is also equipped with a pressure sensor for rough monitoring of gas release from BIOPAN (evacuation) during ascent.

Irradiation by solar light (including UV radiation) is measured by two types of sensors. A radiometer monitors the input of solar electromagnetic radiation over the entire spectrum from UV to IR. Special UV sensors are added to monitor the solar wavelength ranges that are specific to orbital flight. (At the surface of the Earth, there is no incidence of solar UV–C and only a limited incidence of UV-B; see Fig. 11.5.) On the four missions completed so far, the total dose of sunlight has varied from 8.2 kJ cm^{-2} to 20.4 kJ cm^{-2} (Table 11.7). The total dose largely depended on the time when the lid of BIOPAN was open in orbit. This period varied from 7.8 to 14.8 days (Table 11.7). As a consequence, the daily dose varied from 0.99 kJ cm^{-2} to 1.38 kJ cm^{-2}, which corresponds to solar constant hours of 2.02–2.80. On the four missions completed so far, the total dose of specific energy deriving from solar electromagnetic radiation varied from 8.2 kJ cm^{-2} to 20.4 kJ cm^{-2}. Some BIOPAN flights performed measurements of UV–C, UV-B, and UV-A radiation in addition to the solar radiation sensors that integrate over the whole solar spectrum.

Table 11.7 Data of solar irradiation during different BIOPAN missions.

Mission	Lid open	Total dose	Daily dose	
BIOPAN-0	187 h = 7.8 d	8.2 kJ cm^{-2}	1.05 kJ cm^{-2}	2.14 SCh[a]
BIOPAN-1	355 h = 14.8 d	20.4 kJ cm^{-2}	1.38 kJ cm^{-2}	2.80 SCh
BIOPAN-2	239 h = 9.9 d	12.9 kJ cm^{-2}	1.30 kJ cm^{-2}	2.65 SCh
BIOPAN-3	302 h = 12.6 d	12.5 kJ cm^{-2}	0.99 kJ cm^{-2}	2.02 SCh

a SCh = solar constant hours.

The dose received from cosmic radiation is measured in different radiation dosimetry experiments of BIOPAN. Depending on the level of shielding, the experiments in BIOPAN may absorb a radiation dose up to 5.6 Gy per day, which is four orders of magnitude higher than the daily dose received inside the *Foton* capsule or inside the ISS. However, the actual radiation dose absorbed by the experiment samples in BIOPAN is generally much smaller than this value because of their containment. A small change in shielding density may create a dramatic difference. Over the first 0.05 g cm^{-2} of shielding mass, the radiation dose falls steeply down by one order of magnitude, over the first 0.25 g cm^{-2} by two orders of magnitude, and over the first 1 g cm^{-2} by three orders of magnitude.

The microgravity environment was measured by ESA on *Foton-12*, the spacecraft that carried BIOPAN-3. The quasi-steady acceleration level was below 10^{-4} g over all three orthogonal axes.

11.3.2 STONE

The STONE experiments aim to study the physical, chemical, and biological modifications caused by atmospheric entry in rocks, in meteoritic materials, and in their possible embedded organisms. For this purpose, different types of rock, loaded with microorganisms, are mounted in the heat shield of the *Foton* reentry capsule. During reentry into the atmosphere at the end of the two-week flight, the STONE rock samples are subjected to temperature and pressure loads comparable to those experienced by meteorites.

The rationale for the STONE experiments is as follows (see also Chapter 1): most of the 25 000 known meteorites originate from the asteroid belt. There are 18 meteorites that definitely come from the Moon; most are regolith samples, as would be expected when rocks are propelled from planetary surfaces by impact. Then there are the 37 so-called SNC meteorites, which are quite distinct from asteroidal debris and which come from Mars (see Chapter 8), but none of them is a surface sample. Because Mars once had a warm and wet climate, its surface must be covered by both impact-generated regolith and sedimentary rocks deposited by running and/or still water. The sedimentary rocks should comprise detrital deposits as well as chemical sediments such as evaporites. In addition, groundwater can compact loose sediments and regolith by filling the pore spaces with evaporitic minerals. Such consolidated sedimentary hard rocks should be among the Martian meteorites, but they are not. It is possible that they survived their traumatic escape from Mars but not their terrestrial atmospheric entry because of decrepitation in the cementing mineral (very likely to be a sulfate). Therefore, STONE investigated whether such rocks would be destroyed or damaged by atmospheric infall, although *Foton*'s reentry speed of 7.8 km s^{-1} is considerably lower than that experienced by a meteorite (20–70 km s^{-1}).

The STONE experiments take advantage of the accommodation on the heat shield of the reentry capsules of the Russian retrievable carriers of the *Foton* class. They are normally placed around the descent module's hottest point (stagnation point), inserted into the heat shield with specially designed holders that are made of the same ablative material used for the heat shield (Fig. 11.12). Because the rock sample's thermal conductivity is generally of some order of magnitude higher than that of the capsule shielding, a protective layer is placed underneath the exposed sample, between the metallic structure of the capsule and the rock sample itself.

STONE-1 was conducted in 1999 on *Foton-12*. Three different types of rock were fixed into the ablative heat shield of the *Foton* reentry capsule:

- a basalt,
- a dolomite (sedimentary rock), and
- a simulated Martian regolith sample.

The goal was to investigate why all Martian meteorites so far collected on Earth were exclusively of the igneous type, while the surface of Mars consists, apart from igneous rocks, largely of sedimentary materials. The experiment provided a pos-

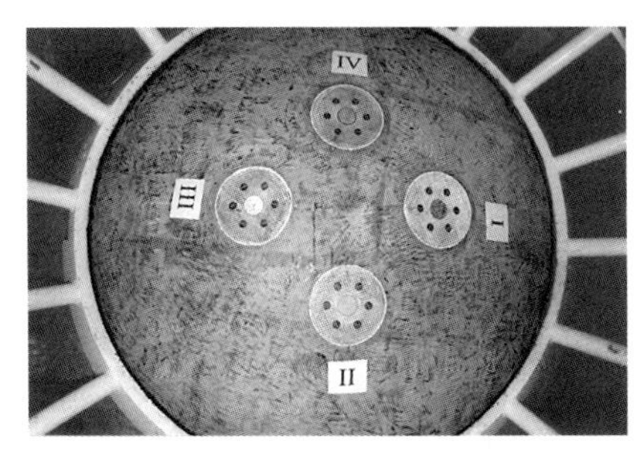

Fig. 11.12 STONE-5 samples installed onto the *Foton M2* heat shield. Picture taken before the mating of the descent capsule with the attitude control module (photo credit: ESA).

sible explanation for this dilemma: it showed that a sedimentary rock (the dolomite) can survive the atmospheric entry process, but after landing, it will easily decompose from the chemical and physical changes it has suffered.

Follow-on experiments were prepared in 2000 (STONE-2), 2001 (STONE-3), and 2002 (STONE-4), but unfortunately none was completed. STONE-2 failed because its carrier was heavily damaged during reentry and touchdown, whereby the rocks were lost. STONE-3 was never launched because of programmatic changes. STONE-4 was carried by the *Foton-M1* capsule, which was lost when the launcher rocket crashed shortly after liftoff.

The STONE-5 experiment was partly a repetition of STONE-1. A sandstone (a sedimentary rock) and, for comparison, a dolerite (an igneous rock) were exposed to the reentry environment. A new aspect of this experiment is the emphasis on biology: both rocks were loaded with microorganisms. This allowed the scientists to investigate the chances for survival of simple life forms transported by meteorites when entering the atmosphere of the Earth. The third rock from STONE-5 was a gneiss (a metamorphic stone) from Antarctica carrying endolithic cyanobacteria (desiccated, vegetative cells).

11.3.3 EXPOSE

The EXPOSE facility, conceived to support astrobiology and radiation biology research in space, was designed and developed by ESA to fly on the external platforms of the ISS, originally on the *Columbus* external balconies. EXPOSE will host experiments prepared by scientists from various European research institutes. Experiment specimens will be located in hundreds of tiny cells, pressurized or vented, protected with windows and filters of various geometries and materials. Most of the selected experiments deal with understanding the influence of the radiation environment, in all its different components, on selected biological samples and organic compounds. Some of them are part of a research program that includes among its targets the understanding of organic matter's survivability of extraterrestrial conditions, such as on Mars.

The EXPOSE development started in 2001. After undergoing a full qualification program, in 2004 the Flight Model passed its flight acceptance review and was delivered to *Columbus* for integration (EXPOSE-E). Because of delays in the launch of *Columbus*, ESA investigated in 2004 the possibility of flying EXPOSE on the Russian segment of the ISS (RS-ISS), mounted on a recently installed external platform (named URM-D), in order to provide European scientists with an earlier and strongly demanded flight opportunity (EXPOSE-R).

11.3.3.1 EXPOSE Facility

The EXPOSE facility includes three experiment trays (Fig. 11.13), each one containing four sample carriers; some of them are sealed and allow accommodation of pressurized containers, while others are vented via a venting system integrated in

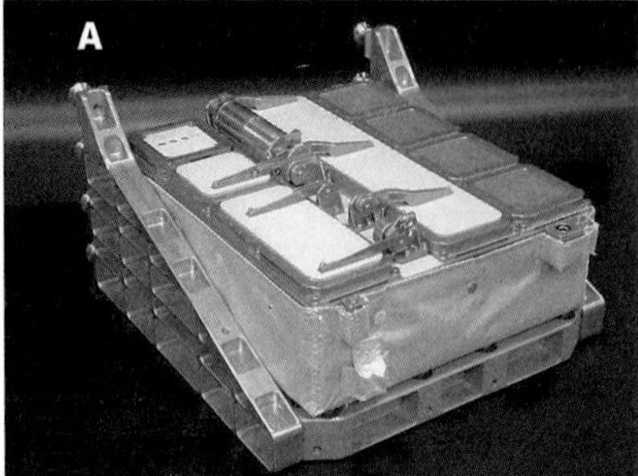

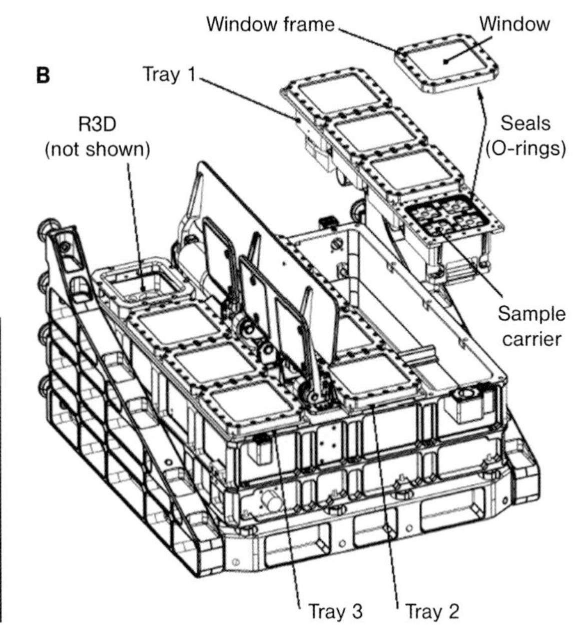

Fig. 11.13 EXPOSE facility, the external facility of the ISS. (A) Photo of EXPOSE-E; (B) drawing of EXPOSE assembly (courtesy of ESA).

the tray. The air channel is sealed with a dedicated valve, which opens after the facility is brought out to the external surface of the ISS by a dedicated control command.

Each sample carrier is structured in two or three layers (for sample exposure to solar radiation and for dark control) and accommodates a variety of sample cells of different size and geometry (from 16 to 64 per layer), each one equipped with different kinds of optical windows, filters, and dosimeters. All biological samples, such as bacterial cells or spores, fungal spores, fern spores, or plant seeds, are deposited in dry layers on quartz or MgF_2 disks. Pressure integrity is provided by means of a rubber gasket between the metallic tray housing and the glass.

The EXPOSE experiments are designed for one to two years' external exposure to outer space. According to the EXPOSE experiment procedure, preparation of the biological samples, deposition on quartz glasses, drying and accommodation in the carriers, placement of the carriers in the trays, and final assembly of the EXPOSE facility will be carried out at ground laboratories.

In addition to the trays, the facility accommodates various sensors that measure parameters characterizing the experiment conditions (solar radiation, temperature, etc.) and an electronic unit that acquires, formats, and transfers these data to the ISS service systems.

The EXPOSE assembly includes the EXPOSE facility and its supporting structure, interfacing with the EUTEF-CEPA (Columbus External Platform Adapter) for

EXPOSE-E or with the external platform (named URM-D) of the RS-ISS for EXPOSE-R.

The EXPOSE-E facility provides common functionalities and services and a controlled environment to the experiment samples (shutters, scientific and housekeeping data acquisition and transmission, thermal control). It includes the following subsystems:

- structure and mechanisms;
- shutters (including lids, hinges, shutter drives);
- trays (including windows, seals, valves, connectors);
- sample carriers (including windows, filters, controls);
- EXPOSE control unit (motor and valves control, data acquisition);
- sensors (temperature, pressure, UV, proximity); four sensor packages (UV-B and radiometer) are positioned on the four facility corners;
- a thermal control system, based on heaters, multilayer insulators (MLI), and coatings;
- an electrical interface (I/F) to EUTEF Data Handling and Power Unit; and
- a mechanical I/F to EUTEF-CEPA (I/F bracket).

The EXPOSE actuators are controlled by software via the experiment control unit micro-controller; they can also be operated via the telemetry/telecommands (TM/TC) link. In particular, the micro-controller provides the following functions:

- acquisition and transmission of housekeeping data and data from eight temperature and five solar sensors;
- acquisition and transmission of R3D data;
- data storage (in case of data link interruption);
- command and control of three motor-driven lids (shutters) by stepper motors;
- command and control of three motor-driven valves for vented experiment compartments;
- thermal control of experiments achieved with two different sets of heaters, thermostat controlled, positioned underneath each experiment tray and on the inner side of the facility frame; and
- overheat protection via automatic lid closure whenever a predefined temperature is exceeded.

11.3.3.2 EXPOSE Experiments

The experiments selected for the flight of EXPOSE-R are listed in Table 11.8. The experiments selected for the flight of EXPOSE-E with *Columbus* are listed in Table 11.9.

Table 11.8 Experiments selected for EXPOSE-R.

Principal investigator	Experiment	Topic of research
C. Cockell Open University (UK)	ROSE-1/ENDO	To assess the impact of extraterrestrial UV radiation on microbial primary producers (algae, cyanobacteria)
R. Mancinelli SETI Institute, NASA Ames (USA)	ROSE-2/OSMO	To assess the protective effects on osmophilic microorganisms enclosed within gypsum-halite crusts
G. Horneck DLR (D)	ROSE-3/SPORES	To assess the protection of spores by meteorite material against space conditions: UV, vacuum, and ionizing radiation
J. Cadet C.E. N. G. (F)	ROSE-4/PHOTO	To determine the photoproducts resulting from exposure of dry DNA samples or bacterial spores to solar UV radiation
N. Munakata University of Tokyo, Japan	ROSE-5/SUBTIL	To determine the mutational spectra of *Bacillus subtilis* spores induced by space vacuum and/or solar UV radiation
G. Rontó Research Lab. for Biophysics, Budapest (HUN)	ROSE-8/PUR	To determine the biologically effective dose of solar extraterrestrial UV radiation by biological dosimetry
A. Brack CNRS- Orleans (F)	AMINO	To study photochemical processing of amino acids in Earth orbit
P. Ehrenfreund Leiden Observatory (NL)	ORGANICS	To study the evolution of organic matter in space
D-P. Häder University of Erlangen (D)	R3D	Radiation dosimetry

Table 11.9 Experiments selected for EXPOSE-E.

Principal investigator	Experiment	Topic of research
G. Horneck DLR (D)	PROTECT	To determine the resistance of spacecraft isolates to outer space for planetary protection purposes
D. Tepfer CNRS, Versailles (F)	SEEDS	To test plant seeds as a terrestrial model for a panspermia vehicle and as a source of universal UV screens
H. Cottin University 5, Paris	PROCESS	To study prebiotic organic chemistry on the ISS
P. Rettberg DLR (D)	ADAPT	To study molecular adaptation strategies of microorganisms to different space and planetary UV climate conditions
S. Onofri Universita'degli studi della Tuscia di Viterbo (I)	LIFE	To study the resistance of lichens and lithic fungi in space conditions
D-P. Häder University of Erlangen (D)	R3D-2	Radiation dosimetry

11.3.3.3 EXPOSE Experiment Hardware

The experiments of EXPOSE (Tables 11.8 and 11.9) are accommodated in three experiment trays. They are equipped with four squared recesses, 77 × 77 × 26 mm each, where the sample carriers are inserted. These are lightweight elements housing a number of sample cells whose quantity, geometry, and characteristics depend on the experiment typology. Each tray hosts four carriers (Fig. 11.14) sealed with optical windows or kept open to the space environment. The trays have a mechanical interface to the structure and comply with the design of the lids. The trays are electrically connected to the structure by matable/dematable connectors.

The sample carriers are designed to accommodate a number of optical filters (neutral density and/or cutoff) to control wavelength and amount of incident light; they are also complemented with passive radiation dosimeters, such as thermoluminescent detectors. Two or three control layers are foreseen in each carrier.

In the flight of EXPOSE-R, tray 1 is dedicated to the AMINO and ORGANICS experiments (Table 11.8). It is the only one not equipped with shutters, and its sample carriers are not covered with windows. It hosts pressurized and vented sample cells, protected with 1- or 2-mm thick MgF_2 windows (8 mm in diameter) (Fig. 11.15A). Both experiments have a dark control layer underneath the exposed set. Peculiarities of this tray include temperature sensors inside some cells (in contact with the inner window) and innovative design concepts for the pressurized cells of the two experiments.

Trays 2 and 3 (Fig. 11.15B, C) are dedicated to the samples of the so-called ROSE (Response of Organisms to Space Environment) consortium, a team of biologists whose experiments are part of a coordinated research effort (Table 11.8). The trays host vented and sealed units, but here the sample carriers are always closed with an 8-mm thick MgF_2 window.

The pressurized compartments carry experiments with a predefined gaseous environment, provided during the integration of the scientific payload. This atmosphere has to be maintained during all mission phases (transportation to the launch site, storage, trays' integration into EXPOSE-R, launch, in-orbit operations, retrieval, landing, and transportation to the de-integration site).

The venting of the other compartments is achieved with motor-driven venting valves that allow depressurization once in orbit and upon reception of a command issued from the ground. Every EXPOSE-R tray is provided with this system, but only specific sample carriers are connected to it.

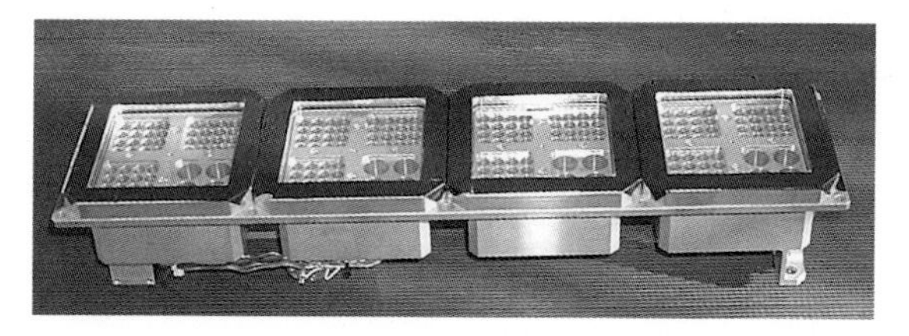

Fig. 11.14 One of the three trays of EXPOSE; photo of the Engineering Model (courtesy of ESA).

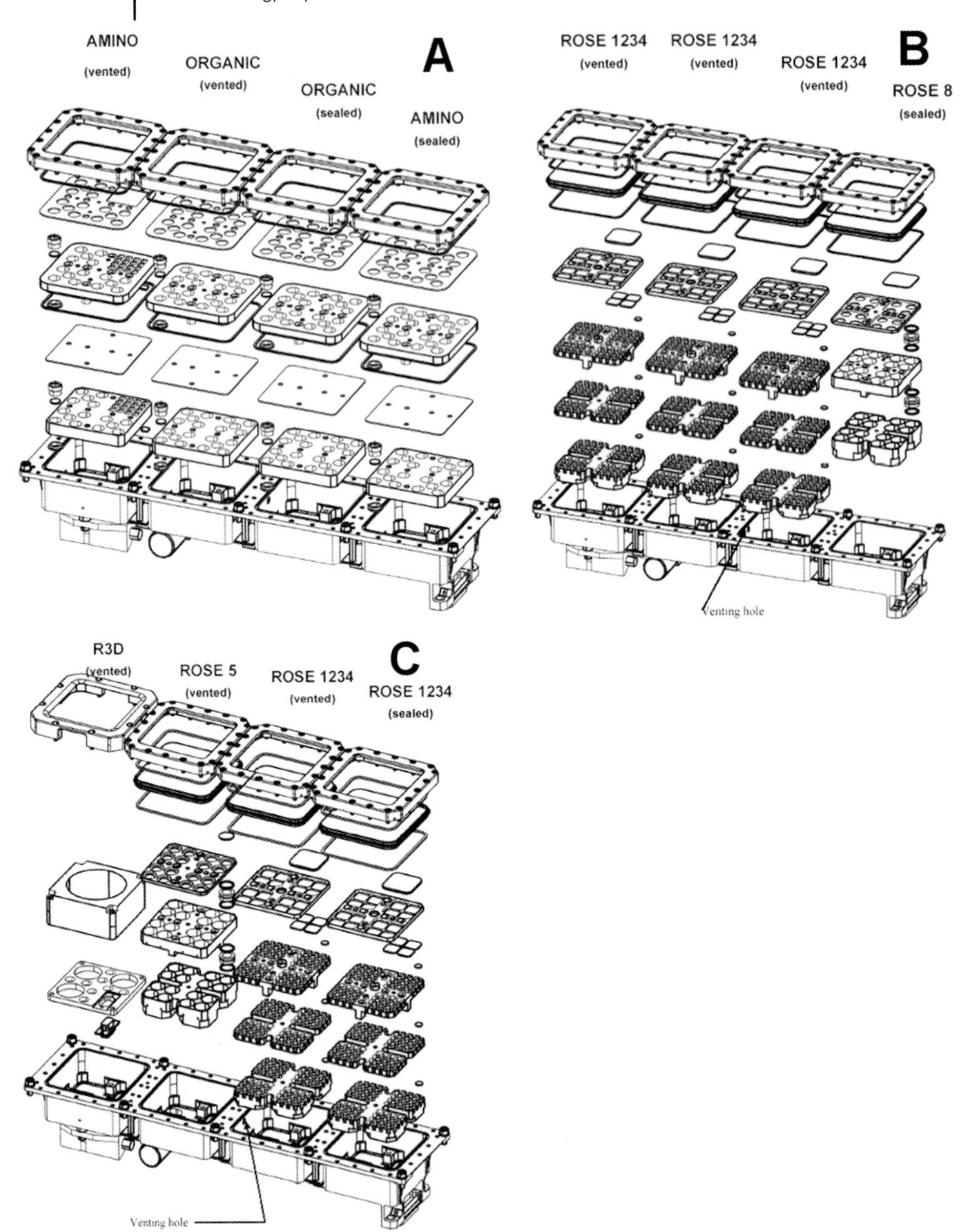

Fig. 11.15 Arrangement of the subunits of the EXPOSE trays according to the different experiment requirements. (A) Tray for the experiments AMINO and ORGANICS; (B) tray for the experiments ROSE 1, 2, 3, 4, and 8; (C) tray for the experiments ROSE 1, 2, 3, 4, 5, and R3D (courtesy of ESA).

Because the only access to the sealed compartments is through the removal of the glass window, integration of the scientific payload has to be done inside a glove box, in sterile or environment-controlled conditions.

11.3.3.4 EXPOSE-R and EXPOSE-E

For EXPOSE-R, the EXPOSE Engineering Model, used for the design and qualification of the EXPOSE Flight Model (called EXPOSE-E), has been upgraded to flight standards and made compatible with the interfaces and the operational aspects of the external platforms of the Russian segment of the ISS (Fig. 11.16). The target launch date is October 2007, with an unmanned *Progress* cargo from Baikonour. EXPOSE-E will fly with *Columbus*.

For both missions, the integration of the facility on the external platforms (and the removal of the experiment samples at exposure conclusion) will be done via extravehicular activity, but the operational scenarios will be quite different. In EXPOSE-R, the experiment trays will be removed at mission completion (planned after one year in space), put in sealed transport containers, and, after a temporary storage inside the RS-ISS, will be downloaded to the Earth onboard a Soyuz reentry capsule. The facility will remain outside the ISS and might then be "refilled" with a new set of trays uploaded with the next *Progress* cargo. Such an "extended" mission is actually envisaged by the astrobiology scientific community in order to use EXPOSE as test bed for experiments, sensor packages, and new technologies to be developed in the frame of the upcoming *ExoMars* mission (see Chapter 12).

EXPOSE-E is planned to stay in orbit for about three years. It will be part of the EUTEF, an assembly of ESA technology demonstration and scientific instruments, and as such, it will have to be dismounted together with it. Because it is the first to have been developed, EXPOSE-E is not provided with removable experiment trays;

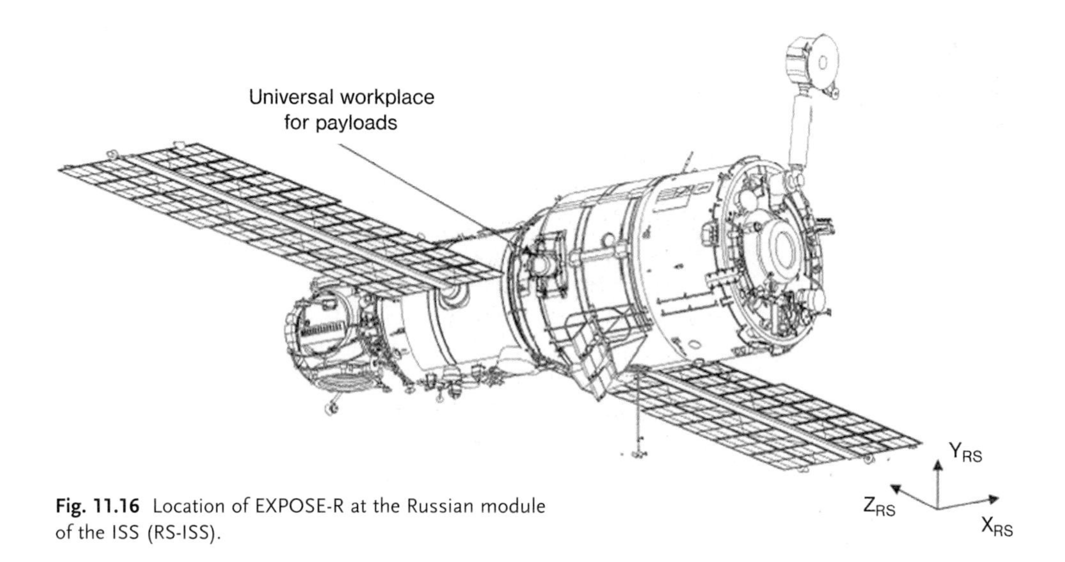

Fig. 11.16 Location of EXPOSE-R at the Russian module of the ISS (RS-ISS).

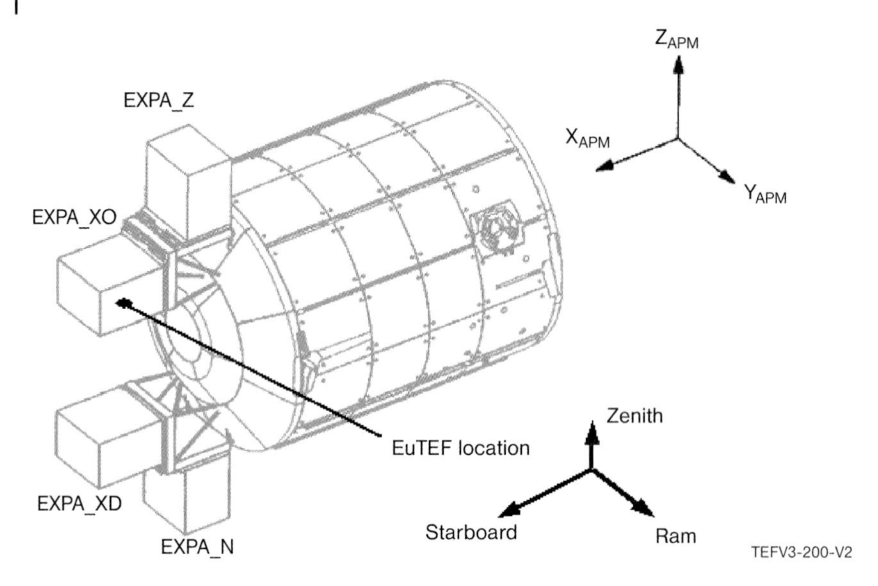

Fig. 11.17 Location of EXPOSE-E at the outer platform EUTEF on the European module *Columbus* of the ISS.

therefore, its flexibility in accommodating new experiment samples and/or upgrading is quite limited.

EXPOSE-E will be mounted on one of the four external balconies of the *Columbus* module, on the ISS (Fig. 11.17). Initially designed in combination with a coarse pointing device that would have allowed a longer direct orientation to the sun, it had to be reconfigured for a less ambitious location on the *Columbus* starboard-oriented external platform, as part of the EUTEF assembly (Fig. 11.17).

11.3.3.5 **Process of Experiment Proposal, Acceptance, Preparation, and Validation**

The experiment preparation process for this kind of research generally starts with a request formulated by the scientists in the form of an unsolicited proposal or in reply to a dedicated Announcement of Flight Opportunity (AO) issued by the ESA. These AOs are published taking into account the interest and demand of scientific research in certain fields as well as the existing flight opportunities in terms of available payloads and retrievable carriers.

Once submitted, the experiment proposals are evaluated and rated by an independent peer review board. The preliminary selected experiments then undergo a detailed engineering assessment on their technical feasibility and on their compatibility with facility characteristics and mission scenarios under the lead of ESA. Finally, the experiment complement is frozen.

After being selected, the experiments undergo a dedicated program of verification and qualification in order to achieve the necessary flight readiness. This normally includes

- selection of samples,
- environmental tests (UV, ionizing radiation, vacuum, etc.),
- flight qualification tests (vibration, thermo-vacuum, thermo-cycling),
- material compatibility tests, and
- experiment sequence tests (pre-mission).

The test results are used as feedback into the experiment hardware design process.

For EXPOSE-R, an Experiment Verification Test Program (EXPOSE-EVT) is established that includes the following elements:

- assessment of the desiccation and UV radiation resistance of the different species of ROSE in order to select suitable organisms for the flight;
- determination of the acceptable temperature ranges;
- assessment of the biocompatibility of the different samples;
- selection of analytical methods for the different biological endpoints to be investigated;
- testing of sample preparation techniques for the space experiment;
- assessment of the resistance to polychromatic UV radiation;
- definition of the optical requirements;
- measurement of dose-effect curves to UV effects in order to choose suitable filter combinations for the flight unit;
- optimization of analytical methods for the different biological endpoints adapted to the small sample volumes required by the space hardware.

11.4 Results from Astrobiology Experiments in Earth Orbit

Since the *Apollo* missions, European scientists have been involved in exobiological or radiation biology experiments in space. Table 11.10 shows the chronologies of ESA missions relevant to astrobiology.

Table 11.10 Missions in low Earth orbit relevant to astrobiology.

Year	Spacecraft	Experiment
1983	*Spacelab 1*	ES029: Response of resistant microbial system (*Bacillus subtilis* spores) to free space
1984–1990	LDEF Facility	Exostack: Long-term survival of spores in space
1992–1993	*EURECA*-ERA	Exposure of different biological systems and complex organic residues to selected parameters of outer space

Year	Spacecraft	Experiment
1993	*Spacelab D2*	UVRAD: Biological responses to extraterrestrial solar ultraviolet radiation and space vacuum
1993–2005	*Foton*/BIOPAN	25 experiments on 5 missions
2003	BIOBOX-5/STS-107	RADCELLS: Effects of cosmic radiation in cells
2004–2005	ISS	MATROSHKA-1: Depth-dose distribution in a human phantom
2006–Future	ISS	EXPOSE-E, EXPOSE-R, MATROSHKA-2

11.4.1
Relevance of Extraterrestrial Organic Molecules for the Emergence of Life

Primary organic chemistry first occurs in the interstellar medium. The chemical processes by which these organic molecules are formed in the interstellar medium are mainly gas-phase ion–molecule reaction paths, which start with the most abundant molecule, hydrogen (H_2), reacting with various ions. The processes include hot and suprathermal reactions, photochemical processes, and surface reactions on interstellar dust grains with successive evaporation. When in the course of pre-stellar evolution a molecular cloud becomes cooler (10–20 K) and denser (10^6 cm^{-3}), trace molecules such as CO, H_2O, CH_3OH, and NH_3 condense onto grain surfaces as ice mantles. In the cyclic evolutionary model for interstellar dust, developed by J. M. Greenberg, interstellar dust grains consist of a silicate core and a mantle of relatively nonvolatile higher-molecular-weight organic compounds. Each complete cycle takes about 10^8 years, and the mean lifetime for an interstellar grain is about 5×10^9 years before it is consumed by star formation. During the EURECA mission of ESA, in the Exobiology Radiation Assembly (ERA), such complex organic residues, produced in the laboratory, were exposed to the full spectrum of solar UV, including vacuum UV radiation, in order to simulate the photolytic processes thought to occur during chemical evolution of interstellar grains. The exposure of the organic residue to solar UV for about six months provided a total fluence of UV that is equivalent to the photon fluence received within about 10 million years in the diffuse clouds of the interstellar medium. Hence, the solar UV-irradiated samples of the EURECA mission were expected to resemble the interstellar organic mantles, which have undergone at least one complete evolutionary cycle. Analysis of the samples after retrieval showed that the UV-irradiated samples had changed in color from yellow to brown, indicating an increase in carbonization. Infrared spectra showed that the 3.4-μm feature of the samples matched quite well with the spectra of the interstellar medium, thereby supporting the cyclic evolutionary model for interstellar dust.

Studies in low Earth orbit have also contributed to answer the question of whether the precursors of life were produced on the primitive Earth or transported to the early Earth via meteorites. The study of micrometeorites collected in the Greenland and Antarctic ice sheets by M. Maurette shows that the Earth captures interplanetary dust as micrometeorites at a rate of about 50–100 tons per day. About 99% of this mass is carried by micrometeorites in the 50–500 μm size range. In the Antarctic micrometeorites, a high percentage of unmelted chondritic micrometeorites from 50 to 100 μm in size have been observed, indicating that a large fraction crossed the terrestrial atmosphere without drastic thermal treatment. In this size range, the carbonaceous micrometeorites represent 80% of the samples and contain 2% of carbon on average. They might have brought about 3×10^{19} g of carbon over a period of 300 million years corresponding to the late terrestrial bombardment phase. This delivery represents more carbon than that engaged in the surface biomass of the Earth, which amounts to about 10^{18} g. Amino acids such as β-amino isobutyric acid have been recently identified in these Antarctic micrometeorites. These grains contain a high proportion of metallic sulfides, oxides, and clay minerals that have catalytic capabilities. In addition to the carbonaceous matter, micrometeorites might have delivered a rich variety of catalysts having perhaps acquired specific crystallographic properties during their synthesis in the microgravity environment of the early solar nebula. They may have functioned as tiny chondritic chemical reactors when reaching oceanic water.

Space environment alters the composition, morphology, and dynamics of these interplanetary dust particles (IDPs) by processes such as thermal annealing, radiation and ion bombardment, or collisions. In the last years, techniques have been developed to collect IDPs in space. The advantage of this type of collection – compared to sampling on Earth – is that the samples are not altered by atmospheric entry effects. Passive aerogel collectors have been exposed on NASAs Long Duration Exposure Facility (LDEF) and the MIR station, in order to collect grains uncontaminated by interaction with Earth's atmosphere or environment, and returned to Earth. Aerogel collectors were used on the NASA *Stardust* mission to collect and return to Earth grains from the interplanetary medium and from the coma of comet Wild 2.

Amino acids like those detected in the Murchison meteorite have been exposed for 10 days to space conditions in Earth orbit during three months onboard MIR (PERSEUS experiment) and for about two weeks within the ESA facility BIOPAN onboard the unmanned Russian satellites *Foton 8* and *Foton* 11. They were exposed to the full spectrum of solar extraterrestrial UV radiation, including vacuum UV, and to space vacuum under different exposure conditions: either as free amino acid or associated with mineral powder. Analysis of the samples after flight showed that exposed amino acids (aspartic acid and glutamic acid) were partially photo-processed during their stay in space. However, decomposition was prevented when the amino acids were embedded in clays (see Chapter 1). The main limitations of these experiments were the relatively weak irradiation of the samples (because of the short flight duration), the non-synchronous orbits, and the absence of an automatic sun pointing device. ESA's EXPOSE facility onboard the ISS offers an interesting opportunity for long-duration exposure of amino acids and other biogenic molecules (see Section 11.3.3.).

11.4.2 Role of Solar UV Radiation in Evolutionary Processes Related to Life

Space experiments using solar UV radiation as the authentic energy source represent new approaches towards our understanding of the evolution of life and potential evolutionary steps on other celestial bodies. They may especially contribute to tackle the following issues:

- the formation and stability of organic molecules in the pre-biotic environment of early Earth;
- the role of the ozone layer in protecting life on Earth; and
- the role of solar UV radiation in the planets and moons of astrobiological interest, such as Mars, Saturn's moon Titan and Jupiter's moon Europa.

11.4.2.1 Efficiency of the Stratospheric Ozone Layer to Protect Our Biosphere

Until recently, assessment of the consequences of increasing environmental UV-B irradiation – caused by a decreasing concentration of stratospheric ozone, the so-called ozone hole – on the biosphere was based merely on model calculations. In an experiment onboard the German *Spacelab D2* mission that used extraterrestrial solar UV radiation as natural radiation source and optical filters, the terrestrial UV radiation climate was simulated at different ozone concentrations down to very low ozone values, thereby using space as a kind of time machine. Biologically effective irradiances as a function of the simulated ozone column thickness were measured with *Bacillus subtilis* spores immobilized in a biofilm (Fig. 11.18). They were compared to the expected irradiance, which was obtained from radiative transfer calculations multiplied with the biofilm action spectrum (an action spectrum gives the spectral efficiency of UV radiation with regard to a certain biological effect, here, cell killing). After the mission, the biofilms were incubated and stained, and

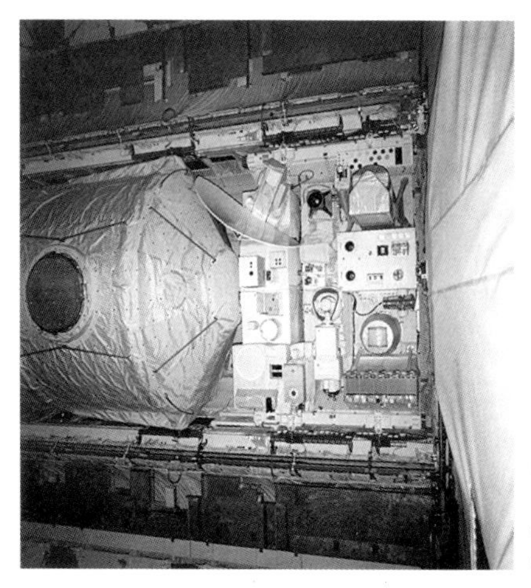

Fig. 11.18 Location of the experiment UV-RAD in the cargo bay of the Space Shuttle during the German *Spacelab D2* mission (photo credit: DLR).

optical densities indicative of the biological activity were determined for each exposure condition by image analysis. It was found that with decreasing (simulated) ozone concentrations, the biologically effective solar UV irradiance strongly increased. The full spectrum of extraterrestrial solar radiation led to an increment of the biologically effective irradiance by nearly three orders of magnitude compared to that of the solar spectrum at the surface of the Earth for average total ozone columns.

Experiments with bacterial endospores, simulating the UV radiation climate of Mars, were performed using the BIOPAN facility of ESA onboard a *Foton* satellite. These experiments tackled the question of the toxicity of the Martian regolith to terrestrial microorganisms exposed to the UV radiation climate of Mars. This information is pivotal for assessing the habitability of the surface of Mars and for planetary protection efforts.

11.4.3
Chances and Limits of Life Being Transported from One Body of Our Solar System to Another or Beyond

The idea of interplanetary transfer of life was originally put forward independently by three scientists: Richter in 1865, Lord Kelvin in 1894, and Svante Arrhenius in 1903. The theory of "panspermia" as postulated by Arrhenius in 1903 holds that reproductive bodies of living organisms can exist throughout the Universe and develop wherever the environment is favorable. This theory implies that during the evolution of the Universe, conditions favorable to the development of life have prevailed at different locations and at different times. For example, early in the development of our Solar System, at a time when conditions on the young Earth may have been hostile to the development of life, conditions conducive to the development of living systems may have existed simultaneously on other bodies either within our Solar System (such as Mars, Venus, Jupiter's moon Europa, or Saturn's moon Titan) or within other solar systems. The panspermia theory does not set a precondition for terrestrial life having arisen on Earth; living organisms could well have arisen elsewhere in the Solar System (or Universe) and been transported to Earth, where they encountered environmental conditions favorable to growth, proliferation, and eventual evolution into more-complex forms. Once established, life on Earth would then be subject to transfer to other celestial bodies.

Since its formulation, the theory of panspermia has been subjected to several criticisms, with arguments such as: (1) it cannot be experimentally tested, (2) it shunts aside the question of the origin of life, and (3) living organisms cannot survive long-time exposure to the hostile environment of space, especially vacuum and radiation. However, recent experimental evidence, summarized below, has led to a reexamination of the feasibility of the notion of interplanetary transfer of living material, particularly microorganisms.

The scenario of interplanetary transfer of life involves three basic hypothetical steps:

- the escape process, which comprises removal to space of biological material that has survived being lifted from the surface to high altitudes;
- an interim state in space, which implies survival of the biological material over time scales comparable with the interplanetary or interstellar passage; and
- the entry process, where a non-destructive deposition of the biological material on another planet is required.

It is not know with certainty whether panspermia is likely to have occurred in the history of the Solar System, or whether it is feasible at all, but additional support to the idea of panspermia is given by a variety of recent discoveries, such as the Martian meteorites (see Chapter 8), the high UV resistance of microorganisms at the low temperatures of deep space, and the recovery of viable spores after about six years of exposure in space (see Chapter 5).

In laboratory experiments that simulate the impact of a meteorite as a feasible process of ejecting material from a planet to reach escape velocity, the survival of microorganisms residing inside of rocks has been proved. Once rocks have been ejected from the surface of their home planet, the microbial passengers face an entirely new set of problems affecting their survival, namely, exposure to the space environment. With space technology, a new tool is available for testing experimentally whether microorganisms are capable of surviving a hypothetical journey from one planet to another. In order to study the survival of resistant microbial forms in

Fig. 11.19 Space Simulation Facility (PSI) at the Institute of Aerospace Medicine, DLR, for the pre-flight testing of the experiments on EURECA and EXPOSE (photo credit: DLR).

the upper atmosphere or in free space, microbial samples have been exposed *in situ* by use of balloons, rockets, or spacecraft, and their responses were investigated after recovery. For this purpose, several facilities were developed, such as the exposure device on *Gemini*, MEED on *Apollo*, ES029 on *SL1*, ERA on EURECA, UV-RAD on *Spacelab D2*, and BIOPAN on *Foton*. These investigations were supported by studies in the laboratory in which certain parameters of space (high and ultrahigh vacuum, extreme temperature, UV radiation of different wavelengths, ionizing radiation) were simulated and the microbial responses were determined (Fig. 11.19).

11.4.3.1 **Effects of Space Vacuum**

Because of its extremely dehydrating effect, space vacuum has been argued to be one of the factors that may prevent interplanetary transfer of life. However, space experiments have shown that up to 70% of bacterial and fungal spores survived short-term (up to 10 days) exposure to space vacuum, even without any protection. The chances of survival in space were increased if the spores were embedded in chemical protectants, such as sugars or salt crystals, or if they were exposed in thick layers. For example, during the LDEF mission, 30% of *B. subtilis* spores survived nearly six years of exposure to space vacuum when embedded in salt crystals, whereas 80% survived in the presence of glucose. Sugar and polyalcohol stabilize the structure of the cellular macromolecules, especially during vacuum-induced dehydration, leading to increased rates of survival. In addition, endolithic micro-organisms in nature are probably distributed in micro-cracks and pore spaces within rock as consortia biofilms embedded in a complex extracellular polysaccharide matrix, which would be expected to provide a degree of protection from extreme vacuum. Recent studies on lichens under simulated space conditions and during the BIOPAN-5 mission have demonstrated the high resistance of these communities to space vacuum.

To determine the protective effects of different meteorite materials, "artificial meteorites" were constructed by embedding *B. subtilis* spores in clay, meteorite dust, or simulated Martian soil. Their survival was determined after exposure to the space environment. Crystalline salt provided sufficient protection for osmophilic microbes to survive at least two weeks in space. For example, a species of the cyanobacterium *Synechococcus* that inhabits gypsum-halite crystals was capable of nitrogen and carbon fixation. Concerning an extreme halophile microorganism, about 5% of a species survived after exposure to the space environment for two weeks within the BIOPAN facility in connection with a *Foton* space flight.

The mechanisms of cellular damage due to vacuum exposure are based on the extreme desiccation imposed. If not protected by internal or external substances, cells in a vacuum experience dramatic changes in such important biomolecules as lipids, carbohydrates, proteins, and nucleic acids. Upon desiccation, the lipid membranes of cells undergo dramatic phase changes from planar bilayers to cylindrical bilayers. The carbohydrates, proteins, and nucleic acids undergo so-called Maillard reactions, i.e., amino-carbonyl reactions, to give products that become cross-linked, eventually leading to irreversible polymerization of the biomolecules. Concomitant with these structural changes are functional changes including altered selective membrane permeability, inhibited or altered enzyme activity, decreased energy production, altered genetic information, etc. Vacuum-induced damage to DNA is especially dramatic because it may lead to death or mutation. During the *Spacelab 1* mission, a 10-fold-increased mutation rate over the spontaneous rate was observed in spores of *B. subtilis* after exposure to space vacuum. This mutagenic effect by vacuum is probably based on a unique molecular signature of tandem double-base changes at restricted sites in the DNA. In addition, DNA strand breaks have been observed to be induced by exposure to space vacuum. Such damage would accumulate during long-term exposure to space vacuum because DNA repair is not active in the desiccated state. Spore survival ultimately depends on the efficiency of DNA repair after rehydration and germination.

The high resistance of bacterial endospores to desiccation is due mainly to a dehydrated core enclosed in a thick protective envelope, the cortex and the spore coat layers, and the protection of their DNA by small proteins whose binding greatly alters the chemical and enzymatic reactivity of the DNA. However, the strategies by which *B. subtilis* spores protect their integrity, including that of DNA, against vacuum damage are not yet fully understood. Non-reducing sugars, such as trehalose or sucrose, generally help to prevent damage to DNA, membranes, and proteins by replacing the water molecules during the desiccation process and thereby preserving the three-dimensional structure of the biomolecules. Although bacterial spores do not naturally accumulate the substances, the addition of glucose to the spores substantially increases the survival rate of spores in space vacuum.

11.4.3.2 Effects of Extraterrestrial Solar UV Radiation

Solar UV radiation has been found to be the most deleterious factor of space, as tested with dried preparations of viruses and bacterial and fungal spores. The reason for this is the highly energetic UV–C and vacuum UV radiation that is directly absorbed by the DNA, as demonstrated by action spectroscopy in space (Fig. 11.5). The full spectrum of extraterrestrial UV radiation kills unprotected spores of *B. subtilis* within seconds, as discovered during the *Spacelab 1* mission. These harmful UV ranges do not reach the surface of the Earth because they are effectively absorbed by the Earth's atmosphere. The photobiological and photochemical effects of extraterrestrial UV radiation are based on the production of specific photoproducts in the DNA that are highly mutagenic and lethal. The most damaging photochemical lesions in the DNA are thymine-containing dimers: cyclobutadipyrimidines, the (6–4) pyrimidine–pyrimidone adducts, and their Dewar valence isomer.

If spores of *B. subtilis* were simultaneously exposed to solar UV radiation and space vacuum, they responded with an increased sensitivity to a broad spectrum of solar UV (>170 nm) as well as to selected wavelengths. Upon dehydration, e.g., in space vacuum, DNA undergoes substantial conformational changes. This conversion in the physical structure leads to altered DNA photochemistry, resulting in a 10-fold increase in UV sensitivity of *B. subtilis* spores in space vacuum as compared to spores irradiated at atmospheric pressure. The photoproducts generated within the DNA of *B. subtilis* spores exposed to UV radiation in vacuum were different from those induced in wet spores; two thymine decomposition products, namely, the *cis-syn* and *trans-syn* cyclobutane thymine dimers, as well as DNA-protein cross-linking were found in addition to the spore photoproduct, the only photoproduct induced by UV in wet spores.

11.4.3.3 Effects of Galactic Cosmic Radiation

Among the ionizing components of radiation in space, the heavy primaries, the so-called HZE particles (see Section 11.1.1.1), are the most biologically effective species. Because of their low flux (they contribute to approximately 1% of the flux of particulate radiation in space), methods have been developed to localize precisely the trajectory of an HZE particle relative to the biological object and to correlate the physical data of the particle relative to the observed biological effects along its path. In the BIOSTACK method (Fig. 11.20), visual track detectors are sandwiched between layers of biological objects in a resting state; one candidate is *B. subtilis* spores. This method allows us to

- localize each HZE particle's trajectory in relation to the biological specimens;
- investigate the responses of each biological individual hit separately, with regard to its radiation effects;
- measure the impact parameter, which is the distance between the particle track and the sensitive target;

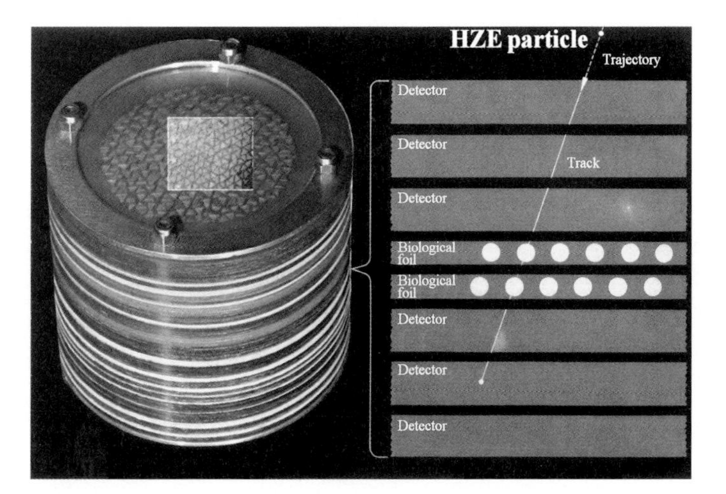

Fig. 11.20 BIOSTACK concept to measure the effects of single heavy ions (HZE particles) of galactic cosmic radiation (GCR) in biological systems in resting state (photo credit: DLR).

- determine the physical parameters (charge [Z], energy [E], and linear energy transfer [LET]); and
- correlate the biological effect with each HZE particle parameter.

Taking the results from several BIOSTACK experiments in space (*Apollo 16*, *Apollo 17*, *Apollo*-Soyuz, *Spacelab*, EURECA) as well as those obtained at accelerators, one can draw the following general conclusions:

- The inactivation probability for spores, centrally hit, is substantially less than 1.
- The effective radial range of inactivation around each HZE particle, the so-called impact factor, extends far beyond the range where inactivation of spores by δ-rays (secondary electron radiation) can be expected.
- This far-reaching effect is less pronounced for ions of low energies (1.4 MeV per mass unit), a phenomenon that may reflect the "thin-down effect" at the end of the ion's path. The dependence of inactivated spores from impact parameter points to a superposition of two different inactivation mechanisms: a short ranged component reaching up to about 1 μm that may be traced back to the δ-ray dose (secondary electrons) and a long-ranged one that extends at least to somewhere between 4 and 5 μm off the particle's trajectory for which additional mechanisms are conjectured, such as shock waves, electromagnetic radiation, or thermophysical events.

The data show that *B. subtilis* spores would survive even a central hit of an HZE particle of cosmic radiation to a certain extent. However, these HZE particles of cosmic radiation are conjectured to set the ultimate limit on the survival of spores in space because they penetrate even thick shielding. Because they interact with the shielding material by creating secondary radiation, with increasing shielding thickness the dose rates go through a maximum. They reach their maximum behind a shielding layer of about 10 cm; behind shielding of about 130 cm, the value is the same as obtained without any shielding, and only for higher shielding thickness does the dose rate reduce significantly. Calculations based on the space experiments have shown that if shielded by 2–3 m of meteorite material, a substantial fraction of a spore population (about 100 from a population of 10^8) would survive the exposure to cosmic radiation, even after several million years in space. The same survival fraction would be reached after about 500 000 years without any shielding and after 350 000 years with 1 m of shielding.

These studies on the biological effects of the HZE fraction of cosmic radiation are of interest for answering astrobiological questions, and they provide important basic information for our attempts to assess the radiation risks to humans in space.

11.4.3.4 Combined Effects of All Parameters of Space

With the LDEF mission, for the first time *B. subtilis* spores were exposed to the full environment of space for an extended period of time (nearly six years), and their survival was determined after retrieval (Fig. 11.21). The samples were separated from space by a perforated aluminum dome only, which allowed access of space vacuum, solar UV radiation including vacuum UV, and most of the components of cosmic radiation. It was found that even in the unprotected samples thousands of spores survived the space journey (from an initial sample size of 10^8 spores). All spores were exposed in multilayer and pre-dried in the presence of glucose. The spore samples had turned from white to yellow during the mission, a phenomenon that is probably due to photochemical processes. It was suggested that all spores in the upper layers were completely inactivated by the high flux of solar UV radiation. With time, they formed a protective crust that considerably attenuated the solar UV radiation for the spores located beneath this layer. Therefore, the survivors probably originated from the innermost part of the samples.

11.4.3.5 Time Scales of Interplanetary Transport of Life

To travel from one planet of our Solar System to another, e.g., from Mars to Earth, by random motion, a mean time of several hundred thousands to millions of years has been estimated for boulder-sized rocks but only two months for microscopic particles. Therefore, the LDEF experiment may be considered a reasonable simulation of an interplanetary transfer event. Because LDEF was in a near-equatorial, low-altitude Earth orbit, during each orbit the satellite traveled about 40 650 km every 90 min. Thus, over the nearly six-year duration of the LDEF experiment, the satellite traveled over 10^9 km, the approximate equivalent of the distance from Earth to Saturn if it had been flying on a straight course rather than in an elliptical orbit. The likelihood of spore survival over the time spans involved in interplanetary transfer is therefore quite high.

For future astrobiological research in space, ESA is developing the EXPOSE facility that is to be attached to external pallets of the ISS for 1–3 years (see Section 11.3.3). EXPOSE will support long-term *in situ* studies of microbes in artificial meteorites, as well as of microbial communities from special ecological niches, such as endolithic and endoevaporitic ecosystems. These experiments on the Responses of Organisms to the Space Environment (ROSE) include the study of

Fig. 11.21 Position of the Exostack experiment on the LDEF of NASA that stayed for nearly 6 years in Earth orbit (photo credit: NASA).

photobiological processes in simulated radiation climates of planets (early Earth, early and present Mars, and the role of the ozone layer in protecting the biosphere from harmful UV-B radiation) as well as studies of the probabilities and limitations for life to be distributed beyond its planet of origin. Results from the EXPOSE experiments will provide us with a better understanding of the processes regulating the interactions of life with its environment.

A hypothetical scenario of interplanetary transfer of life requires that the microorganisms finally be deposited from space to a target planet. When captured by a planet with an atmosphere, most meteorites are subjected to very high temperatures during landing. However, because the fall through the atmosphere takes only a few seconds, the outermost layers form a kind of heat shield and the heat does not reach the inner parts of the meteorite. During entry, the fate of the meteorite strongly depends on its size. Large meteorites may break into pieces; however, these may still be large enough to remain cool inside until hitting the surface of the planet. Medium-sized meteorites may obtain a melted crust, whereas the inner part still remains cool. Micrometeorites of a few micrometers in size may tumble through the atmosphere without being heated at all above 100 °C. Therefore, it is quite possible that a substantial number of microbes can survive the landing process on a planet. A first attempt to study the fate of microorganisms embedded inside a small meteorite was performed within the STONE experiment (Fig. 11.12) during the flight of the *Foton* satellite in 2005 (see Section 11.3.2).

11.4.4
Radiation Dosimetry in Space

Measurements of radiation exposure in Earth orbit have been performed during manned and robotic space flights at various altitudes, orbital inclinations, durations, periods during the solar cycle, and mass shielding (Fig. 11.22). Because of

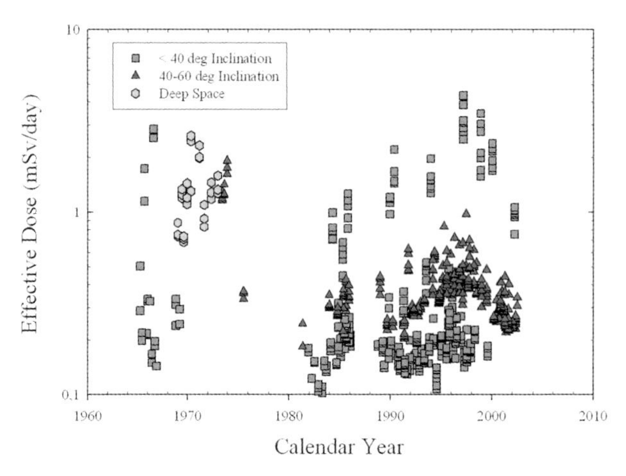

Fig. 11.22 Effective doses measured in low Earth orbit missions and missions to the Moon (courtesy of G. Reitz, DLR).

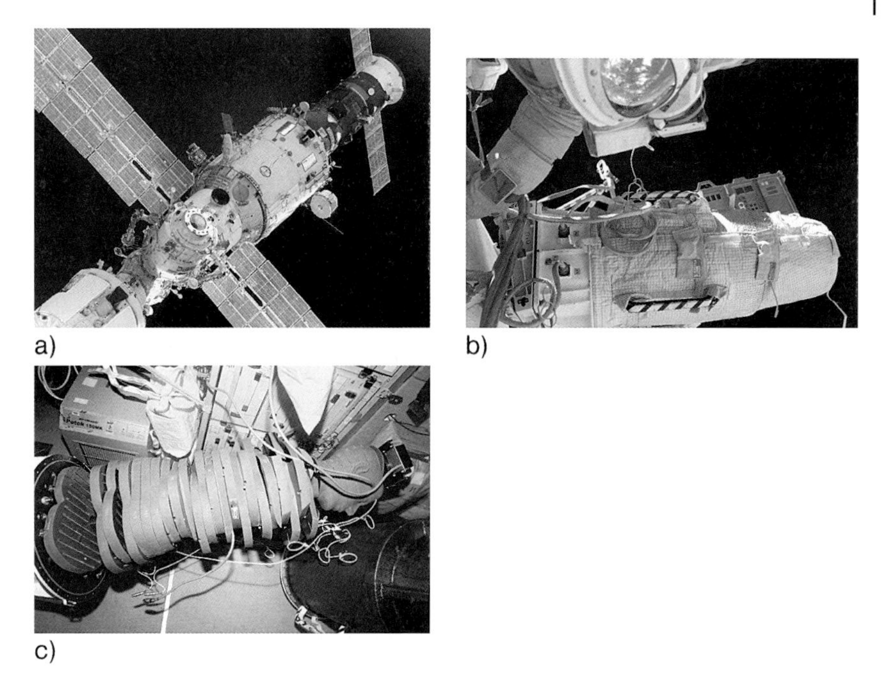

Fig. 11.23 Human phantom MATROSHKA (a) attached to the outer platform of the Russian module of the ISS during its one-year flight in 2005; (b) mounting during an extravehicular activity; (c) during disassembly inside the Russian segment of the ISS (photo credit: ESA, courtesy of G. Reitz, DLR).

the complex mixture of radiations occurring in space, comprising sparsely ionizing components (photons, electrons, pions, muons, and protons) and densely ionizing components (heavy ions, neutrons, and nuclear disintegration stars) (see Section 11.1.), different dosimetry systems have been applied that specifically respond to the quality of the radiation under consideration. However, most dosimetric systems used provided data on the "surface" or skin dose only. For radiation protection purposes, it is important to assess the depth-dose distribution within the human body and especially at the most radiation-sensitive organs, such as the brain, the blood-forming organs, and the gonads. Therefore, human phantoms equipped with different dosimetric systems at the sites of sensitive organs are required. The anthropomorphic phantom MATROSHKA is one such example; it was exposed for one year to the radiation in space outside of the ISS (Fig. 11.23) in order to determine the depth-dose distribution of radiations within the human body during EVAs. It consists of a human phantom upper torso composed of 33 slices equipped with passive and active radiation dosimeters at the sites of different organs.

11.5 Future Development and Applications of Exposure Experiments

The features of EXPOSE allow passive exposure of samples to space with continuous monitoring of the environmental conditions during the mission, in particular of UV radiation, broad solar spectrum, temperature, and residual pressure. As previously described (see Section 11.3.3), predefined thermal conditions are guaranteed during exposure by use of

- electrical heaters positioned under the trays and inside the frame structure;
- thermal blankets around the facility and around its support structure;
- low-conductive interfaces (Ti bolts/washers) between the facility and EUTEF-CEPA; and
- coatings on the exposed surfaces with low absorbivity–high emissivity characteristics.

Temperature should be maintained within specific ranges (+5 to +15 °C or –20 to –5 °C) despite the environmental conditions. The strategy aims to avoid "crossing" of the freezing point. These thermal requirements are generally originated by the radiation biology experiments, where biological samples sensitive to large temperature variations are exposed to space radiation.

On the contrary, astrobiology experiments try to simulate as closely as possible the conditions experienced by organic matter when it is traveling with comets or asteroids or when exposed to extraterrestrial conditions. In this case, low temperatures, far below 0 °C, are desired, while maintaining a full exposition to the solar UV radiation. This is why a future evolution of a facility like EXPOSE should be provided with some cooling devices, either passive, such as heat pipes and radiators conductively linked with the experiment trays, or semi-active, such as thermoelectric junctions (Peltier) linked to heat exchangers.

The onboard sensor packages, including active radiation measurement devices capable of correlating input data with time (a prototype, Radfet, flew with BIOPAN-5 in May 2005), could be improved. In addition, Sun pointing devices and sample compartments with embedded temperature sensors are part of future development.

EXPOSE might be used in the near future as a "test bed" for experiments in support of the upcoming planetary exploration missions. The design and development of a mobile exobiology package (Pasteur) to be operated on the Martian surface (*ExoMars* mission) (see Chapter 12) will most probably require testing of materials, electronic components, and mechanical devices that will have to be operated continuously when exposed for a long time to extraterrestrial conditions (low pressure or vacuum, large temperature variations, high radiation, electric and magnetic fields, etc.).

Looking at the near-term plans of ESA for Mars exploration, it is clear that the first missions (*ExoMars* Pasteur, *Mars Sample Return*) will bring to the Martian surface a set of instrumentation sterilized of terrestrial microorganisms and cleansed of Earth organic agents, in an attempt to avoid any contamination when measuring and investigating the Martian soil characteristics. These planetary protection issues are being investigated at the moment (see Chapter 13). These studies may require the help of an exposure platform like EXPOSE for validation, qualification, and testing of materials or technical devices.

The second phase of the exploration program of ESA foresees that life and organic matter will be brought to Mars in support of human activities on Mars. In this case, EXPOSE or EXPOSE-like facilities may be used for the understanding of non-terrestrial environment effects on organic matter. ISS external platforms provide unique environmental conditions, different from those of Earth and Mars, that allow testing and verifying the effects of specific parameters of outer space, applied singly or in combination.

11.6 Further Reading

11.6.1 Books and Articles in Books

Bücker, H., Horneck, G. Studies on the effects of cosmic HZE particles in different biological systems in the Biostack experiments I and II, flown on board of *Apollo 16* and *17*. In *Radiation Research*, Nygaard, O. F., Adler, H. I., Sinclair, W. K. (Eds.), Academic Press, New York, pp. 1138–1151, 1975.

ESA, *The ESA Radiation Handbook*, ESA PSS-01–609 Issue 1, 1993.

ESA, *Proceedings of the 2nd European Symposium on the Utilization of the International Space Station*, ESTEC, Noordwijk, The Netherlands, 16–18 November 1998, ESA SP-433, 1999, with the following articles:
- Brack, A., Ehrenfreund, P., Otroshenko, V., Raulin, F. From interstellar chemistry to terrestrial life: exposure experiments in Earth orbit, pp. 455–458.
- Horneck, G., Wynn-Williams, D. D., Mancinelli, R. L., Cadet, J., Munakata, N., Ronto, G., Edwards, H. G. M., Hock, B., Wänke, H., Reitz, G., Dachev, T., Häder, D. P., Brillouet, C. Biological experiments on the EXPOSE facility of the International Space Station ISS, pp. 459–468.

Feuerbacher, B., Stoewer, H. (Eds.), *Utilization of Space – Today and Tomorrow*, Springer, Berlin Heidelberg, 2006

Horneck, G., Brack, A. Study of the origin, evolution and distribution of life with emphasis on exobiology experiments in Earth orbit. In *Advances in Space Biology and Medicine*, Vol. 2, Bonting, S. L. (Ed.), JAI Press, Greenwich, CT, pp. 229–262, 1992.

Horneck, G. The Biostack concept and its application in space and at accelerators: studies on *Bacillus subtilis* spores. In *Biological Effects and Physics of Solar and Galactic Cosmic Radiation*, Part A, Swenberg, C. E., Horneck, G., Stassinopoulos, E. G. (Eds.) Plenum Press, New York, pp. 99–115, 1993.

Horneck, G., Baumstark-Khan, C., Reitz, G. Space microbiology: effects of ionizing radiation on microorganisms in space. In *The Encyclopedia on Environmental Microbiology*, Bitton, G. (Ed.), Wiley, New York, pp. 2985–2996, 2002.

Innocenti, L., Mesland, D. (Eds.) *EURECA Scientific Results*, Adv. Space Res. Vol. 16, No. 8, Pergamon, Oxford, 1995, especially the following articles:
- Dose, K., Bieger-Dose, A., Dillmann, R., Gill, M., Kerz, O., Klein, A., Meinert, H., Nawroth, T., Risi, S. Stridde, C. ERA experiment "Space Biochemistry" pp. (8)119–(8)129.
- Horneck, G., Eschweiler, U., Reitz, G., Wehner, J., Willimek, R., Strauch, K. Biological responses to space: results of the experiment "Exobiological Unit" of ERA on EURECA I. pp. (8)105–(8)118.

Maurette, M. Micrometeorites and the Mysteries of our Origins, Springer, Berlin Heidelberg New York, 2006.

Moore, D., Bie, P., Oser, H. (Eds.) *Biological and Medical Research in Space*, Springer, Heidelberg, 1996, especially the following articles:
- Kiefer, J., Kost, M., Schenk-Meuser, K. Radiation biology, pp. 300–367.
- Horneck, G. Exobiology, pp. 365–431.

Seibert, G. A *World without Gravity –Research in Space for Health and Industrial Processes*, ESA SP-1251, ESA-ESTEC, Noordwijk, The Netherlands, 2001, available online: http://www.esa.int/esapub/sp/sp1251/sp1251web.pdf.

11.6.2
Articles in Journals

Barbier, B., Chabin, A., Chaput, D., Brack, A. Photochemical processing of amino acids in Earth orbit, *Planet. Space Sci.* **1998**, *46*, 391–398.

Brack, A., Baglioni, P., Borruat, G., Brandstätter, F., Demets, R., Edwards, H. G. M., Genge M., Kurat, G., Miller, M. F., Newton, E. M., Pillinger, C. T., Roten, C.-A., Wasch E. Do meteoroids of sedimentary origin survive terrestrial atmospheric entry? The ESA artificial meteorite experiment STONE, *Planet. Space Sci.* **2002**, *50*, 763–772.

Demets, R., Schulte, W., Baglioni, P. The past, present and future of BIOPAN, *Adv. Space Res.* **2005**, *36*, 311–316.

Facius, R., Reitz, G., Schäfer, M. Inactivation of individual Bacillus subtilis spores in dependence on their distance to single cosmic heavy ions, *Adv. Space Res.* **1994**, *14 (10)*, 1027–1038.

Ferrini, G., Colangeli, L., Palumbo, P., Westphal, J. A., Borg, J. Capture experiments of meteoroids and space debris in LEO: recent results and future possibilities, *Microgravity and Space Station Utilisation* **2001**, *2*, 2–4.

Greenberg, J. M. Approaching the interstellar grain organic refractory component. *Astrophys. J.* **1995**, *455*, L177.

Horneck, G. Responses of *Bacillus subtilis* spores to space environment: Results from experiments in space, *Origins Life Evol. Biosph.* **1993**, *23*, 37–52.

Horneck, G. Exobiological experiments in Earth orbit, *Adv. Space Res.* **1998**, *22(3)*, 317–326.

Horneck, G. European activities in exobiology in Earth orbit: results and perspectives, *Adv. Space Res.* **1999**, *23*, 381–386.

Horneck, G., Bücker, H., Reitz, G., Requardt, H., Dose, K., Martens, K. D., Menningmann, H. D., Weber, P. Microorganisms in the space environment, *Science* **1984**, *225*, 226–228.

Horneck, G., Bücker, H., Reitz, G. Long-term survival of bacterial spores in space, *Adv. Space Res.* **1994**, *14 (10)*, 41–45.

Horneck, G., Rettberg, P., Rabbow, E., Strauch, W., Seckmeyer, G., Facius, R., Reitz, G., Strauch, K., Schott, J. U. Biological dosimetry of solar radiation for different simulated ozone column thicknesses, *J. Photochem. Photobiol. B: Biol.* **1996**, *32*, 189–196.

Mancinelli, R. L., White, M. R., Rothschild, L. J. BIOPAN survival I: exposure of the osmophiles *Synechococcus sp.* (Nageli) and *Haloarcula sp.* to the space environment. *Adv. Space Res.* **1998**, *22(3)*, 327–334.

Mileikowsky, C., Cucinotta, F., Wilson, J. W., Gladman, B., Horneck, G., Lindegren, L., Melosh, J., Rickman, H., Valtonen, M., Zheng, J. Q. Natural transfer of viable microbes in space, Part 1: From Mars to Earth and Earth to Mars, *Icarus* **2000**, *145*, 391–427.

Nicholson, W. L., Munakata, N., Horneck, G., Melosh, H. J., Setlow, P. Resistance of *Bacillus* endospores to extreme terrestrial and extraterrestrial environments, *Microb. Mol. Biol. Rev.* **2000**, *64, 548–572.*

Obe, G., Johannes, I., Johannes, C., Hallmann, K., Reitz, G., Facius, R. Chromosomal aberrations in blood lymphocytes of astronauts after long-term space flights, *Int. J. Radiat. Biol.* **1997**, *72, 726–734.*

Reitz, G., Beaujean, R., Heckeley, N., Obe, G. Dosimetry in the space radiation field,. *Clin. Investig.* **1993**, *71*, 710–717.

Rettberg, P., Horneck, G. Biologically weighted measurement of UV radiation in space and on Earth with the biofilm technique, *Adv Space Res.* **2000**, *26, 2005–2014.*

11.6.3
ESA Online Archives

Erasmus Experiment Archive, all ESA experiments in microgravity with details and publications: spaceflight.esa.int/eea

ESA ISS Spaceflight Users Web site: single-stop shop for all scientific utilization–related issues: spaceflight.esa.int/users

European User Guide to Low Gravity Platforms: The reference document for Scientific and Technical Users: http://spaceflight.esa.int/users/index.cfm?act=default.page&level=1c&page=2130

Web Streaming Network: all Astrobiology Course Network lectures are available in QuickTime (300 kbps) as streaming video on-demand with synchronized slides: streamiss.spaceflight.esa.int

Erasmus Virtual Campus E-Learning tool: all Astrobiology Course Network lectures are available within the Erasmus User Centre iLinc tool, user ID and password available on request: isslinc.spaceflight.esa.int/virtual

ESA Space Radiation and Effects: http://space-env.esa.int/index.html

11.7
Questions for Students

Question 11.1

The space environment in low Earth orbit is quite similar to that of deep space. Describe the main differences between the two environments.

Question 11.2

Describe the main current flight opportunities for astrobiology experiments offered by the European Space Agency and the scientific aims pursued.

Question 11.3

What are the effects of UV radiation (extraterrestrial and terrestrial) on living systems? Describe the physical, chemical, and biological interactions and the spectral dependence of the effects.

Question 11.4

Describe the interaction between the Earth's magnetosphere, the incoming galactic cosmic radiation, and the radiation originating in the Sun.

Question 11.5

Describe an astrobiology experiment for one of the two experiment carriers of the European Space Agency (EXPOSE or BIOPAN), its scientific objectives, and how to realize them.

12
Putting Together an Exobiology Mission: The *ExoMars* Example

Jorge L. Vago and Gerhard Kminek

In this chapter, the different steps towards the making of an exobiological mission to Mars are presented, exemplified by the ExoMars project of the European Space Agency (ESA). The chapter starts with a short description of the exobiological activities at ESA that have led to the development of the ExoMars concept. Then, the scientific rationale and objectives of the mission are covered, followed by a description of the instruments and engineering design required to reach the objectives. It is pointed out that besides a strong science case and a feasible engineering concept, programmatic and financial boundaries play a major role in shaping a mission.

12.1
Background of the *ExoMars* Mission

12.1.1
Searching for Life on Mars

The term *exobiology*, in its broadest definition, denotes the study of the origin, evolution, and distribution of life in the Universe. It is well established that life arose very early on the young Earth. Fossil records show that life had already attained a large degree of biological sophistication 3.4 billion years ago (see Chapter 1). Since then, it has proven extremely adaptable, colonizing the most disparate ecological niches, from the very cold to the very hot, and spanning a wide range of pressure and chemical conditions (see Chapter 5). For organisms to have emerged and evolved, water must have been readily available on our planet. Life as we know it relies, above all else, upon liquid water. Without it, the metabolic activities of living cells are not possible. In the absence of water, life either ceases or slips into quiescence.

Mars today is cold, desolate, and dry. Its surface is oxidized and exposed to sterilizing and degrading ultraviolet (UV) radiation. Low ambient temperature and

Complete Course in Astrobiology. Edited by Gerda Horneck and Petra Rettberg

ISBN: 978-3-527-40660-9

pressure preclude the existence of liquid water – except, perhaps, in localized environments, and then only episodically. Nevertheless, numerous features, such as large channels, dendritic valley networks, gullies, and sedimentary rock formations, suggest the past action of surface liquid water on Mars – and lots of it. The sizes of Martian outflow channels imply immense discharges, exceeding any floods known on Earth (see Chapter 8).

Mars's observable geological record spans approximately 4.5 billion years. From the number of superposed craters, the oldest terrain is believed to be about 4 billion years old and the youngest possibly less than 100 million years. Most valley networks are ancient (3.5–4.0 billion years old), but as many as 25–35 % may be more recent. Today, water on Mars is stable only as ice at the poles, as permafrost in widespread underground deposits, and in trace amounts in the atmosphere. From a biological perspective, past liquid water itself motivates the question of life on Mars. If Mars's surface was warmer and wetter for the first 500 million years of its history, perhaps life arose independently there, at more or less the same time as it did on Earth.

An alternative pathway may have been the transport of terrestrial organisms embedded in meteoroids, delivered from Earth to Mars. Yet another hypothesis is that life may have developed within a warm, wet, subterranean environment. In fact, given the discovery of a flourishing biosphere 1 km or more below Earth's surface, a similar vast microbial community may be active on Mars, having long ago retreated into that ecological niche following the disappearance of a more benign surface environment. The possibility that life may have evolved on Mars during an earlier period when water existed on its surface, and that organisms may still exist underground, marks the planet as a prime candidate to search for life beyond Earth.

12.1.2
Exobiology Research at ESA

Exobiology activities at ESA started in the 1980 s with the preparation of experiments for the Exobiology and Radiation Assembly (ERA). ERA flew in 1992, onboard the *European Retrievable Carrier* mission, and was active for almost a year. It provided results on the exposure of invertebrates, microorganisms, and organic molecules to long-term space conditions, such as UV radiation, cosmic radiation, and vacuum (see Chapter 11).

Other experiments were conducted using BIOPAN, a facility externally attached to the Russian *Foton* retrievable satellite. BIOPAN's upper shell is a motor-driven lid that opens when in orbit to expose its samples to space. At the end of its 10-day mission, the lid is closed. To withstand the extreme heat of reentry, the entire BIOPAN structure is protected by an ablative heat shield; upon landing, the specimens can be retrieved and examined. Five flights took place, in 1992, 1994, 1997, 1999, and 2005. Microbes, seeds, and organic molecules were subjected to the harsh low Earth orbit environment in different manners, such as with and without radiation protection, in space vacuum, or in the presence of a simulated atmosphere. The response of the samples was determined. It was found that unprotected bacterial spores were completely or nearly totally inactivated by the UV radiation. Thin layers of clay, rock, or meteorite material were successful in UV shielding only when they were in direct contact with the spores. Thus, concerning a possible scenario for the interplanetary transfer of life, the BIOPAN data suggest that small rock ejecta of a few centimeters in diameter may provide sufficient protection for organisms to survive the space journey. However, micron-sized grains, as invoked in some panspermia theories, would most likely prove inadequate (see Chapter 11).

Meteorites may be natural vehicles for transporting resistant life forms across space. Hence, also on *Foton*, suitable meteorite analogues, the STONE experiments, were subjected to the searing environment of spacecraft reentry. The first three rock samples were fixed to the *Foton* capsule's heat shield and recovered for study upon landing in 1999. The goal was to investigate why, among the known meteorites believed to have come from Mars, none is of sedimentary origin. Can sedimentary rocks survive reentry? Are they altered beyond recognition by their passage through the Earth's atmosphere? STONE provided valuable results on the physical and chemical modifications undergone by sedimentary rocks during atmospheric infall. The 2005 mission contained four rock specimens, this time also including microorganisms. The goal of this work was to simulate a meteorite's atmospheric impact and to observe to what degree the embedded microorganisms were affected (see Chapter 11).

From 2007 onwards, the ESA facility EXPOSE will be mounted on an external payload site of the International Space Station (ISS). Carefully controlled parameters, such as space vacuum and well-defined wavebands of solar UV and cosmic radiation, will act on the samples, which can be combined with chemical and/or physically protective agents. This will help to elucidate whether, and to what extent, meteoritic material may offer enough protection for life to remain viable after a long permanence in space. It will also allow the study of long-term survivability and damage/repair mechanisms operating in microorganisms under space conditions. Finally, EXPOSE will shed light on space chemistry in the Solar System in relation to the origin of life. Organic molecules of biological interest, such as amino acids, peptides, and nucleic acids, will be exposed to characterize any variations in their stability and reactivity. Additionally, powders of clay, meteorite, and terrestrial rock will be used to model the mineral fraction present in meteoroid and interstellar dust to understand their effect as filters or as potential catalysts (see Chapter 11).

Other ESA initiatives that will contribute to our knowledge of important prebiotic chemical processes are the cometary mission *Rosetta* and the *Huygens* probe of the *Cassini–Huygens* mission. *Rosetta* will be the first mission to orbit and land on a comet. For this, the comet 67P/Churyumov-Gerasimenko has been chosen. It will collect essential information to help us understand the formation and evolution of our Solar System. *Rosetta* will also help us to determine whether comets could have contributed to the origin of life on Earth by seeding our planet with complex organic molecules through impacts. Light, volatile substances carried by comets also may have played an important role in the formation of Earth's oceans and atmosphere (see Chapter 3).

ESA's *Huygens* probe, traveling to Titan aboard NASA's *Cassini* spacecraft, successfully completed its mission in 2005 (see Chapter 9). Many scientists consider that the present composition of Titan's atmosphere – mainly nitrogen and methane – might closely resemble that of early Earth, before life began on our planet. Throughout its 2.5-h descent, *Huygens* made a detailed study of Titan's atmosphere and characterized its surface in the proximity of the landing site. Ultraviolet radiation from the Sun breaks methane molecules apart to produce a thick layer of smog at middle altitudes. An organic rain of methane and nitrogen containing aerosols falls steadily onto the satellite's surface, creating an Earth-like terrain that includes extended river networks. The results of *Huygens* reveal the uniqueness of Titan in the Solar System as a planetary-scale laboratory for studying prebiotic chemistry.

The Russian *Mars '96* mission consisted of an orbiter, two landers, and two penetrators to perform subsurface measurements. It was launched in November 1996, but because of a failure in the rocket's upper stage, it fell back to Earth. European scientists had contributed many instruments to *Mars '96*. With no possibility of a Russian reflight, in 1997, within the Space Science Program of ESA, work was started on the design and development of the first ESA spacecraft to visit another planet: *Mars Express*. *Mars Express*, comprising an orbiter and the *Beagle 2* lander, was launched in 2003 using a *Soyuz* rocket. Presently in operation, the mission addresses a wide variety of scientific objectives, concentrating mainly on surface geology and mineralogy; subsurface structure; and atmospheric circulation, composition, and long-term evolution (see Chapter 8).

Regarding exobiology, the orbiter payload is able to identify signatures of water in liquid, solid, and vapor form. In particular, the subsurface sounding radar altimeter MARSIS has delivered data to construct underground water distribution maps to depths of a few kilometers. Other *Mars Express* instruments continue to break scientific ground with important discoveries. Among these are the volcanic and glacial structures observed by the high-resolution stereo camera (HRSC); the detection of trace amounts of methane in the Martian atmosphere by the planetary Fourier spectrometer (PFS), which some scientists believe to have a biogenic origin; and the identification of water-altered minerals by the visible and infrared mineralogical mapping spectrometer OMEGA. A further task of *Mars Express* is to identify life-friendly geological regions that could become candidate landing sites for *ExoMars*.

Regrettably, the *Beagle 2* lander failed. It was meant to undertake a detailed chemical and morphological study of its landing site and to look for water in the soil, in rocks, and in the atmosphere. It would have sampled material from protected niches – subsurface and rock interiors– with a mole and a rock grinder/corer mounted on a small robotic arm. *Beagle 2* was designed to investigate the occurrence of carbonate minerals, to determine the samples' isotopic fractionation, and to search for trace atmospheric species.

12.1.2.1 The ESA Exobiology Science Team Study

As a logical progression from its activities in low Earth orbit (LEO), in 1997 ESA created an Exobiology Science Team. Its objective was to conduct a state-of-the-art survey of exobiology research and to formulate recommendations for the future search for life in the Solar System. The full findings were published in 1999, in the ESA Special Publication ESA SP-1231, the so-called "Red Book Report."

The main recommendations were that Mars should constitute ESA's primary goal and that efforts should be directed mainly to the search for extinct life. The team identified three fundamental requirements:

- That the landing area be of high exobiological interest. This has not been the case in past missions. Locations rich in sedimentary deposits and relatively free from wind-blown dust should be targeted.
- That samples be collected at different sites, with a rover containing a drill to reach well into the soil and surface rocks. This means that mobility and subsurface access are required.
- That an integral set of measurements be performed on each sample and on the place it is obtained from.

The team suggested the following instruments for an exobiology package: a microscope for general examination of the samples at a resolution of 3 μm, plus a close-up camera with 50 μm resolution; an infrared or Raman spectrometer for identifying minerals and organic molecules; an alpha proton X-ray spectrometer (APXS) for establishing the samples' atomic composition; a Mössbauer spectrometer for studying iron mineral compositions and oxidation states; a gas chromatograph–mass spectrometer (GC-MS) assembly for isotopic, organic, and inorganic molecular determination and for chirality measurements; and an oxidants sensor.

During 1999–2000, two parallel phase-A studies were undertaken to examine the feasibility of accommodating the instrument package proposed by the Exobiology Science Team in a *Surveyor*-class lander. At the time, NASA had very ambitious plans for the exploration of Mars, with missions to be launched every two years. ESA saw a potential for scientific cooperation through the contribution of one or more payload elements to a future U. S. mission. The outcome of those industrial studies was a preliminary design concept for what was called the Exobiology Multi-User Facility (EMF).

12.1.2.2 The 1999 Exobiology Announcement of Opportunity

In view of a possible collaboration with NASA, in 1999 ESA issued an Announcement of Opportunity (AO) requesting proposals for exobiology experiments to be performed on Mars using the EMF. No specific flight opportunity was identified at the time. The agency would provide the infrastructure needed for the various instruments: mechanical, control, power, thermal, and communications. Further, it would furnish a drill unit and a sample distribution and preparation system. The investigators were to propose the scientific instruments.

Sadly, the unfortunate demise of NASA's *Mars Polar Lander* and *Mars Climate Orbiter* put the joint-mission scenario on hold. NASA undertook a critical review of its Mars exploration program. This resulted in a revised sequence, with fewer and less frequent missions than previously envisioned. All landers after the twin 2003 Mars Exploration Rovers (MER) *Spirit* and *Opportunity*, dedicated to the study of surface mineralogy, were postponed to 2009 and beyond. In view of these events, the conditions for participating in a U. S. endeavor, as defined in the 1999 exobiology AO, were no longer realistic. ESA, therefore, decided to take the initiative in creating its own mission to search for life on Mars.

12.1.3 The AURORA and the ELIPS Program of ESA

In November 2001, exobiology activities at ESA received a boost at the Ministerial Conference in Edinburgh, Scotland, when the European ministers approved funding for two important new programs: the exploration program AURORA and the European Life and Physical Sciences in Space (ELIPS) Program. The goal of the AURORA program is to formulate and implement a European long-term plan for the robotic and human exploration of the Solar System, particularly of those bodies holding promise for life. The ELIPS program complements AURORA by supporting exobiology and ISS research in LEO.

To prepare for the human exploration of Mars, the AURORA program must first develop the necessary technologies by conducting a number of robotic missions. These missions must resolve important scientific questions connected to exobiology, planetary protection, and hazards to human missions to Mars. For the early exploration phase, ESA assessed a range of possible missions in cooperation with scientists. This resulted in the selection of the first two missions in the AURORA program:

- *ExoMars*: An exobiology mission for performing *in situ* analysis to search for traces of past and present life on Mars and to study the environment in preparation for future human missions.
- Mars Sample Return (MSR): This challenging mission will return to Earth a small capsule carrying samples from the Martian surface. It requires a Mars orbiter, accommodating the Earth return and reentry capsule, and a descent module/ Mars ascent vehicle. It is envisioned to implement this mission in an international collaboration effort.

The approval of the AURORA and ELIPS programs signaled a strong commitment by the member states to continue supporting exobiology research, and to ensure the further consolidation of Europe's role as an important partner in planetary exploration.

12.1.3.1 The 2003 Pasteur Call for Ideas

During 2002, at its Concurrent Design Facility, ESA carried out a study to define the foundations for the first AURORA mission: *ExoMars*. This work resulted in a preliminary mission architecture concept and helped to estimate the level of resources that would be available to perform surface science on the Red Planet. With that information in hand, in early 2003, ESA issued its Pasteur Call for Ideas. Scientists were invited to propose instruments for the Pasteur payload and investigations to be performed with the *ExoMars* rover.

The scientific community's response was extremely encouraging: nearly 600 investigators from 260 universities, research institutions, and companies expressed their interest in participating in *ExoMars*. In all, 50 proposals were received. The proposing teams consisted of international, multidisciplinary groups of investigators. Thirty countries were represented, a demonstration that interest in exobiology research is shared across national borders and that scientists favor international collaboration. During September 2003, all instrument proposals were reviewed for scientific merit by a panel of independent experts drawn from the international scientific community.

In March 2004, 40 scientists from the Pasteur-selected teams (representing approximately 400 investigators) gathered at ESA/ESTEC for a full week. They were requested to assign priorities to the measurements needed to accurately identify signs of past or present life on Mars and to characterize surface hazards to humans. Three working groups were formed, on life detection, geological context, and environment information.

On the basis of that prioritization, the science team recommended the following science exploration scenario for *ExoMars*. It calls for mobility, access to the subsurface, and research at multiple scales, starting with a visual/spectroscopic assessment of the geological environment around the rover, progressing to smaller-scale investigations through the study of interesting surface rocks using a suite of contact instruments, and culminating with the collection of appropriate samples to be analyzed by the instruments inside the rover's laboratory.

During the second half of 2004, two Rover-Pasteur phase A studies were conducted in consultation with the scientists and ESA, with the goal of proposing well-integrated concepts for the Pasteur payload and the rover that would be able to realize the *ExoMars* science objectives. These activities were concluded in February 2005.

Intense discussions seeking to contain the overall mission cost in preparation for the 2005 ESA Ministerial Conference resulted in a revised *ExoMars* mission concept: a *Soyuz* version, carrying the rover and a small station but no orbiter. The mass allocation for the rover and Pasteur was substantially reduced from that considered in the phase A studies, and a new element was introduced – the geophysics and environment package (GEP), which could constitute a better platform for some of the Pasteur environment measurements.

To address these payload issues, 40 scientists from the Pasteur-selected teams gathered once again at ESA/ESTEC for the second Pasteur working group meeting during September 2005. Also present were investigators from the GEP community and from ESA's advisory bodies, delegations, and NASA representatives. The participants recommended a payload of 12.5 kg for the rover. They stressed that *ExoMars*, with its subsurface drill, would provide a unique opportunity to effectively search for life on Mars. Having reduced the rover's instrument mass from 24 kg to 12.5 kg, they underlined that the recommended payload is considered the minimum necessary to do the job properly. The meeting concluded with a strong request by the scientists that the proposed 12.5 kg Pasteur payload for life detection be implemented in its entirety onboard the rover. Equally firmly was stressed the need to include the Pasteur environment instruments in the mission and to confirm the implementation of the GEP station in *ExoMars*.

12.1.3.2 Approval of the *ExoMars* Mission

The *ExoMars* mission was approved at the ESA Ministerial Conference in Berlin, Germany, in December 2005. At the time of this writing, mid-2006, the project is halfway through its design phase. Development of the manufacturing phase is expected to begin in early 2008.

12.2 *ExoMars* Science Objectives

Establishing whether life ever existed on Mars, or is still active today, is one of the outstanding scientific questions of our time. It is also a necessary prerequisite to prepare for future human exploration endeavors. To address this important goal, ESA plans to launch the *ExoMars* mission.

The *ExoMars* mission's scientific objectives are:

- to search for signs of past and present life on Mars;
- to characterize the water/geochemical environment as a function of depth in the first 2 m of the subsurface;
- to study the surface environment and identify hazards to future human missions; and
- to investigate the planet's deep interior to better understand Mars's evolution and habitability.

12.2.1 Searching for Signs of Life

In attempting to define an effective strategy to search for carbon-based life on Mars, a useful approach is initially to ponder separately the issues of past and present life detection and, subsequently, to look for a common thread that may suggest a way to address both cases.

12.2.1.1 Extinct Life

If life ever arose on the Red Planet, it may have happened when Mars was warmer and wetter, during its initial half-billion to one billion years. Conditions then were probably similar to those on early Earth. Therefore, one may reasonably expect that, likewise, life emerged on early Mars and that microbes quickly became a global phenomenon. Nevertheless, there is inevitably a large measure of chance involved in finding convincing evidence of ancient life forms.

On our planet's surface, the permanent presence of running water, atmospheric oxygen, and life itself quickly erases all traces of any exposed, dead organisms. The only opportunity to detect them is to find their biosignatures encased in a protective environment, i.e., in suitable rocks. However, high-temperature metamorphic processes and plate tectonics have resulted in the reforming of most ancient terrains, and thus it is very difficult to find rocks on Earth older than 3 billion years in good condition. The ensuing chemical and isotopic degradation of many putative bacterial fossils makes their reliable identification far from trivial.

A further complication is that a range of inorganic processes is known to result in mineral structures closely resembling primitive biological shapes. This issue lies at the heart of a heated debate among paleobiologists. Two recent examples that have attracted much attention are the early Archean rock specimens (Archean is the earliest part of the Precambrian era on Earth, approximately 3.8–2.5 billion years ago) obtained from the Pilbara region of Western Australia, claimed to contain Earth's oldest fossils to date (see Chapter 1), and the Martian meteorite ALH84001, whose alleged fossil microorganisms were seen worldwide in 1996. Although resembling very small bacteria found on our planet, there is still considerable controversy as to the nature of these structures (see Chapter 8). The difficulty is that, in essence, one is looking for the exiguous remnants of minuscule, unshelled, uncompartmentalized beings whose fossilized forms can be sufficiently simple to be confused with tiny mineral precipitates. It is therefore doubtful that the living origin of ancient candidate microfossils may be accurately established on the basis of their morphology alone. Although important, comparative anatomy by itself cannot be relied upon to provide sufficient proof.

Another useful clue may lie in the isotopic signature of carbon. Many life processes favor the assimilation of the light isotope, ^{12}C, over that having an extra neutron, ^{13}C. This gives rise to a higher concentration of ^{12}C in living cells relative to the one found in the surrounding dead environment. For instance, the enzymatic uptake of carbon during photosynthesis can result in a ^{12}C:^{13}C ratio of about 91–94, whereas the terrestrial ratio taken as standard for abiotic material is just 89. Consequently, provided that they can be isolated, carbon residues stemming from previously living matter may be recognized by their ^{12}C enrichment. However, for a useful interpretation of isotope biosignatures, a detailed understanding of the sources and sinks, as well as their temporal evolution, is crucial.

Some compounds synthesized by living organisms are so stable that they can last for billions of years after the parent organisms have died and decomposed. It is not the whole molecule that survives, but rather the backbone of carbon atoms with its distinctive geometry. Typical examples are amino acids, the lipids that comprise cell walls, and some pigments, such as chlorophyll, which absorbs light to power photosynthesis in plants. These telltale molecules are very common on our planet and can constitute very reliable biomarkers. Identifying one of them could prove as informative as finding a dinosaur bone.

Regrettably, a major problem with the study of biomarkers is that many decompose when exposed to temperatures greater than 200 °C. On Earth, most Archean rocks have been heated beyond this value. Mars, on the other hand, has not suffered such widespread tectonic activity. Long-lasting volcanism seems to have concentrated mainly on large provinces, such as the Tharsis and Elysium regions. This would imply that rock formations from the earliest period of Martian history, which have not been exposed to high-temperature recycling, may exist. Consequently, well-preserved ancient biomarkers may still be accessible for analysis.

Some of life's most important molecular building blocks, namely, amino acids and sugars, can exist in left- and right-handed configurations called enantiomers (from the Greek *enantios*, denoting "opposite" or "opposing"), which, like a pair of gloves, are mirror images of one another (Fig. 12.1) (see also Chapter 3). On Earth all living organisms use one enantiomer only: left-handed in the case of amino acids and right-handed for sugars. This property is called homochirality, a compound word derived from Greek, meaning "same handedness." Homochirality is essential for an efficient metabolism. Key life processes, such as protein synthesis and gene transcription, rely on amino acids and sugars having the correct spatial conformation to "shake hands" at the molecular level with their counterparts. Conversely, synthetic chemicals prepared in the laboratory exhibit equal abundances of both right- and left-handed enantiomers. Such a mixture is said to be racemic. Homochirality probably constitutes the most reliable indicator of the biological versus abiotic origin of organic molecules. Surely, testing for homochirality becomes crucial when searching for life. However, as with the previous methods outlined, this biosignature unfortunately will be lost when the sample is exposed to high temperatures or wet conditions for extended periods.

In summary, the best chance to find signatures of ancient life on Mars is in the form of chemical biomarkers and fossil communities, either preserved underground or encased within surface rocks. A few life-detection methods – by no means exhaustive – were presented to illustrate how important it is to use complementary techniques that, combined, give more credence to the proposition of a sample's biological potential. Several independent lines of evidence are required to construct a compelling case. *ExoMars* must therefore pursue a holistic search strategy, attacking the problem from multiple angles, including geological

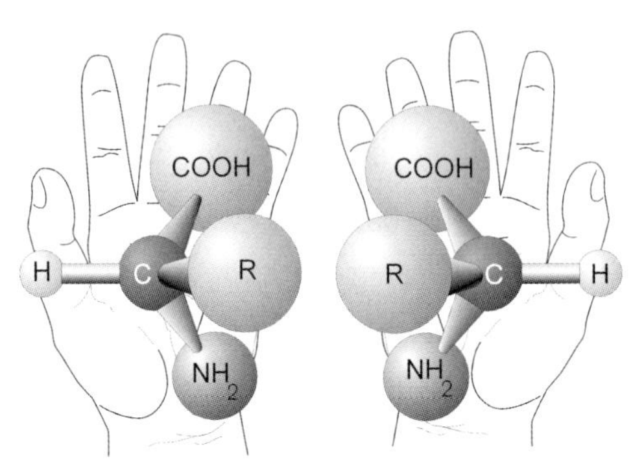

Fig. 12.1 Sketch demonstrating the chirality of amino acids (image credit: J. L. Bada).

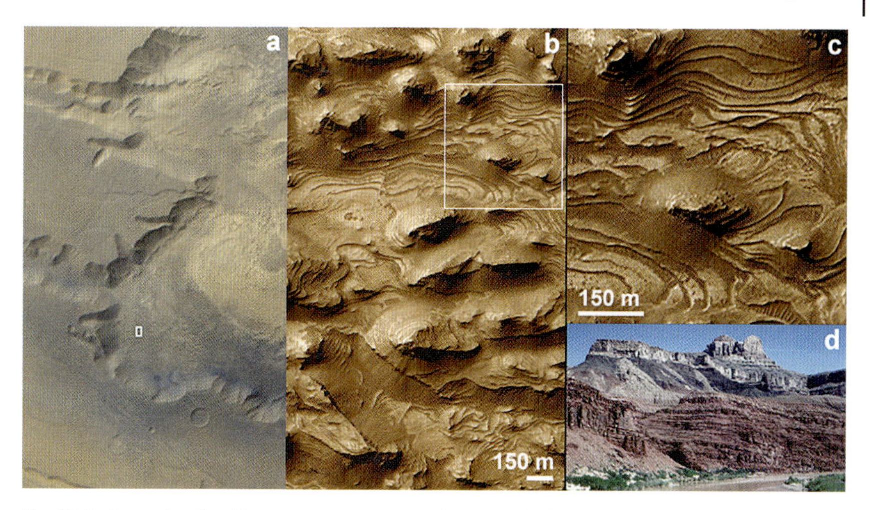

Fig. 12.2 Example of sedimentary outcrops on Mars. (a) Liquid-carved terrain in southwestern Candor Chasma (6.4° S, 77.1° W); (b) high-resolution image corresponding to the small white rectangle in (a); (c) detailed view of the layers or beds whose individual height is estimated to be about 10 m; (d) for comparison, similar features in the Grand Canyon (image credit: NASA/JPL/MalinSSS).

and environmental investigations (to characterize potential former habitats), visual examination of samples (morphology), and spectrochemical composition analyses.

It is important to note that exposure of fossil-bearing rocks to thermal reformation greatly degrades any biogenic morphological, chemical, or isotopic signatures. Thus, geologically recent volcanic regions and lava flows should be avoided when searching for traces of extinct life.

Because liquid water is a prerequisite for active life, good candidate locations to look for microfossils are terrains occupied by long-lasting bodies of water during Mars's early history, e.g., within ancient lacustrine or marine sedimentary rocks that accumulated rapidly, where subsequent diagenesis did not obliterate the original texture and compositional evidence – isotopic, organic, and mineralogical – of the deposition environment. (The term *diagenesis* refers to the physical and chemical changes occurring in sediments between the time of deposition and petrification.) Some promising sites have been identified through the analysis of images taken by the cameras of the current spacecraft orbiting Mars, such as *Mars Global Surveyor, Odyssey,* and *Mars Express.* One such example is presented in Fig. 12.2.

Figure 12.2 a shows a section of what may have been a lake in southwestern Chandor Chasma (6.4° S, 77.1° W), a region within the Valles Marineris system. Figure 12.2 b is a high-resolution image, corresponding to the small, white rectangle in Fig. 12.2 a. It reveals that the floor at this location contains numerous eroded, layered outcrops. Particularly interesting is the large number and uniformity of the layers or beds. They can be appreciated in more detail in Fig. 12.2 c. The thickness of the individual layers is estimated to be approximately 10 m. They have smooth upper surfaces and are hard enough to form sharp, cliff-like edges, which suggest that the beds may consist of indurated fine particles. (*Induration* means the hardening or cementing process that consolidates sediments to form sedimentary rocks.) On Earth, such regular layers signal episodic interruptions in the material deposition process. Patterns like these are usually associated with sediments accumulated in dynamic aqueous environments over very long periods, spanning several hundred million years. The similarities to terrestrial water sedimentation patterns are striking. For comparison, Fig. 12.2 d displays sedimentary rocky outcrops in the Grand Canyon. The layers in the foreground are known as the Dox Formation; they were deposited in rivers and floodplains more than a billion years ago.

It is trapped within old, exposed sedimentary material and evaporitic deposits that the record of ancient Martian life, if it ever existed, is likely to be preserved.

12.2.1.2 Extant Life

In 1976, NASA's twin *Viking* landers conducted the first *in situ* measurements focusing on the detection of organic compounds and life on Mars. The *Viking* biology package contained three experiments, all looking for signs of metabolism in soil samples. One of them, the labeled-release experiment, produced very provocative results (see Chapter 5). If other information had not been available as well, these data could have been interpreted as proof of biological activity. However, theoretical modeling of the Martian atmosphere and regolith chemistry hinted at the existence of powerful oxidants, which could more or less account for the results of the three biology package experiments. The biggest blow was the failure of the *Viking* gas chromatograph–mass spectrometer (GC-MS) to find evidence of organic molecules at the parts-per-billion (ppb) level. Although not intended as a life-detection experiment, the GC-MS measurements were regarded to be inconsistent with a biochemical explanation for the labeled-release data. It must be stressed that the *Viking* GC-MS was not designed to search for living cells, but rather for the gaseous pyrolytic degradation products of organic substances. Hence, this instrument could have missed significant amounts of non-volatile organic compounds. Recent work suggests that the sensitivity of the *Viking* GC-MS may have been insufficient to detect the organics released by heat from up to 30 million bacteria per gram of soil. This number reflects the concentration of microorganisms detected, e.g., in permafrost samples on Earth. The interpretation that the labeled-release results may be due to the action of highly reactive oxidants in the regolith is still debated; nevertheless, with few exceptions, the majority of the scientific community has concluded that the *Viking* results do not demonstrate the presence of life.

Numerous attempts have been made in the laboratory to simulate the reactions observed by the *Viking* biological package. While some have reproduced certain aspects of the data, none has succeeded entirely. An oxidant experiment was part of the Russian *Mars '96* scientific payload. Alas, a launcher failure meant the loss of that opportunity to characterize the reactivity of the Martian soil. Incredibly, almost 30 years after the *Viking* missions, the crucial chemical oxidant hypothesis remains untested. *ExoMars* will therefore include a powerful instrument to study oxidants and their relation to organics distribution on Mars.

Undoubtedly, the present environment on Mars is exceedingly hostile for the widespread proliferation of surface life: it is simply too cold and dry, not to mention the large doses of UV radiation. Notwithstanding these hazards, basic organisms may still flourish in protected places: deeply underground, at shallow depths in especially benign environments, or within rock cracks and inclusions.

Perhaps a good first step is to consider Earth ecosystems with conditions approximating those of the Red Planet. In this regard, it is the frigid desert of the Antarctic dry valleys (77° 45' S) that bears the closest resemblance to the Martian environment today. This region has temperatures varying between −15 °C and 0 °C in the summer, and as low as −60 °C during the winter, with a relative humidity of 16–75 %. The melting of the infrequent snow coverage on rocks is the main source of water for life there.

The primary producers are photosynthetic endolithic microbial communities dominated by cryptoendolithic lichens. These microorganisms colonize a narrow zone a few millimeters beneath the surface of rocks (Fig. 12.3). This habitat provides a favorable microclimate and is well protected from the harsh outside environment (strong winds, temperature fluctuations, desiccation, and UV radiation). Cryptoendolithic communities are found only in weathered or porous rocks, because only these types of rocks offer the necessary substrate to colonize their interior, the permeability for the uptake of liquid water and moisture, and the translucent property required for the photosynthetic primary producer. Usually, endoliths grow only on the faces of rocks where the highest insolation is received: in Antarctica, north-facing or horizontal. Water is provided to the rock by blowing snow or frost, which melts into the rock when it is warmed by sunlight. During the summer, freeze–thaw transitions are common. The endoliths are wetted either by equilibration with the high-humidity air in the rock or by direct moisture uptake after snow/frost melt.

A similar kind of microhabitat could be available on Mars. From the *Viking* mission we know that surface frost is widespread in the morning hours at certain locations; in addition, the *Pathfinder* landing site contained many weathered rocks. Cryptoendolithic communities colonizing the interstitial space in Martian rocks would be protected from the harsh UV radiation. Even after the extinction of these

Fig. 12.3 Cryptoendolithic microorganisms from the McMurdo Dry Valleys, Antarctica. These cold-adapted communities live in favorable microclimates, just beneath the surface of porous rocks facing the Sun (image credit: Russ Kinne, NSF).

communities, certain stable biomarkers, such as pigments, might be preserved in the cold and dry environment for long periods. Life could have escaped the deteriorating climatic conditions on the surface of Mars by finding refuge in habitats that are very similar to those colonized by cryptoendolithic communities in Antarctica. Finding out whether this happened will be part of the investigations performed by the *ExoMars* mission.

12.2.1.3 Search for Life: Conclusions

The strategy to find traces of past biological activity rests on the assumption that any surviving signatures of interest will be preserved in the geological record, in the form of buried/encased remains, organic materials, and fossil communities. Similarly, because current Martian surface conditions are hostile to most known organisms, also when looking for signs of extant life, the search methodology should focus on investigations in protected niches: underground, in permafrost, or within surface rocks. Therefore, the same sampling device and instrumentation can adequately serve both types of studies. The biggest difference is due to location requirements. When searching for extinct life, the interest lies in areas occupied by ancient bodies of water over many thousands of years. For extant life, the emphasis is on water-rich environments close to the surface and accessible to our sensors today. For the latter, the presence of permafrost alone may not be enough. Permafrost in combination with a sustained heat source, probably of volcanic or hydrothermal origin, may be necessary. Such warm oases can only be identified by an orbital survey of the planet. In the next few years, a number of remote sensing satellites, such as ESA's *Mars Express* and NASA's *Mars Reconnaissance Orbiter* (MRO), will determine the water–ice boundary across Mars and may help to discover any such warm spots. If they do exist, they would be prime targets for missions such as *ExoMars*.

On Earth, microbial life quickly became a global phenomenon. If the same explosive process occurred on the young Mars, the chances of finding evidence of past life may be good. Even more interesting would be the discovery and study of life forms that have successfully adapted to modern Mars, such as the cryptoendolithic communities discussed in Section 12.2.1.2. However, this presupposes the prior identification of geologically suitable, life-friendly locations where it can be demonstrated that liquid water still exists, at least episodically throughout the year. For these reasons, the "Red Book" science team advised ESA to focus mainly on the detection of extinct life, but also to build enough flexibility into the mission design to be able to target sites with potential for present life.

12.2.2 Hazards for Human Operations on Mars

Before astronauts can be sent to Mars, one must understand and control any risks that may pose a threat to a mission's success. Some of these risks will be assessed with *ExoMars*.

Ionizing radiation is probably the single most important limiting factor for human interplanetary flight. To evaluate its danger and to define efficient mitigation strategies, it is desirable to incorporate radiation-monitoring capabilities during cruise, orbit, and surface operations on precursor robotic missions to Mars.

Another physical hazard may result from the basic mechanical properties of the Martian soil. Dust particles will invade the interior of a spacecraft during surface operations, as shown by the *Apollo* missions to the Moon. Dust can pose a threat to astronauts on Mars, and even more so under microgravity during the return flight to Earth. Characteristics of the soil, such as the size, shape, and composition of individual particles, can be studied with dedicated *in situ* instrumentation. However, a more in-depth assessment, including a toxicity analysis, will require the return of a suitable Martian sample.

Reactive inorganic substances could constitute chemical hazards on the Martian surface. Free radicals, salts, and oxidants are very aggressive in the presence of humidity, e.g., in the lungs and eyes. In addition, toxic metals, organics, and pathogens are potential hazards. As with dust, chemical hazards present in the soil might contaminate the interior of a spacecraft during surface operations. They could affect the health of astronauts and the operation of equipment. Many potential inorganic and organic chemical hazards may be identified with the *ExoMars* search-for-life instruments.

12.2.3 Geophysics Measurements

The processes that have determined the long-term habitability of Mars depend on the geodynamics of the planet and on its geological evolution and activity. Important issues still need to be resolved: What is Mars's internal structure? Is there any volcanic activity on Mars? The answers may allow us to extrapolate into the past, to estimate when and how Mars lost its magnetic field and the importance of volcanic outgassing for the early atmosphere.

ExoMars will carry a geophysics and environment package (GEP), accommodated on the descent module.

12.3 *ExoMars* Science Strategy

The mission strategy to achieve the scientific objectives of *ExoMars* is as follows:

- To land on, or be able to reach, a location of high exobiology interest for past and/or present life signatures, i.e., access to the appropriate geological environment.
- To collect scientific samples from different sites, using a rover carrying a drill system capable of reaching well into the soil and surface rocks. This requires mobility and access to the subsurface.

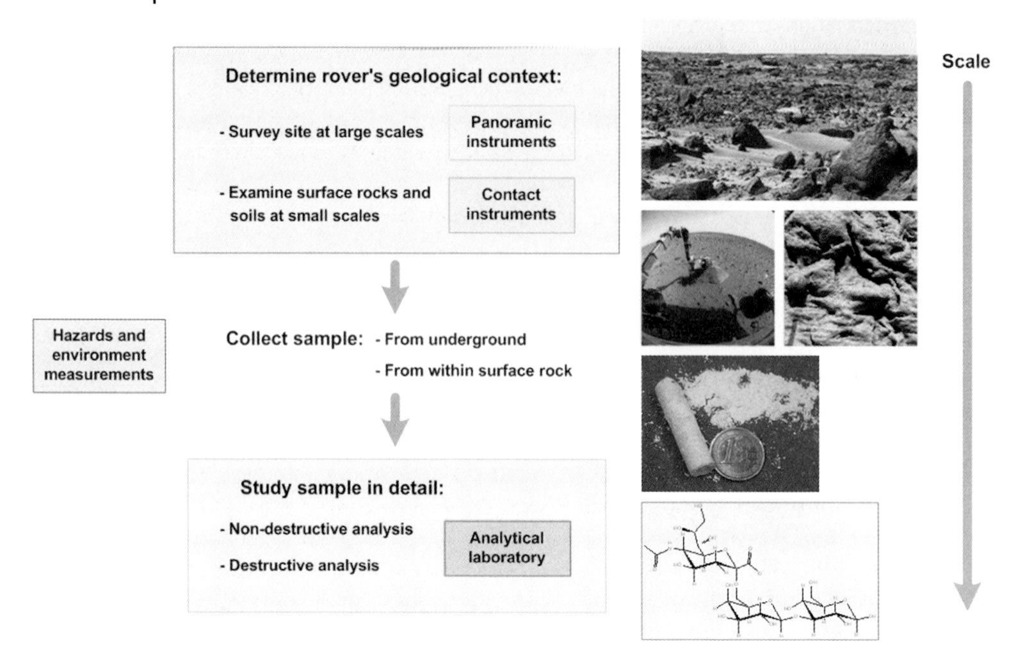

Fig. 12.4 The *ExoMars* surface science exploration scenario. The rover will conduct measurements at multiple scales, starting with a panoramic assessment of the geological environment, progressing to smaller-scale investigations on interesting surface rocks using a suite of contact instruments, and culminating with the collection of well-selected samples to be studied by the rover's analytical laboratory (image credit: ESA).

- To conduct at each site an integral set of measurements at multiple scales (Fig. 12.4), beginning with a panoramic assessment of the geological environment, progressing to smaller-scale investigations on interesting surface rocks using a suite of contact instruments, and culminating with the collection of well-selected samples to be studied by the rover's analytical laboratory.
- To characterize the environment and the geophysical parameters relevant for life, the hazards to humans, and the planetary evolution.

To arrive at a clear and unambiguous conclusion on the existence of past or present life at rover sites, it is essential that the search instrumentation be able to provide mutually reinforcing lines of evidence, while minimizing the opportunities for alternative interpretations. It is imperative that all instruments be carefully designed so that none becomes a weak link in the chain of observations. Performance limitations in an instrument intended to confirm the results obtained by another should not generate confusion and discredit the whole measurement.

The science strategy for the Pasteur payload is therefore to provide a self-consistent set of instruments that can obtain reliable evidence for or against the existence of a range of biosignatures at each search location. This will be achieved through the detection of organic compounds and oxidants, as well as by determining the geological context, mineralogy, and geochemistry of the samples being analyzed.

12.4 *ExoMars* Mission Description

The baseline mission scenario, shown in Fig. 12.5, consists of a spacecraft composite with a carrier and a descent module, launched by a *Soyuz* 2 b rocket from Kourou, Europe's spaceport in French Guiana. It will follow a two-year "delayed trajectory" designed to reach Mars after the statistical dust storm season. The carrier will release the descent module from the hyperbolic arrival path. The descent module will enter the Martian atmosphere, land (using either bouncing [non-vented, as in MER] or non-bouncing [vented] airbags), and deploy the rover and GEP. In the baseline mission, data relay for the rover will be provided by a NASA orbiter. The main characteristics of the *ExoMars* mission are listed in Table 12.1.

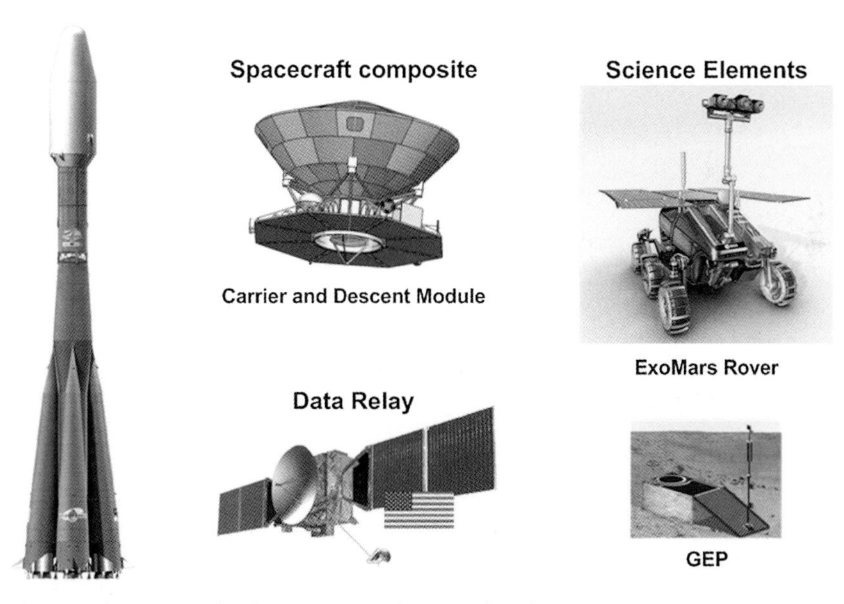

Fig. 12.5 The *ExoMars* "baseline" mission architecture, based on a "carrier plus descent module" composite launched by a *Soyuz* 2 b rocket. The descent module is released from the hyperbolic arrival trajectory and delivers the rover to the Martian surface (image credit: ESA).

Table 12.1 Main characteristics of the *ExoMars* mission.

Area	Type	Specification
Technical data	Spacecraft	Carrier plus descent module including rover and GEP[c] Data relay provided by NASA
	Launch	May–June 2013, from Kourou on Soyuz 2 b (backup 2015)
	Arrival at Mars	June 2015 (backup 2017)
	Landing on Mars	Direct entry, from hyperbolic trajectory, after the dust storm season. Latitudes between 15° S and 45° N, all longitudes; Altitude: ≤ 0 m, relative to the MGS/MOLA[a] zero level
Science	Rover + Pasteur	Mass 170–190 kg, includes drill system, SPDS[b], and instruments (8 kg); Lifetime ≥ 180 sols
	GEP[c]	Mass ≤ 20 kg; includes instruments (4–5 kg); Lifetime ≥ 1 year
Ground segment	Mission control/mission operations	European Space Observation Center (Germany)
	Rover operation on Mars surface	Rover Operations Center (Italy)
	GEP[c] operations	To be decided

a *Mars Global Surveyor*/Mars Orbiter Laser Altimeter
b Sample preparation and distribution system
c Geophysics and environment package

An alternative configuration, based on an Ariane 5 ECA launcher, may be implemented depending on programmatic, technical, and financial considerations (Ariane 5 ECA is the latest version of the Ariane 5 launcher. In this option (Fig. 12.6), the carrier is replaced by an orbiter that provides end-to-end data relay for the surface elements. The orbiter will also carry a science payload to complement the results from the rover and the GEP and provide continuity to the great scientific discoveries of *Mars Express*.

ExoMars is a search-for-life mission targeting regions with high life potential. It has therefore been classified as planetary protection category IVb/c (see Chapter 13). This, coupled with the mission's ambitious scientific goals, imposes challenging requirements regarding sterilization and organic cleanliness.

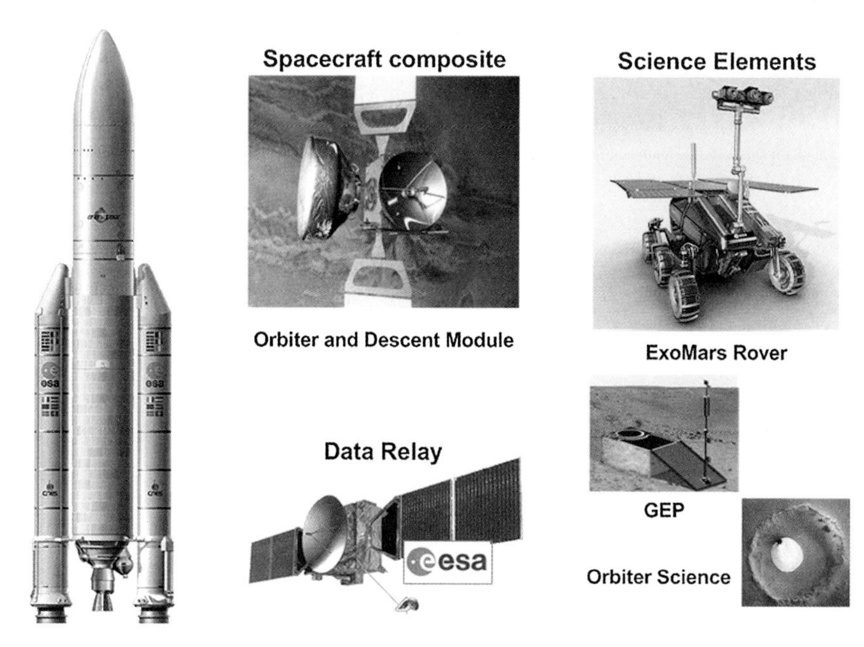

Fig. 12.6 The *ExoMars* "orbiter" option architecture, based on an "orbiter plus descent module" composite launched by an Ariane 5 ECA rocket. The descent module delivers the rover to the Martian surface, and the orbiter includes a science payload (image credit: ESA).

12.4.1
The *ExoMars* Rover

The *ExoMars* rover will have a nominal lifetime of 180 sols (approximately 6 months, 1 sol = 24 h and 39 min). During this period, it will ensure a regional mobility of several kilometers, relying on solar array electrical power (Fig. 12.7). The Pasteur model payload (see Tables 12.2 and 12.3) contains:

- panoramic instruments (cameras, a ground-penetrating radar, and an IR spectrometer);
- contact instruments for studying surface rocks and soils (close-up imager and Mössbauer spectrometer);
- a subsurface drill capable of reaching depths down to 2 m and of collecting specimens from exposed bedrock;
- a sample preparation and distribution system; and
- the analytical laboratory.

The latter includes a microscope, an oxidation sensor, and a variety of instruments for characterizing the organic substances and the geochemistry in the collected samples.

Table 12.2 Recommended Pasteur exobiology instruments.

Instrument	Function
Panoramic instruments	To characterize the rover's geological environment, both on the surface and in the subsurface. Typical scales span from kilometers to meters, with sub-centimeter resolution for close targets.
Panoramic camera system	Two wide-angle stereo cameras plus one high-resolution camera; to characterize the rover's geological environment. Very important for scientific target selection.
Infrared (IR) spectrometer	For the remote identification of water-related minerals and for scientific target selection.
Ground penetrating radar	To establish the subsurface stratigraphy down to 3-m depth and to help plan the drilling strategy.
Contact instruments	To investigate surface rocks and soils. Among the scientific interests at this scale are macroscopic textures, structures, and layering and bulk mineralogical and elemental content. This information will be fundamental for collecting samples for more detailed analysis. The preferred solution is to deploy the contact instruments using an arm-and-paw arrangement, as in *Beagle 2*. Alternatively, in case of mass limitations, they could be accommodated at the base of the subsurface drill.
Close-up imager	To visually investigate scientific targets at the centimeter range with sub-millimeter resolution.
Mössbauer spectrometer	To study the mineralogy of Fe-bearing rocks and soils.
Raman-LIBS[a] external optical heads	To determine the geochemistry of rocks and soils. These are external heads connected to the instruments inside the analytical laboratory.
Facility-support equipment	These essential systems are devoted to the acquisition and preparation of samples for detailed investigations in the analytical laboratory. They must follow specific acquisition and preparation protocols to guarantee the optimal survival of any organic molecules and volatiles in the samples. The mission's ability to break new scientific ground, particularly for signs-of-life investigations, depends on these two instruments.

Instrument	Function
Subsurface drill	Capable of obtaining samples from 0 to 2 m depth, where organic molecules and volatiles might be well preserved. It also integrates temperature sensors and an IR spectrometer for borehole mineralogy studies.
SPDS[b]	Receives a sample from the drill system, prepares it for scientific analysis, and presents it to all analytical laboratory instruments. A very important function is to produce particulate material while preserving the organic and water content.
Analytical laboratory	To conduct a detailed analysis of each collected sample. The first step is a visual and spectroscopic inspection. If the sample is deemed interesting, it is ground up and the resulting particulate material is used to search for organic molecules and to perform more-accurate mineralogical investigations.
Microscope (VIS + IR)	To examine the collected samples to characterize their structure and composition at the grain-size level. These measurements will also be used to select sample locations for further detailed analyses by the Raman-LIBS spectrometers.
Raman-LIBS	To determine the geochemistry of the collected samples.
X-ray diffractometer	To determine the true mineralogical composition of a sample's crystalline phases.
Urey (Mars Organics + Oxidant Detector)	The Mars Organics Detector searches for amino acids, nucleobases, and polyaromatic hydrocarbons in the collected samples with extremely high sensitivity (ppt). Can also function as front-end to the GC-MS. The Mars Oxidant Instrument determines the chemical reactivity of oxidants and free radicals in the soil and atmosphere.
GC-MS	To conduct a broad-range, very high-sensitivity search for organic molecules in the collected samples; also for atmospheric analyses.
Life-Marker Chip	Antibody-based instrument to detect organic molecules with very high specificity.

Mass (without drill and SPDS): 12.5 kg.
a LIBS: laser-induced breakdown spectroscopy.
b SPDS: sample preparation and distribution system.

Fig. 12.7 Artist's representation of the *ExoMars* rover's egress from the descent module (image credit: ESA).

Table 12.3 Recommended Pasteur environment instruments.

Instrument	Function
Environment instruments	To characterize possible hazards to future human missions and to increase our knowledge of the Martian environment.
Dust suite	Determines the dust grain size distribution and deposition rate. It also measures water vapor with high precision.
UV spectrometer	Measures the UV radiation spectrum.
Ionizing radiation	Measures the ionizing radiation dose reaching the surface due to cosmic rays and solar particle events.
Meteorological package	Measures pressure, temperature, wind speed and direction, and sound.

Mass: 1.9 kg.

A key rover element is the drill. The reason for the 2-m requirement is the need to obtain pristine sample material for analysis. The tenuous atmosphere of Mars allows damaging space radiation to penetrate into the uppermost meters of the planet's subsurface, affecting the long-term preservation of organic molecules that might be associated with extinct life or meteoritic infall. Figure 12.8 shows the surviving fraction of amino acids versus depth after simulated exposures of 0.5, 1, and 3 billion years to ionizing radiation in the Martian subsurface. With the present detection limit of 0.01 ppb per amino acid, reductions of up to 10^{-6} can be tolerated (dashed line in Fig. 12.8), beyond which amino acid signatures become undetectable. This result indicates that when one is searching for biomarkers of Martian life that became extinct more than 3 billion years ago, it is necessary to access the subsurface in the range of 2 m.

On Mars, any exposed organic biomarker or life form is readily degraded by UV radiation and by reactive oxidant species resulting from surface UV-induced photochemistry. Although no clear proof regarding the presence of reactive oxidants has yet been established, radiochemical models of the Martian atmosphere and measurements from the *Viking* experiments suggest their presence on the Martian surface. Depending on the modeling parameters, the calculated range for the oxidant extinction depth (i.e., the depth at which the modeled concentration of the oxidizing agents approaches zero because of their interaction with the subsurface environment) is between centimeters and several meters. However, these estimates refer to global mean values. The actual penetration of putative oxidants will depend on local conditions, such as soil/rock porosity, which can limit their diffusion. To minimize the effects of oxidants, the *ExoMars* drill must be able to penetrate and obtain samples from well-consolidated (hard) formations, such as sedimentary rocks and evaporitic deposits, at depths beyond 1.5 m. Additionally, it is essential to avoid loose dust deposits distributed by eolian transport. While driven by the wind, this material was intensely processed by UV radiation, ionizing radiation, and potential oxidants in the atmosphere and on the surface of Mars. Any organic biomarkers will be highly degraded in these samples.

The *ExoMars* rover must monitor and control torque, thrust, penetration depth, and temperature at the drill bit. Grain-to-grain friction in a continuous rotary drill can generate a heat wave in the sample, destroying the organic molecules that *ExoMars* seeks to detect. The drill therefore has a variable cutting protocol that can dissipate heat in a science-safe manner. The drill's full 2-m extension is achieved by assembling four sections: one drill tool rod, with an internal shutter and sampling

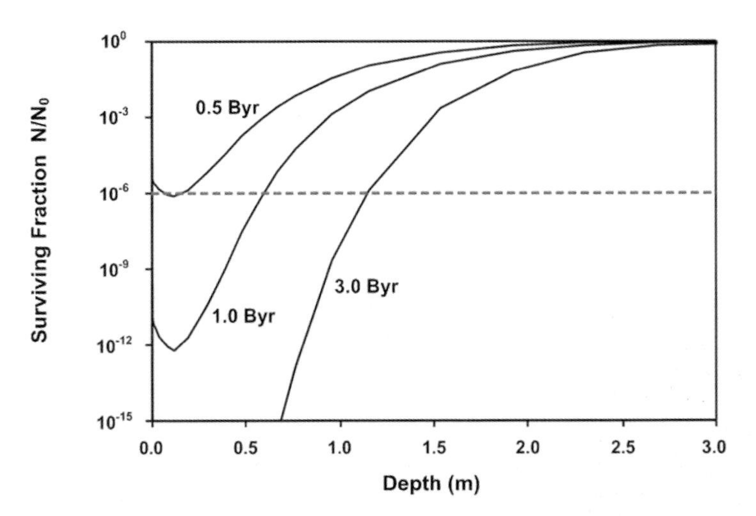

Fig. 12.8 Surviving fraction of amino acids versus depth in the Martian subsurface after simulated exposures of 0.5, 1, or 3 billion years to ionizing radiation. The dashed line at a survival rate of 10^{-6} gives the present detection limit of 0.01 ppb per amino acid, beyond which amino acid signatures become undetectable. Therefore, when searching for biomarkers of Martian life that became extinct more than 3 billion years ago, it is necessary to access the subsurface in the range of 2 m. Adapted from Kminek and Bada (2006).

collection capability, plus three extension rods (Fig. 12.9). The drill is also equipped with an IR spectrometer for mineralogy studies inside the borehole.

The analysis sequence within Pasteur's analytical laboratory is presented in Fig. 12.10. The drill system delivers the acquired sample to the sample preparation and distribution system (SPDS). A detailed optical/spectral inspection is performed using the microscope, the Raman spectrometer, and the laser-induced breakdown spectrometer. After these data have been studied, the rover control center instructs the SPDS to grind the sample. Powder material is then distributed to the remaining instruments. The X-ray diffractometer determines the sample's mineralogy. The Mars Oxidant Instrument characterizes chemical species in the Martian soil and measures their reactivity. The Mars Organics Detector searches for trace levels of specific organic molecules, such as amino acids, nucleotides, and polycyclic aromatic hydrocarbons with unprecedented sensitivity (a factor of 10^4 greater than the *Viking* GC-MS package; see Section 12.2.1.2). The GC-MS conducts a broad search for organic molecules. Finally, the Life-Marker Chip looks for biomarkers using recombinant antibody technology.

If any organic compounds are detected on Mars, it should be possible to demonstrate that they were not brought from Earth. The only way to reliably prove that the rover is free from contaminants is to perform an initial measurement run using a blank sample. It is not sufficient to do this at the instrument level, as the contamination could still come from the sampling device. For this reason, the *ExoMars* rover will carry three blank calibration samples that are accessible to the drill system. Upon landing, one of the first scientific actions will be to use the drill to collect and pass a blank sample to the analytical laboratory and perform a full investigation. The result of this investigation should indicate "no life" and "no organics." Failure to obtain this first negative reading may invalidate any later search-for-life findings. The second blank sample serves a different purpose. It

Fig. 12.9 Artist's view of the *ExoMars* rover with the drill obtaining a sample from the Martian subsurface (image credit: ESA).

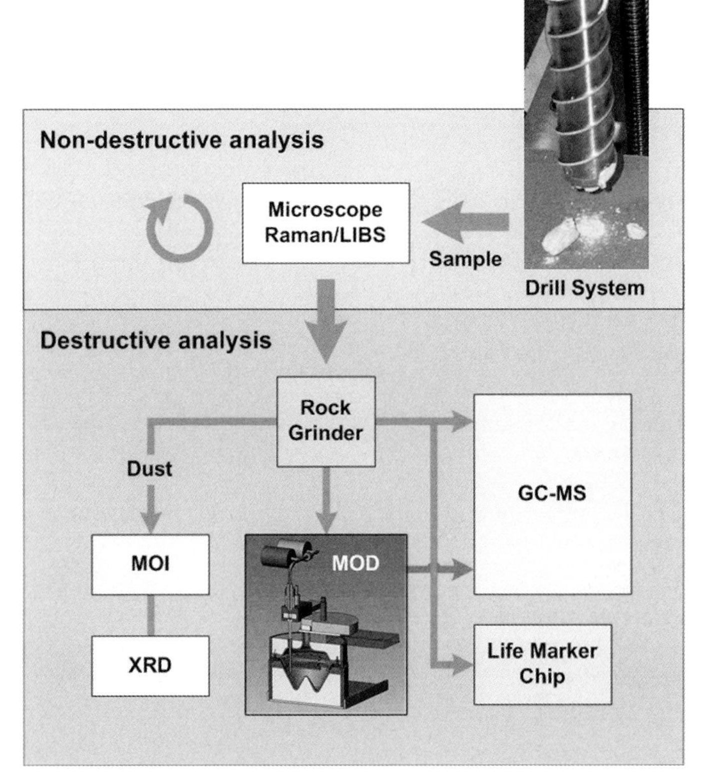

Fig. 12.10 Analysis sequence within Pasteur's analytical laboratory (image credit: ESA).

provides a means to directly assess the impact of sample cross-contamination. In case organics are detected, a blank sample can be introduced into the measuring chain to quantify *in situ* how much the organic levels drop from those measured using material from the previous sample. The third blank sample is a reserve.

12.5
Outlook and Conclusions

NASA's highly successful twin MER missions were conceived as robotic geologists. They have demonstrated the past existence of long-lasting, wet environments on Mars. Their results have persuaded the scientific community that mobility is a must-have requirement for all future surface missions. Recent discoveries from *Mars Express* have revealed multiple deposits containing salt and clay minerals that can form only in the presence of liquid water. This reinforces the hypothesis that ancient Mars may have been wetter, and possibly warmer, than it is today. NASA's 2009 *Mars Science Laboratory* will study surface geology and organics, with the goal of identifying habitable environments. ESA's 2013 *ExoMars* is the next logical step. It will have instruments to investigate whether life ever arose on the Red Planet. It will also be the first mission combining mobility and access to subsurface locations where organic molecules may be well-preserved, thus allowing for the first time an investigation of Mars's third dimension: depth. This alone is a guarantee that the mission will break new scientific ground. Finally, the many technologies developed for this project will allow ESA to prepare for international collaboration on future missions, such as *Mars Sample Return*.

12.6 Further Reading

12.6.1 Books and Articles in Books

Brack, A, Fitton, B., Raulin, F. (Scientific Coordinators) *Exobiology in the Solar System & The Search for Life on Mars*, Report from the ESA Exobiology Team Study 1997–1998, Wilson A. (Ed.), SP-1231, ESA Publications Division, ESTEC, Noordwijk, The Netherlands, 1999.

Godwin, R., *Viking '75 Report,* in: *Mars – The NASA Mission Reports,* Apogee Books, Burlington, Ontario, 2000.

Tokano, T. (Ed.) *Water on Mars and Life,* Springer, Berlin, Heidelberg, 2005.

Walter, M. (Ed.) *The Search for Life on Mars,* Perseus Books, Cambridge, Massachusetts, 1999

Wilson, A. (Ed.) *Mars Express – The Scientific Payload,* ESA SP-1240, ESA Publications Division, ESTEC, Noordwijk, The Netherlands, 2004.

12.6.2 Articles in Journals

Allwood, A. C., Walter, M. R., Kamber, B. S., Marshall, C. P., Burch, I. W. Stromatolite reef from the early archaen era of Australia, *Nature* **2006**, *441, 714–718, doi:10.1038/nature04764.*

Anderson, R. C., Dohm, J. M., Golombek, M. P., Haldemann, A. F.C., Franklin, B. J., Tanaka, K. L., Lias, J., Peer, B. Primary centers and secondary concentrations of tectonic activity through time in the western hemisphere of Mars, *J. Geophys. Res.* **2001**, *106*(E9), 20563–20585.

Bada, J. L., Sephton, M. A., Ehrenfreund, P., Mathies, R. A., Skelley, A. M., Grunthaner, F. J., Zent, A. P., Quinn, R. C., Josset, J.-L., Robert, F., Botta, O., Glavin, D. P. Life on Mars: New strategies to detect life on Mars, *Astron. Geophys.* **2005**, *46*(6), 26–27, doi:10.1111/j.1468–4004.2005.46626.x.

Baker, V. R. Icy Martian misteries, *Nature* **2003**, *426*, 779–780.

Baker, V. R. Water and the Martian landscape, *Nature* **2001**, *412*, 228–236.

Bell III, J. F., McSween, H. Y., Jr., Crisp, J. A., Morris, R. V., Murchie, S. L., Bridges, N. T., Johnson, J. R., Britt, D. T., Golombek, M. P., Moore, H. J., Ghosh, A., Bishop, J. L., Anderson, R. C., Brückner, J., Economou, T., Greenwood, J. P., Gunnlaugsson, H. P., Hargraves, R. M., Hviid, S., Knudsen, J. M., Madsen, M. B., Reid, R., Rieder, R., Soderblom, L. Mineralogic and compositional properties of Martian soil and dust: Results from Mars Pathfinder, *J. Geophys. Res.* **2000**, *105*(E1), 1721–1755.

Bell, J. Tip of the Martian iceberg? *Science* **2002**, *297*, 60–61.

Bibring, J.-P., Langevin, Y., Mustard, J. F., Poulet, F., Arvidson, R., Gendrin, A., Gondet, B., Mangold, N., Pinet, P., Forget, F., and the OMEGA team. Global mineralogical and aqueous Mars history derived from

OMEGA/Mars Express data, an ESA study for the search for life on Mars, *Science* **2006**, *312*. 5772, 400–404, doi:10.1126/science.1122659.

Biemann, K., Oro, J., Toulmin III, P., Orgel, L. E., Nier, A. O., Anderson, D. M., Simmonds, P. G., Flory, D., Diaz, A. V., Rushneck, D. R., Billier, J. E., Lafleur, A. L. The search for organic substances and inorganic volatile compounds in the surface of Mars, *J. Geophys. Res.* **1977**, *82*, 4641–4658.

Boynton, W. V. Feldman, W. C., Squyres, S. W., Prettyman, T., Brückner, J., Evans, L. G., Reedy, R. C., Starr, R., Arnold, J. R., Drake, D. M., Englert, P. A. J., Metzger, A. E., Mitrofanov, I., Trombka, J. I., d'Uston, C., Wänke, H., Gasnault, O., Hamara, D. K., Janes, D. M., Marcialis, R. L., Maurice, S., Mikheeva, I., Taylor, G. J., Tokar, R., Shinohara, C. Distribution of hydrogen in the near surface of Mars: Evidence for subsurface ice deposits, *Science* **2002**, *297*, 81–85.

Brack, A., Clancy, P., Fitton, B., Hofmann, B., Horneck, G., Kurat, G., Maxwell, J., Ori, G. G., Pillinger, C., Raulin, F., Thomas, N., Westall, F. An Integrated Exobiology Package for the search for life on Mars, *Adv. Space Res.* **1999**, *23*(2), 301–308.

Carr, M. H., Head, J. W.III. Oceans on Mars: An assessment of the observational evidence and possible fate, *J. Geophys. Res.* **2003**, *108*(E5), 5042, doi:10.1029/2002JE001963.

Carr, M. H., Malin, M. C. Meter-Scale characteristics of Martian channels and valleys, *Icarus* **2000**, *146*, 366–386.

Clark, B. C. Surviving the limits to life at the Martian surface, *J. Geophys. Res.* **1978**, *103*(E12), 28545–28555.

Cockell, C. S. Catling, D. C., Davis, W. L., Snook, K., Kepner, R. L., Lee, P., McKay, C. P. The ultraviolet environment of Mars: Biological implications past, present, and future, *Planet. Space Sci.* **2000**, *48*, 343–359.

Cockell, C. S., Raven, J. A. Zones of photosynthetic potential on Mars and the early Earth, *Icarus* **2004**, *169*, 300–310.

Craddock, R. A., A. D. Howard, A. D. The case for rainfall on a warm, wet early Mars, *J. Geophys. Res.* **2002**, *107*(E11), 5111, doi:10.1029/2001JE001505.

Dohm, J. M., Anderson, R. C., Baker, V. R., Ferris, J. C., Hare, T. M., Strom, R. G., Rudd, L. P., Rice, J. W. Jr., Casavant, R. R., Scott, D. H. System of gigantic valleys northwest of Tharsis, Mars: Latent catastrophic flooding, northwest watershed, and implications for Northern Plains ocean, *Geophys. Res. Lett.* **2000**, *27*(21), 3559–3562.

Dohm, J. M., Anderson, R. C., Baker, V. R., Ferris, J. C., Rudd, L. P., Hare, T. M., Rice, J. W. Jr., Casavant, R. R., Strom, R. G., Zimbelman, J. R., Scott, D. H. Latent outflow activity for Western Tharsis, Mars: Significant flood record exposed, *J. Geophys. Res.* **2001**, *106*(E6), 12301–12314.

Dohm, J. M., Ferris, J. C., Baker, V. R., Anderson, R. C., Hare, T. M., Strom, R. G., Barlow, N. G., Tanaka, K. L., Klemaszewski, J. E., Scott, D. H. Ancient drainage basin of the Tharsis region, Mars: Potential source for outflow channel systems and putative oceans or paleolakes, *J. Geophys. Res.* **2001**, *106*(E12), 32943–32958.

Doran, P. T., Wharton, R. A., Jr., Des Marais, D. J., McKay, C. P. Antarctic paleolake sediments and the search for dxtinct life on Mars, *J. Geophys. Res.* **1998**, *103*(E12), 28481–28493.

Edgett, K. S. The sedimentary rocks of Sinus Meridiani: Five key observations from data acquired by the Mars Global Surveyor and Mars Odyssey Orbiters, *Mars* **2005**, *1*, 5–58, doi:10.1555/mars.2005.0002.

Edwards, H. G. M., Moody, C. D., Jorge Villar, S. E., Wynn-Williams, D. D. Raman spectroscopic detection of key biomarkers of cyanobacteria and lichen symbiosis in extreme Antarctic habitats: Evaluations for Mars lander missions, *Icarus* **2005**, *174, 560–571, doi:10.1016/j.icarus.2004.07.029.*

Edwards, H. G. M., Moody, C. D., Newton, E. M., Jorge Villar, S. E., Russel, M. J. Raman Spectroscopic analysis of cyanobacterial colonisation of hydromagnesite, a putative Martian extremophile, *Icarus* **2005**, *175, 372–381, doi:10.1016/j.icarus.2004.12.006.*

Farmer, J. D., Des Marais, D. J. Exploring for a record of ancient Martian life, *J. Geophys. Res.* **1999**, *104*(E11), 26977–26995.

Formisano, V., Atreya, S., Encrenaz, T., Ignatiev, N., Giuranna, M. Detection of methane in the atmosphere of Mars, *Science* **2004**, *306*, 5702, 1758–1761, doi:10.1126/science.1101732.

Gendrin, A., Mangold, N., Bibring, J.-P., Langevin, Y., Gondet, B., Poulet, F., Bonello, G.,

Quantin, C., Mustard, J., Arvidson, R., Le-Mouélic, S. Sulfates in Martian layered terrains: The OMEGA/Mars Express view, *Science* **2005**, *307, 5715, 1587–1591, doi:10.1126/science.1109087.*

Gibson, E. K. Jr., McKay, D. S., Thomas-Keprta, K., Romanek, C. S. The case for relic life on Mars, *Scientific American* **1997**, *277*(6), 58–65.

Gillespie, A. R., Montgomery, D. R., Mushkin, A. Planetary science: Are there active glaciers on Mars? *Nature* **2005**, *438*, E9, doi:10.1038/nature04357.

Glavin, D. P., Schubert, M., Botta, O., Kminek, G., Bada, J. L. Detecting pyrolysis products from bacteria on Mars, *Earth and Planetary Science Letters* **2000**, *5721*, 1–5.

Greeley, R., Kuzmin, R. O., Nelson, D. M., Farmer, J. D. Eos Chasma, Mars: Regional setting for a potential landing site for astrobiology, *J. Geophys. Res.* **2003**, *108*(E12), 8083, doi:10.1029/2002JE002014.

Haberle, R. M., McKay, C. P., Schaeffer, J., Cabrol, N. A., Grin, E. A., Zent, A. P., Quinn, R. On the possibility of liquid water on present-day Mars, *J. Geophys. Res.* **2001**, *106*, E10, 23317–23326.

Hartmann,W.K., Thorsteinsson, T., Sigurdsson, F. Martian hillside gullies and Icelandic analogues, *Icarus* **2002**, *162*, 259–277.

Head, J. W., Mustard, J. F., Kreslavsky, M. A., Milliken, R. E., Marchant, D. R. Recent ice ages on Mars, *Nature* **2003**, *426*, 797–802.

Head, J. W., Neukum, G., Jaumann, R., Hiesinger, H., Hauber, E., Carr, M., Masson, P., Foing, B., Hoffmann, H., Kreslavsky, M., Werner, S., Milkovich, S., van Gasselt, S., and The HRSC Co-Investigator Team. Tropical to mid-latitude snow and ice accumulation, flow and glaciation on Mars, *Nature* **2005**, *434*, 346–351, doi:10.1038/nature03359.

Head, J. W., G. Neukum, G., Jaumann, R., Hiesinger, H., Hauber, E., Carr, M., Masson, P., Foing, B., Hoffmann, H., Kreslavsky, M., Werner, S., Milkovich, S., van Gasselt, S., and The HRSC Co-Investigator Team. Planetary science: Are there active glaciers on Mars? (Reply), *Nature* **2005**, *438*, E10, doi:10.1038/nature04358.

Herkenhoff, H. E., Squyres, S. W., Arvidson, R., Bass, D. S., Bell III, J. F., Bertelsen, P., Ehlmann, B. L., Farrand, W., Gaddis, L., Greeley, R., Grotzinger, J., Hayes, A. G., Hviid, S. F., Johnson, J. R., Joliff, B., Kinch, K. M., Knoll, A. H., Madsen, M. B., Maki, J. N., McLennan, S. M., McSween, H. Y., Ming, D. W., Rice, J. W., Jr., Richter, L., Sims, M., Smith, P. H., Soderblom, L. A., Spanovich, N., Sullivan, R., Thompson, S., Wdowiak, T., Weitz, C., Whelley, P. Evidence from Opportunity's Microscopic Imager for water on Meridiani Planum, *Science* **2004**, *306*, 1727–1730.

Horneck, G. The microbial world and the case for Mars, *Planet. Space Sci.* **2000**, *48*, 1053–1063.

Jakosky, B. M., Phillips, R. J. Mars' volatile and climate history, *Nature* **2001**, *412*, 237–244.

Janhunen, P. Are the Northern Plains of Mars a frozen ocean?, *J. Geophys. Res.* **2002**, *107*(E11), 5103, doi:10.1029/2000JE001478.

Kargel, J. S. Proof for water, hints for life? *Science* **2004**, *306*, 1689–1691.

Klein, H. P. The search for life on Mars: What we learned from Viking, *J. Geophys. Res.* **1978**, *103*(E12), 28463–28466.

Klein, H. P. The Viking biological experiments on Mars, *Icarus* **1978**, *34*, 666–674.

Kminek, G., Bada, J. L., Botta, O., Glavin, D. P., Grunthaner, F. MOD: An organic detector for the future robotic exploration of Mars, *Planet. Space Sci.* **2000**, *48*, 1087–1091.

Kminek, G., Bada, J. L. The effect of ionising radiation on the preservation of amino acids on Mars, *Earth Planet. Sci. Lett.* **2006**, *245, 1–5.*

Komatsu, G., Ori, G. G. Exobiological implications of potential sedimentary deposits on Mars, *Planet. Space Sci.* **2000**, *48*, 1043–1052.

Krasnapolsky, V. A. Some problems related to the origin of methane on Mars, *Icarus* **2006**, *180*, 359–367, doi:10.1016/j.icarus.2005.10.015.

Lepland, A., van Zuilen, M. A., Arrhenius, G., Whitehouse, M. J., Fedo, C. M. Questioning the evidence for Earth's earliest life–Akilia revisited, *Geology* **2005**, *33, 1, 77–79, doi:10.1130/G20890.1.*

Leovy, C. Weather and climate on Mars, *Nature* **2001**, *412, 245–249.*

Magnani, P. C., Ylikorpi, T., Cherubini, G., Olivieri, A. Deep Drill (DeeDri) for Mars application, *Planet. Space Sci.* **2004**, *52*, 79–82, doi:10.1016/j.pss.2003.08.023.

Mahaney, W. C., Dohm, J. M., Baker, V. R., Newsom, H. E., Malloch, D., Hancock,

R. G. V., Campbell, I., Sheppard, D., Milner, M. W., Morphogenesis of Antarctic paleosols: Martian analogue, *Icarus* **2001**, *154*, 113–130.

Malin, M. C., Carr, M. H. Groundwater formation of Martian valleys, *Nature* **1999**, *397*, 589–591.

Malin M. C., Edgett, K. S. Sedimentary rocks of early Mars, *Science* **2000**, *290*, 1927–1937.

Mangold, N., Quantin, C., Ansan, V., Delacourt, C., Allemand, P. Evidence for precipitation on Mars from dendritic valleys in the Valles Marineris area, *Science* **2004**, *305*, 78–81.

Mars, M., Malin, C., Edgett, K. S. Evidence for persistent flow and aqueous sedimentation on early Mars, *Science* **2003**, *302*, 1931–1934.

McKay, D. S., Gibson, E. K. Jr., Thomas-Keprta, K. L., Vali, H., Romanek, C. S., Clemmett, S. J. Chillier, X. D. F., Maechling, C. R., Zare, R. N. Search for past life on Mars: Posible relic biogenic activity in Martian meteorite ALH84001, *Science* **1996**, *273*, 924–930.

Milliken, R. E., Mustard, J. F., Goldsby, D. L. Viscous flow features on the surface of Mars: Observations from high-resolution Mars Orbiter Camera (MOC) images, *J. Geophys. Res.* **2003**, *108*(E6), 5057, doi:10.1029/2002JE002005.

Mitrofanov, I., Anfimov, D., Kozyrev, A., Litvak, M., Sanin, A., Tret'yakov, V., Krylov, A., Shvetsov, V., Boynton, W., Shinohara, C., Hamara, D., Saunders, R. S. Maps of subsurface hydrogen from the High-Energy Neutron Detector, Mars Odyssey, *Science* **2002**, *297*, 78–81.

Moore, J. M., Howard, A. D. Large alluvial fans on Mars, *J. Geophys. Res.* **2005**, *110*, E04005, doi:10.1029/2004JE002352.

Mustard, J. F. A wet and altered Mars, *Nature* **2002**, *417*, *234–235*.

Neukum, G., Jaumann, R., Hoffmann, H., Hauber, E., Head, J. W., Basilevsky, A. T., Ivanov, B. A., Werner, S. C., van Gasselt, S., Murray, J. B., McCord, T., and The HRSC Co-Investigator Team. Recent and episodic volcanic and glacial activity on Mars revealed by the High Resolution Stereo Camera, *Nature* **2004**, *432*, 971–979, doi:10.1038/nature03231.

Ori, G. G., Marinangeli, L., Komatsu, G., Martian paleolacustrine environments and their geological constrains on drilling operations for exobiological research, *Planet. Space Sci.* **2000**, *48*, 1043–1052.

Patel, M. R., Bérces, A., Kolb, C., Lammer, H., Rettberg, P., Zarnecki, J. C., Selsis, F. Seasonal and diurnal variation in Martian surface UV irradiation: Biological and chemical implications for the Martian regolith, *Int. J. Astrobio.* **2003**, *2*(1), 21–34.

Piccardi, G., Plaut, J. J., Biccari, D., Bombaci, O., Calabrese, D., Cartacci, M., Cicchetti, A., Clifford, S. M., Edenhofer, P., Farrel, W. M., Federico, C., Frigeri, A., Gurnett, D. A., Hagfors, T., Heggy, E., Herique, A., Huff, R. L., Ivanov, A. B., Johnson, W. T. K., Jordan, R. L., Kirchner, D. L., Kofman, W., Leuschen, C. J., Nielsen, E., Orosei, R., Pettinelli, E., Phillips, R. J., Plettemeier, D., Safaeinili, A., Seu, R., Stofan, E. R., Vannaroni, G., Watters, T. R., Zampolini, E. Radar soundings of the subsurface of Mars, *Science* **2005**, *310*, 1925–1928.

Pondrelli, M., Baliva, A., Di Lorenzo, S., Marinangeli, L., Rossi, A. P. Complex evolution of paleolacustrine systems on Mars: An example from the Holden Crater *J. Geophys. Res.* **2005**, *110*, E04016, doi:10.1029/2004JE002335.

Poulet, F., Bibring, J. P., Mustard, J. F., Gendrin, A., Mangold, N., Langevin, Y., Arvidson, R. E., Gondet, B., Gomez, C., and the Omega Team. Phyllosilicates on Mars and implications for early Martian climate, *Nature* **2005**, *438*, 623–627, doi:10.1038/nature04274.

Quinn, R. C., Zent, A. P., Grunthaner, F. J., Ehrenfreund, P., Taylor, C. L., Garry, J. R. C. Detection and characterization of oxidizing acids in the Atacama desert using the Mars Oxidation Instrument, *Planet. Space Sci.* **2005**, *53*, 1376–1388, doi:10.1016/j.pss.2005.07.004.

Rull Pérez, F., Martinez Frias, J. Raman spectroscopy goes to Mars, *Spectroscopy Europe* **2006**, *18*(1), 18–21.

Schulze-Makuch, D., Irwin, L. N., Lipps, J. H., LeMone, D., Dohm, J. M., Fairén, A. Scenarios for the evolution of life on Mars, *J. Geophys. Res.* **2005**, *110*, E12S23, doi:10.1029/2005JE002430.

Sims, M. R., Cullen, D. C., Bannister, N. P., Grant, W. D., Henry, O., Jones, R., McKnight, D., Thompson, D. P., Wilson,

P. K. The Specific Molecular Identification of Life Experiment (SMILE), *Planet. Space Sci.* **2005**, *53*, 781–791, doi:10.1016/j.pss.2005.03.006.

Skelley, A. M., Scherer, J. R., Aubrey, A. D., Grover, W. H., Ivester, R. H. C., Ehrenfreund, P., Grunthaner, F. J., Bada, J. L., Mathies, R. A. Development and evaluation of a microdevice for amino acid biomarker detection and analysis on Mars, *Proc. Natl. Acad. Sci.* **2005**, *102*(4), 1041–1046, doi:10.1073/pnas.0406798102.

Squyres, S. W., Arvidson, R. E., Bell III, J. F., Brückner, J., Cabrol, N. A., Calvin, W., Carr, M. H., Christensen, P. R., Clark, B. C., Crumpler, L., Des Marais, D. J., dÚston, C., Economou, T., Farmer, J., Farrand, W., Folkner, W., Golombek, M., Gorevan, S., Grant, J. A., Greeley, R., Grotzinger, J., Haskin, L., Herkenhoff, K. E., Hviid, S., Johnson, J., Klingelhöffer, G., Knoll, A. H., Landis, G., Lemmon, M., Li, R., Madsen, M. B., Malin, M. C., McLennan, S. M., McSween, H. Y., Ming, D. W., Moersch, J., Morris, R. V., Parker, T., W. Rice, J. W., Jr., Richter, L., Rieder, R., Sims, M., Smith, M., Smith, P., Soderblom, L. A., Sullivan, R., Wänke, H., Wdowiak, T., Wolff, M., Yen, A. The Opportunity Rover's Athena science investigation at Meridiani Planum, Mars, *Science* **2004**, *306*, 1698–1703.

Squyres, S. W., Grotzinger, J. P., Arvidson, R. E., Bell III, J. F., Calvin, W., Christensen, P. R., Clark, B. C., Crisp, J. A., Farrand, W. H., Herkenhoff, K. E., Johnson, J. R., Klingelhöffer, G., Knoll, A. H., McLennan, S. M., McSween, H. Y., Jr., Morris, R. V., Rice, J. W., Jr., Rieder, R., Soderblom, L. A. *In-situ* evidence for an ancient aqueous environment at Meridiani Planum, Mars, *Science* **2004**, *306*, 1709–1714.

Tung, H.C., Bramall , N. E., Price, P. B. Microbial origin of excess methane in glacial ice and implications for life on Mars, *PNAS* **2005**, *102*, 51, 18292–18296, doi:10.1073/pnas.0507601102.

Wang, A., Haskin, L. A., Squyres, S. W., Jolliff, B. L., Crumpler, L., Gellert, R., Schröder, C., Herkenhoff, K., Hurowitz, J., Tosca, N. J., Farrand, W. H., Anderson, R., Knudson, A. T. Sulfate deposition in subsurface regolith in Gusev Crater, Mars, *J. Geophys. Res.* **2006**, *111, E02S17, doi:10.1029/2005JE002513.*

Wang, A., Korotev, R. L., Jolliff, B. L., Haskin, L. A., Clumper, L., Farrand, W. H., Herkenhoff, K. E., de Souza, P., Jr., Kusack, A. G., Hurowitz, J. A., Tosca, N. J. Evidence of phyllosilicates in Wooly Patch, an altered rock encountered at West Spur, Columbia Hills, by the Spirit Rover in Gusev Crater, Mars, *J. Geophys. Res.* **2006**, *111, E02S16, doi:10.1029/2005JE002516.*

Wentworth, S. J., Gibson, E. K., Velbel, M. A., McKay, D. S. Antarctic Dry Valleys and indigenous weathering in Mars meteorites: Implications for water and life on Mars, *Icarus* **2004**, *174, 383–395, doi:10.1016/j.icarus.2004.08.026.*

Westall, F., Brack, A., Hofmann, B., Horneck, G., Kurat, G., Maxwell, J., Ori, G. G., Pillinger, C., Raulin, F., Thomas, N., Fitton, B., Clancy, P., Prieur, D., Vassaux, D. An ESA Study for the Search for Life on Mars, *Planet. Space Sci.* **2000**, *48*, 181–202.

Williams, R. M. E., Malin, M. C. Evidence for late stage fluvial activity in Kasei Valles, Mars, *J. Geophys. Res.* **2004**, *109*, E06001, doi:10.1029/2003JE002178.

Wynn-Williams, D. D., Edwards, H. G. M. Antarctic ecosystems as models for extraterrestrial surface habitats, *Planet. Space Sci.* **2000**, *48*, 1065–1075.

Wynn-Williams, D. D., Edwards, H. G. M. Proximal analysis of regolith habitats and protective biomolecules *in situ* by Laser Raman Spectroscopy: Overview of terrestrial Antarctic habitats and Mars analogs, *Icarus* **2000**, *144, 486–503.*

Wynn-Williams, D. D., Edwards, H. G. M., Newton, E. M., Holder, J. M. Pigmentation as a survival strategy for ancient and modern photosynthetic microbes under high ultraviolet stress on planetary surfaces, *Internat. J. Astrobi.* **2002**, *1*(1), 39–49.

Zent, A. P. On the thickness of the oxidized layer of the Martian regolith, *J. Geophys. Res.* **1998**, *103*(E13), 31491–31498.

Zent, A. P., McKay, C. P. The chemical reactivity of the Martian soil and implications for future missions, *Icarus* **1994**, *108*, 146–157.

12.7 Questions for Students

Question 12.1

When searching for traces of past life on Mars, describe why it is important to have access to "protected" environments – protected from what? Which sort of soils should a sampling device be able to obtain sample material from (hard, soft, porous, dust) and why?

Question 12.2

Assuming that the level of organics/organisms in the Martian soil is very small, describe a strategy to maximize your chances for a positive detection. In a couple of paragraphs, describe what you think is important.

13
Astrobiology Exploratory Missions and Planetary Protection Requirements

Gerda Horneck, André Debus, Peter Mani, and J. Andrew Spry

This chapter describes the legal and scientific issues of planetary protection requirements required for each space mission within our Solar System. Planetary protection is based on an international treaty to prevent the introduction of microbes from the Earth to another celestial body or vice versa, whether this occurs intentionally or unintentionally. Based on this treaty, a concept of contamination control has been elaborated by the Committee of Space Research (COSPAR) with regard to specific classes of mission–target combinations, and it is recommended that each space-faring organization follow it. The intention of planetary protection is twofold: first, to protect the planet being explored and to prevent jeopardizing search-for-life studies, including those for precursors and remnants, and second, to protect the Earth from the potential hazards posed by extraterrestrial matter carried on a spacecraft returning from another celestial body. Missions to different bodies in our Solar System that are of astrobiological interest are considered with regard to planetary protection requirements.

13.1
Rationale and History of Planetary Protection

Space activities started about 50 years ago, in 1957, when the first manmade object reached outer space and broadcasted its signals around the world. With *Sputnik* – the name of this first satellite – a long-desired vision of mankind became reality, namely, the ability to leave the boundaries of our Earth and enter space. In parallel with the rapid development of space activities that followed this first enterprise (1961, first cosmonaut in space; 1966, first safe robotic landing on the Moon with *Surveyor 1*; 1969, first human landing on the Moon), concern was raised about possible contamination of the Moon and planets by terrestrial microorganisms.

Complete Course in Astrobiology. Edited by Gerda Horneck and Petra Rettberg

ISBN: 978-3-527-40660-9

This concern was formulated by the National Academy of Sciences (NAS) of the United States as early as 1958:

> *"The NAS of the USA urges that scientists plan lunar and planetary studies with great care and deep concern so that initial operations do not compromise and make impossible forever after critical scientific experiments. For example, biological and radioactive contamination of extraterrestrial objects could easily occur unless initial space activities be carefully planned and conducted with extreme care. The NAS will endeavor to plan lunar and planetary experiments in which the Academy participates so as to prevent contamination of celestial bodies in a way that would impair the unique and powerful scientific opportunities that might be realized in subsequent scientific exploration."*

Of course, such efforts to prevent contamination of the Moon and planets can be efficient only if they are followed globally by all space-faring nations.

It took until 1967, with 10 years of meetings and consultations after the first launch of a satellite, for the requirement for planetary protection to be incorporated into Article IX of the Outer Space Treaty of the United Nations (Treaty on Principles Governing the Activities of States in the Exploration and Use of Outer Space, including the Moon and Other Celestial Bodies). The text reads as follows:

> *"States Parties to the Treaty shall pursue studies of outer space, including the Moon and other celestial bodies, and conduct exploration of them so as to avoid their harmful contamination and also adverse changes in the environment of the Earth resulting from the introduction of extraterrestrial matter, and where necessary, shall adopt appropriate measures for this purpose."*

This treaty has been signed and ratified by practically all space-faring nations. It provides that any of its parties can request consultation with any other party concerning the latter's activities or experiments that might be harmful to the use of outer space. This treaty is the foundation on which all subsequent recommendations and guidelines of "planetary protection" (or "planetary quarantine," a term used until 1975) are based.

The rationale for planetary protection is the supposition that life could be a universal or planetary phenomenon that emerges at a certain stage of either cosmic or planetary evolution, if the right environmental physical and chemical requirements are present (see Chapter 1). Therefore, although the Earth is the only planet known to harbor life, there may be many more habitable or even inhabited celestial bodies in our Solar System or beyond (see Chapter 6). With space exploration, we possess for the first time the tools required to reach candidate moons and planets in our Solar System by means of spacecraft and to search *in situ* for signatures of indigenous life, either extinct or extant. However, the introduction – by means of orbiters, entry probes, or landing vehicles – and possible proliferation of terrestrial life forms on other planets could entirely destroy the opportunity to examine the target planets or moons in their pristine conditions. From this concern by the scientific community, the concept of planetary protection has evolved. Its intent is twofold:

- To protect the planet or moon being explored and to preserve its environment from terrestrial biological contamination in order to prevent jeopardizing search-for-life studies, including those for precursors or remnants; this process is called "forward contamination prevention."
- To protect the Earth and its biosphere and human population from the potential hazards posed by extraterrestrial matter carried on a spacecraft returning from another celestial body; this process is called "backward contamination prevention."

In addition, planetary protection is concerned with the preservation of the properties of the extraterrestrial samples themselves, as far as they are to be subjected to exobiological investigations, either *in situ* at the surface of the extraterrestrial body or when brought back to Earth. If correct and reliable preservation of the extraterrestrial samples (or of parts of them) is not ensured, resulting in possible contamination by terrestrial organisms or a change in the properties of the sample material, one might draw false conclusions about the presence of extraterrestrial life forms or biohazards. The consequence of such mistreatment of extraterrestrial samples could be either a false-positive result, requiring additional but unnecessary planetary protection measures, or a false-negative result, giving rise to free distribution of potentially hazardous samples. False-negative results could endanger a human exploratory mission to a planet under consideration if the environment is interpreted as harmless when in reality it is hazardous.

13.2 Current Planetary Protection Guidelines

As previously noted, under the 1967 Outer Space Treaty of the United Nations, the exploration of the Solar System needs to be in compliance with a set of planetary protection constraints, for both outbound and inbound missions. These constraints impose specific requirements on spacecraft design and construction, such as the introduction of sterilization procedures, integration under ultra-clean or even sterile conditions, microbiological and cleanliness control, and the use of highly reliable systems in order to avoid crash landings. In the particular case of sample return missions, the potential introduction of sample quarantine is also added. Such requirements are required solely for planetary protection purposes and need to be considered right at the beginning of the planning of a mission to the planets, moons, or small bodies (comets or asteroids) of our Solar System.

The current planetary protection guidelines were developed by the scientific community under the auspices of the Committee on Space Research (COSPAR) through its Panel on Planetary Protection. COSPAR was formed in 1958 by the

International Council of Scientific Unions to coordinate worldwide space research. In this context, COSPAR also assumed responsibility for the contamination problem. COSPAR is an observer at the United Nations Committee on Peaceful Uses of Outer Space and reports to the United Nations through its periodic Scientific Assemblies.

Concerning Planetary Protection, the current COSPAR policy is as follows:

> *"Although the existence of life elsewhere in the Solar System may be unlikely, the conduct of scientific investigations of possible extraterrestrial life forms, precursors, and remnants must not be jeopardized. In addition, the Earth must be protected from the potential hazard posed by extraterrestrial matter carried by a spacecraft returning from another planet. Therefore, for certain space mission/target planet combinations, controls on contamination shall be imposed, in accordance with issuances implementing this policy."*

This notation reflects that the current COSPAR policy is based on a scientific interpretation of the Outer Space Treaty and that topics such as protection of the environment and ethical considerations are not covered. However, it is important to note that all space-faring nations are bound to comply with the COSPAR policy and, consequently, with its recommendations. Hence, the planetary protection guidelines of COSPAR shall be included in the relevant planetary protection documentation of the various space agencies.

The first planetary protection guidelines (developed by COSPAR in 1959 and initially called COSPAR Objectives for Planetary Quarantine) were based on relevant information about the probability of the survival and release of organisms contained either in or on exposed surfaces of spacecraft, about the surface and atmosphere characteristics of the planet under consideration, and about the probable distribution and growth of the types of organisms involved. The first requirements for missions to Mars and other planets set by COSPAR, in 1964, imposed a maximum probability of 10^{-4} to find one viable microorganism per spacecraft (or one viable microorganism per 10 000 spacecrafts) and stated that the probability of any unsterilized hardware crashing must be less than 3×10^{-5}. Later, in 1969, the basic objective of the planetary quarantine was to achieve a probability of no more than 1×10^{-3} for contamination during the period of biological exploration. This period was assumed to be 20 years for Mars, starting from 1969 and extending through the end of 1988, and was assumed to include approximately 100 missions. These numbers demonstrate the optimistic view that was taken at the beginning of Solar System exploration. However, in this period, the only successful landings on Mars were the two *Viking* missions, and a further three landers were sent to Venus (see Section 13.4).

The early COSPAR concept of probability of contamination of a planet of biological interest was replaced in 1989 by a concept of contamination control. This was to be elaborated specifically for certain space mission–target planet combinations, such as orbiters, landers, or sample return missions. The current planetary protection recommendations, formulated by COSPAR through its Panel on Planetary Protection and in agreement with its own policy, separately consider the *target* planet or body (including Solar System bodies encountered as flybys during the mission) on one hand and the *type* of mission on the other:

- Concerning the extraterrestrial body to be explored, one has to consider that, for instance, the risk of contaminating Mars is not the same as that of contaminating Venus, because a landing mission to the latter planet, which has an ambient surface temperature of more than 400 °C, will already be sufficiently sterilized during the landing process.
- Concerning the type of mission (orbiter, flyby, lander, or sample return), it is obvious that the contamination risk is not the same for a spacecraft orbiting around a planet as for a probe landing on its surface.

The current position is described in Table 13.1. This information is updated periodically by COSPAR in line with the best scientific advice.

According to the scientific interest and mission goals for the different targets in our Solar System, COSPAR's planetary protection guidelines group missions into five categories (see Table 13.1).

13.2.1 Category I Missions

Category I missions include any mission (orbiters, flyby, or landers) to a target body that is not of direct interest for understanding the process of chemical evolution, origin of life, or other questions of astrobiology. The target bodies include the planets Mercury and Venus, the Moon, and undifferentiated, metamorphosed asteroids and the Sun. However, other celestial bodies may be added to this list depending on the scientific results obtained in the future. In category I, no protection of the planets or bodies is warranted, and therefore no planetary protection requirements are imposed on the spacecraft and mission. Examples include *Smart I* (targeting the moon) and *Genesis* (outbound – not encountering any Solar System bodies).

13.2.2 Category II Missions

Category II refers to all types of missions (orbiters, flyby, or landers) to target bodies where there is significant interest relative to the process of chemical evolution, but where there is only a remote chance that contamination carried by a spacecraft could jeopardize future exploration. Target planets and other bodies include the giant planets Jupiter, Saturn, Uranus, and Neptune; the remote dwarf planet Pluto with its moon Charon; comets; carbonous chondrite asteroids (these asteroids contain a substantial amount of organic material); and the objects of the

Table 13.1 Current COSPAR planetary protection categories with regard to missions to Solar System bodies and types of missions.

Category	Category I	Category II	Category III	Category IV	Category V
Type of mission	Any but Earth return	Any but Earth return	No direct contact (flyby, some orbiters)	Direct contact (lander, probe, some orbiters)	Earth return
Target body	See text[a]	See text[a]	See text[a]	See text[a]	See text[a]
Degree of concern	None	Record of planned impact probability and contamination control measures	Limit on impact probability; passive bioload control	Limit on probability of non-nominal impact; limit on bioload (active control)	If restricted Earth return: No impact on Earth or Moon; returned hardware sterile; containment of any sample.
Range of requirements	None	Documentation only: PP plan[b]; prelaunch report; boostlaunch report; postencounter report; end-of-mission report	Documentation (category II plus): contamination control; organics inventory; implementing procedures (trajectory biasing, clean room, bioload reduction, as necessary)	Documentation (category II plus): Pc[c] analysis plan; microbial reduction plan; microbial assay plan; organics inventory; implementing procedures (trajectory biasing, clean room, bioload reduction, partial sterilization of contacting hardware, bioshield, monitoring of bioload via bioassay)	Outbound: Same category as target body/outbound mission. Inbound: If restricted Earth return: Documentation (category II plus): Pc analysis plan; microbial reduction plan; microbial assay plan; trajectory biasing; sterile or contained returned hardware; continual monitoring of project activities; project advanced studies/research. If unrestricted Earth return: None

a See Section 13.4.
b PP: Planetary protection.
c Pc: Probability of contamination.

Kuiper Belt. Again, other bodies may be added following new scientific results. In category II, the planetary protection recommendations require relatively simple documentation. The main requirement is to prepare a planetary protection plan before the mission describing a brief pre- and post-launch analysis and whether it is planned to impact on the target – or whether this may happen accidentally. In both cases, a detailed description of impact strategies is required. Furthermore, reports are required after the encounter with the celestial body and at the end of mission. These reports should give further details on the location of impact if such an event occurred. Reporting of planned impact probability and contamination control measures is also required. Mission examples include *Venus Express, Cassini–Huygens* (which had flybys at Venus, Earth, and Jupiter before arriving at the Saturnian system), and *Stardust* (which encountered the comet Wild 2).

13.2.3 Category III Missions

Category III comprises certain types of missions (mostly flyby or orbiters) to a target planet or body that – according to the current understanding – is of interest with regard to chemical evolution and/or the origin of life or for which scientific opinion considers that a potential contamination with terrestrial microorganisms could jeopardize future biological experiments, e.g., those searching for signatures of indigenous life. Our neighbor planet Mars and the Jovian moon Europa are classified in this category. However, depending on future scientific results, other bodies, especially other moons of the giant planets, may be included in this category. It is important to note that an impact with the target body is not foreseen for the missions of category III. However, if the probability of accidental impact is significant, an inventory of the bulk constituent organics is required. In addition to a more extensive documentation than required in category II, for category III it is necessary to assemble and test the spacecraft in a clean room, to monitor the bioburden (number of terrestrial microorganisms present) and reduce if necessary, and to include trajectory biasing. For instance, for orbital or flyby missions sent toward Mars, the spacecraft crash probability must be less than 1% during the 20 years following the launch and 5% for the 50 years following the launch. For the Jovian moon Europa, the crash probability must be less than 0.01%. Example missions include all the Mars orbiters (*Mars Global Surveyor, Mars Odyssey, Mars Express, Mars Reconnaissance Orbiter*).

13.2.4 Category IV Missions

Category IV missions are those missions of the utmost astrobiological interest. They include mostly entry probe or lander missions to a target planet of interest with regard to chemical evolution and/or origin of life or for which scientific opinion provides a significant chance that contamination by terrestrial microorganisms could jeopardize future biological experiments. As in category III, the celestial targets are Mars and Jupiter's moon Europa, but in this case, missions to the surface or subsurface are considered. For missions to Mars, depending on the landing site and the scientific purpose of the mission, the following subcategories are defined.

- Category IVa: Mission with lander systems not carrying instruments for the investigation of extant Martian life.
- Category IVb: Mission with lander systems carrying instruments for the investigation of extant Martian life.
- Category IVc: Landing mission to a special region, defined as a region within which terrestrial microorganisms are likely to propagate or a region that is interpreted to have a high potential for extant Martian life forms. Given current understanding, this category concerns regions where liquid water is present or may occur. Examples include subsurface access in an area and to a depth where the presence of liquid water is probable, penetrations into the polar caps, or areas of potential hydrothermal activity. At the time of writing (2006), uncertainty over what exactly constitutes a Martian "special region" is a subject of discussion within the Mars exploration community. Clear guidance on this issue is required for the ongoing mission set; some clarification can be expected from within the community within the next few months and years. With regard to Jupiter's moon Europa, the probability of contaminating Europa with one single viable microorganism must be less than 0.01 %.

For category IV missions, more detailed planetary protection documentation is required than for category III missions. The planetary protection requirements include bioassays to enumerate the bioload, a probability of contamination analysis, an inventory of the bulk constituent organics, and an increased number of implementation procedures. The implementation procedures required include trajectory biasing, clean-room assembly, bioburden reduction and monitoring, and, possibly, partial sterilization of the directly contacting hardware and a bioshield for that hardware, depending on the subcategory considered. *Pathfinder*, *Beagle 2*, and the Mars Exploration Rover missions are all examples of category IV missions.

13.2.5 Category V Missions

Category V includes all missions within our Solar System that return to the Earth. The concern for these missions is the protection of the Earth and the Moon, which are considered a single system (preventing back contamination). In this context, the Moon is to be protected from back contamination to retain freedom from planetary protection requirements on Earth–Moon travel.

Return missions from all Solar System bodies are included in category V. Those classified as "safe for Earth return" do not have constraints on handling or mission design. This is a risk-based decision, determined by scientific opinion when

- there was never liquid water in or on the target body,
- metabolically useful energy sources were never present,
- there was never sufficient organic matter (or CO_2 or carbonates and an appropriate source of reducing equivalents) in or on the target body to support life,
- the target body has been subjected to extreme temperatures (e.g., >160 °C),
- there is or was sufficient radiation for biological sterilization of potential life forms, and
- there has been a natural influx to Earth (e.g., via meteorites) of material equivalent to a sample returned from the target body.

Missions from all other planets and moons of our Solar System are classified as "restricted for Earth return" and must follow the additional category V requirements. The highest degree of concern is expressed by the absolute prohibition of impact upon return, the need for any returned hardware that directly contacted the target body to be sterile (contact chain breaking if sterilization is not possible), and the need for containment of any non-sterilized sample collected and returned to Earth. In general, this concern is reflected in a range of requirements that encompass those of category IV plus a continuing monitoring of project activities. Special research is required before the mission in order to develop suitable and efficient remote sterilization procedures and reliable containment techniques.

The return elements of *Genesis* and *Stardust* are examples of "unrestricted-Earth-return" category V missions. No restricted-Earth-return missions have flown since the development of the category-based planetary protection approach.

13.2.6
Future Development of Planetary Protection Guidelines

It should be stressed that the COSPAR Planetary Protection Guidelines are a living document that is steadily adapted to new findings on the habitability of the bodies of our Solar System and new knowledge about terrestrial microbiology. Especially in view of the current and planned landing activities on Mars (see Chapter 12), with robotic and finally human visits, the planetary protection guidelines are currently under review. For this purpose, the European Space Agency (ESA) and the National Aeronautics and Space Administration (NASA) have established Planetary Protection Advisory Groups gathering scientists from fields such as astrobiology, planetology, microbiology, hygiene, and sterilization technology. The priority task of these groups is to advise the agencies concerning efficient planetary protection measures for samples returned from Mars and for human exploratory missions, e.g., those landing on Mars. For human mission concepts, no planetary protection guidelines have been developed by COSPAR to date.

13.3
Implementation of Planetary Protection Guidelines

Planetary protection considerations must be implemented very early in the development of a space mission to one of the bodies of our Solar System, because they may influence the design and integration of the final spacecraft and its operations. The initial starting point is the determination of the relevant category of the space mission–target planet combination (Table 13.1) and whether some sort of bioload control may be necessary. Usually the first formal recognition of a requirement is in the Planetary Protection Plan. Here, the planetary protection requirements and how the project plans to meet them are described. After the Planetary Protection Plan is approved, generally by the planetary protection officer of the responsible space agency, a subsequent document, the Planetary Protection Implementation Plan, will give the detailed "how to" once the design is more clearly understood.

This approach can differ significantly from mission to mission, according to the mission goals, complexity, and budgets (both funding and mass–volume budgets). However, planetary protection requirements need to be factored into the design process at an early stage (pre–phase A stage) of any mission to ensure compatibility between the design and the COSPAR requirements. For smooth implementation of planetary protection overall, it is important to ensure that iterations of the design and planetary protection implementation proceed in an integrated fashion, such that changes in design are assessed for their impact on planetary protection and vice versa, ensuring that the planetary protection requirements are met by the spacecraft and instruments finally selected to fly. To train spacecraft engineers in planetary protection issues, ESA and NASA regularly organize planetary protection courses for representatives of the space industry.

As mentioned above, the first step is determining the category of the space mission–target planet combination (Table 13.1). Of special concern with regard to preventing forward contamination are missions of categories III and IV. Depending on the categorization, the following measures need to be implemented during assembly until the final integration of a space system:

- The bioburden should be reduced by defined and validated sterilization techniques.
- Assembly of biocleaned space systems should take place in a clean room with biological cleanliness control measures and bioload control.
- In order to avoid recontamination of the system after sterilization, a cover to the space system as bioshield should be provided; typically this will be jettisoned in flight.
- Terminal sterilization after final integration (through the bioshield using typically dry heat) should take place on a case-by-case basis, such as for category IVb or IVc missions to Mars, e.g., the *Viking* mission.

Most of the implementation procedures, as currently used, were originally developed by NASA and are outlined in NASA documents. For example, the current planetary protection protocols for microbial examination, cleaning, and sterilization of space hardware are largely based on *Viking* planetary protection state of knowledge (i.e., from the 1970 s). Revision is currently underway because our knowledge about microbial diversity and adaptive strategies has immensely increased since the *Viking* era (see Chapters 5 and 12) and because several international groups have now gained experience in microbial sampling from spacecraft (e.g., from *Viking, Pathfinder, Mars Polar Lander, Beagle 2,* Mars Exploration Rovers, *Smart-1, Rosetta*). In a series of workshops, international space agencies are currently revisiting these planetary protection protocols with the aim of amending

- the bioload measurements, by adding modern molecular biology analysis methods, and
- the sterilization methods, by using modern methods such as those from treatment of medical devices or pharmaceuticals.

13.3.1 Bioload Measurements

Monitoring the bioload of a spacecraft is essential for determining the compliance of a project with the planetary protection guidelines issued by COSPAR. There are several reasons for knowing the number and type of microorganisms in and around a spacecraft with regard to planetary protection issues:

- The data are the basis for determining appropriate sterilization or cleaning procedures.
- The data will help us to understand the survival and growth potential of microorganisms on the target planet or moon.
- If searching for signatures of life – especially extant life forms – is the goal of the mission, there is a need to know what type of terrestrial microorganisms may be present on the spacecraft.
- For sample return missions, it is required to know the bioload of the outbound spacecraft in order to avoid false-positive results upon return to Earth.

The standardized, validated, culture-based microbial assay used by planetary protection practitioners for determining the microbial bioload is given in the NASA document NPR 5340. For the sampling of microorganisms on the surface of spacecraft, its subunits, or equipment within a spacecraft assembly facility, the standard method includes either swabbing of multiple small areas (5×5 cm^2) on and near the probe with sterile cotton swabs (Fig. 13.1) or wiping larger areas (1 m^2) with a sterile wipe. The microorganisms are detached from the swabs/wipes by vortexing and sonication in distilled water or buffer, and part of the suspension is spread on agar plates of a defined growth medium and aerobically incubated at 32 °C or 37 °C. The colony-forming units (CFU) are counted every 24 h, and the

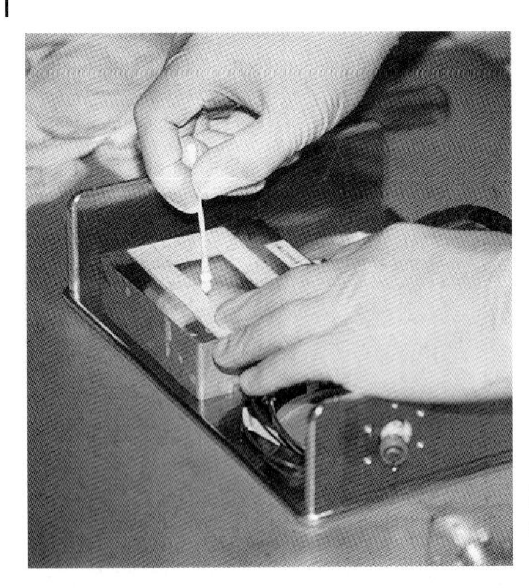

Fig. 13.1 Swabbing of an area (5×5 cm^2) of a spacecraft subunit of the *Mars '96* mission (photo credit: CNES).

overall bioload is estimated from the final count after 72 h of growth. This procedure is based on a cultivation approach, allowing the detection of organisms able to grow under "standard" laboratory conditions.

In order to determine the fraction of bacterial spores in the samples, a portion of the suspension is heat shocked at 80 °C for 15 min before plating and incubation. This process, generally known as pasteurization, leads to inactivation of most of the vegetative bacterial cells, whereas bacterial spores and also cells of heat-tolerant microorganisms survive the heat treatment. It was found that spore formers constitute the dominant fraction of microorganisms isolated from the surfaces of various spacecraft when the heat-shock method is used. These spore-forming microbes are of particular concern in the context of planetary protection, because in many species their endospores can be highly resistant to a variety of environmental extremes, including certain sterilization procedures and the harsh environment of outer space or planetary surfaces (see Chapter 5). Among these environmental isolates, strains have been identified that excel by an elevated resistance to various physical and chemical conditions, such as ionizing and UV radiation and desiccation and oxidative stress, as compared to spores of related laboratory strains. It is speculated that the laboratory strains may have lost their resistance because of prolonged exposure to artificial laboratory conditions.

In a study to further understand the spectrum of organisms present in spacecraft environments, selected microbial species isolated from several spacecraft (*Mars Odyssey, Smart, Rosetta* lander) and their assembly facilities were identified using 16S rDNA sequencing. This molecular biology method uses genomic DNA isolated from the cultures under investigation as template for amplification by polymerase chain reaction (PCR). The phylogenetic relationship of the microorganism is then determined by comparison of the individual 16S rDNA sequences to other already-existing sequences in the public database. Among 45 *Bacillus* spore lines screened, 19 isolates showed high resistance to UV–C irradiation after exposure to 1 000 J m^{-2} at UV–C of 254 nm. Using separate UV bands at the Mars-simulated solar constant including UV-A (315–400 nm), UV-(A+B) (280–400 nm), and full UV spectrum (200–400 nm), it was found that spores of *Bacillus* species isolated from spacecraft surfaces were more resistant to the Martian UV radiation climate than a laboratory strain, *B. subtilis* 168. Such studies are important because dust suspended in the Martian atmosphere can attenuate UV–C while allowing both UV-A and UV-B to reach the surface. Among all *Bacillus* species tested, spores of *B. pumilus* SAFR-032 showed highest resistance to all three spectral ranges of simulated Martian UV, compared to the other strains of the same species. In a further study, representatives of these highly resistant spore formers will be tested under the conditions of space and simulated Martian surface radiation for more than one year in the EXPOSE facility of ESA, which will be attached to the International Space Station (see Chapter 11).

Microorganisms in laboratory culture have very specific growth requirements, e.g., with regard to temperature, oxygen availability, and pH value (see Chapter 5). Therefore, in order to determine the microbial diversity of the bioload of a spacecraft or spacecraft assembly facility by laboratory methods, the samples need to be divided and subjected to a palette of different growth conditions with regard to pH, temperature, oxygen, and pasteurization (Fig. 13.2). In the ESA study "Determination of the microbial diversity of spacecraft assembly facilities" the group of Rettberg obtained the following bioload results for the space missions *Smart-1* and *Rosetta*. The vast majority of colonies, about 96.7%, were identified after growth at a temperature of 30 °C. Only 3.2% grew at 5 °C and only about 0.1% (six in total) at 55 °C. Seventy-two percent of the colonies were formed under aerobic conditions and 28% under anaerobic conditions. Seventy-nine percent of the colonies were found on agar plates with a pH of 7.0, 21% on agar plates with a pH of 9.0, and no colony at a pH of 4.5. The preference of one-fifth of the microorganisms for an alkaline environment (pH 9) could be the result of a selection due to detergents with a high pH that are used for routine cleaning of that particular clean room. In this study, about 66% of the colony-forming units originated from pasteurized aliquots of collected samples. Unlike sterilization, pasteurization is not intended to kill all microorganisms in the sample. Instead, pasteurization aims to achieve a reduction in the number of viable organisms (e.g., in milk processing). The viability of thermotolerant or thermophilic microorganisms (for definition, see Chapter 5) as well as of especially robust dormant microbial forms such as bacterial spores is not impaired by pasteurization. Many isolates from pasteurized samples could be identified as either *Bacillus* species or as *Micrococcus* species. Because *Micrococcus* does not form spores, it is not strictly correct to describe all survivors of the pasteurization process as spores (as is currently assumed in the planetary protection guidelines).

In summary, it can be noted that a large fraction of the microorganisms (58–75%) isolated from spacecraft or space assembly facilities is considered to be indigenous to humans and was brought into the facility by the workers assembling the spacecraft. The majority of these isolates belong to the *Staphylococcus* or *Micrococcus* groups. This fraction of the overall bioload could be reduced by appropriate training of the personnel and controlling of the working procedures. The remaining fraction (25–42%) of the bioload either is associated with soil and

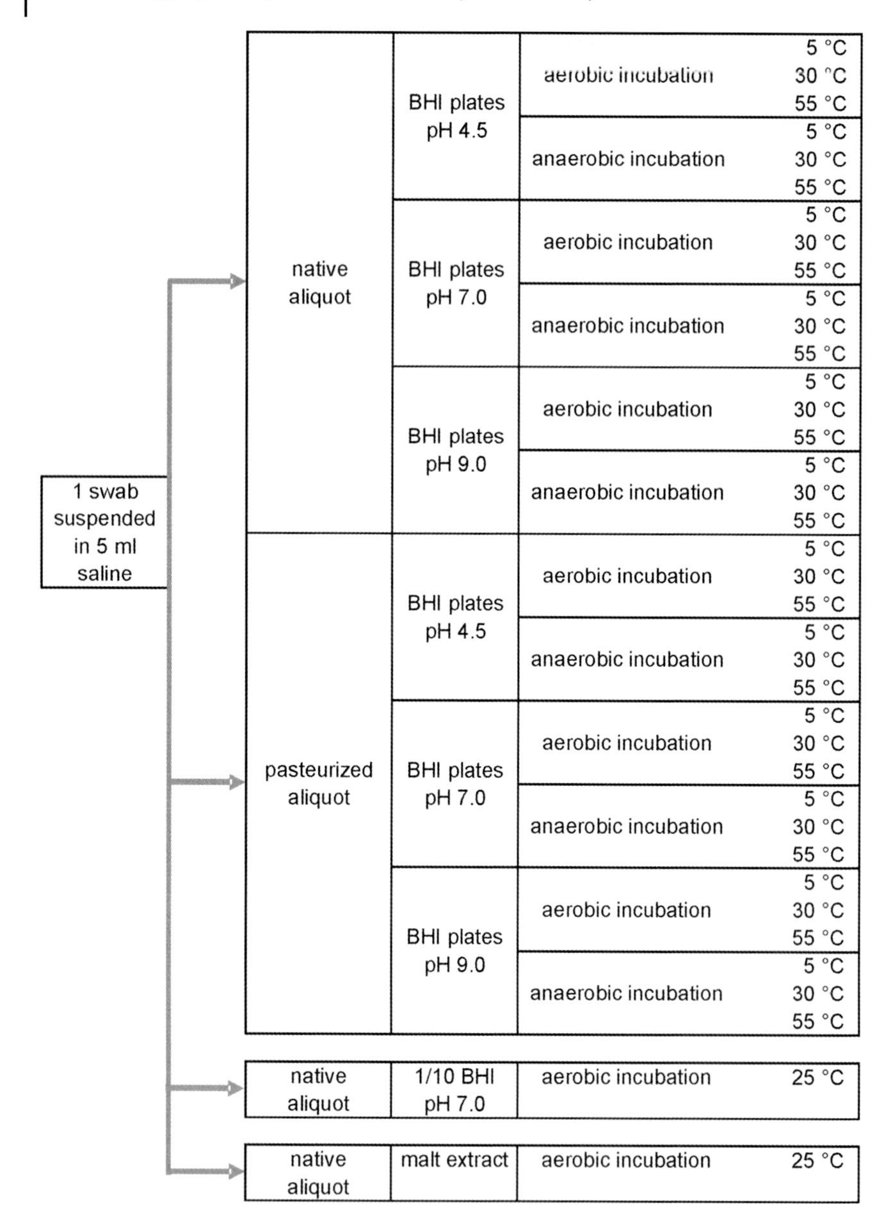

Fig. 13.2 Plating scheme with a broad range of cultivation conditions for the detection of microorganisms, isolated from spacecraft or spacecraft assembly facility within the ESA project MiDiv (Rettberg et al. 2006).

dust (*Bacillus* group) or originates from the specific environment and is determined mainly by the clean-room class of the spacecraft assembly facility.

So far, most work on bioload control has been restricted to cultivable microorganisms. However, it is well known that many microorganisms found in environmental samples are not readily cultivable in the laboratory. Use of new technologies, e.g., fluorescent DNA staining of bacterial cells or epifluorescence microscopy, has shown that the number of microorganisms in aquatic environments is 100–1000 times greater than that determined by cultivation techniques. These so-called non-cultivable microorganisms (we probably have not yet found the correct cultivation conditions) are assumed to represent the majority of terrestrial microorganisms.

Although these organisms cannot be cultured today, they can still be detected and described by analyzing their biomolecules. One way to bypass the cultivation limitation is to focus on the detection of specific DNA sequences as marker molecules for microorganisms present. PCR is a powerful molecular tool to detect even minute amounts of specific unknown DNA sequences. This technique is now being applied to detect and identify very low numbers of microorganisms present in environmental samples, such as those from spacecraft or their assembly facilities. In addition, when the right target gene is chosen for analysis, this technique may be used for classifying the taxa or even species of the non-cultivable organisms. Highly conserved genes such as those of the 16S rRNA are suitable targets for phylogenetic identification of prokaryotes. A major advantage of this approach is that a large databank of sequenced 16S rDNA is already available.

Other techniques to track non-cultivable microorganisms include Live/Dead fluorescence analysis, the ATP bioluminescence assay as an indicator of live cells, or assays for lipopolysaccharides (e.g., the limulus amebocyte lysate [LAL] assay) as membrane building blocks for detecting microorganisms of the Bacteria domain. However, each of these assays has limitations that mediate against being used as a direct replacement for the classical spore assay. For example, the LAL method does not detect gram-positive bacteria or representatives of the Archaea domain, because of the specific membrane components detected by the assay (see Chapter 5).

13.3.2 Bioburden Reduction

It is normal aerospace practice for spacecraft to be assembled in clean rooms. These rooms have specific air filtration installations to keep the concentration of particles low. They are used for processing space hardware requiring a high degree of cleanliness. Planetary protection requirements go over and above these spacecraft-specific requirements and are dictated by the category of the mission.

Research in the *Viking* era showed that, depending on the class of the clean room, a lower limit of background surface bioload density can be assumed (Table 13.2). Below this level, further reductions are not possible because of the erratic and non-uniform nature of microbial contamination and the ever-present risk of cross-contamination by the activities of clean-room operators and operations.

Table 13.2 Lower limit of surface bioload density in clean rooms of different classes.

Room classification	Surface bioload density
Uncontrolled manufacturing	1×10^5 spores m^{-2}
Class 100 000 clean room with normal controls	1×10^4 spores m^{-2}
Class 100 000 clean room with stringent controls	1×10^3 spores m^{-2}
Class 10 000 clean room with normal controls	5×10^2 spores m^{-2}
Class 10 000 clean room with stringent controls	5×10^1 spores m^{-2}

Category I and category II missions generally do not need special measures of bioload reduction. Category III missions are orbital or flyby missions sent to targets of high astrobiological interest, such as Mars or Jupiter's moon Europa, and require bioburden reduction if the mission cannot meet the trajectory biasing requirements (i.e., crash probability of <1% for 20 years after launch and <5% for 50 years after launch). In this case, the total bioload of the spacecraft must be reduced to a level lower than 500 000 bacterial spores. This requirement includes exposed external and internal surfaces as well as mated and encapsulated bioburden.

Category IV missions are surface-landing missions to Mars or Jupiter's moon Europa that require sterilization or biocleaning of lander-exposed surfaces (summarized in Table 13.3) in order to comply with the following levels:

- For systems landing on the surface of Mars but not carrying instruments for the investigations of putative extant Martian life forms, the level of contamination shall be less than 300 bacterial spores m^{-2} and less than 300 000 bacterial spores per lander concerning exposed external and internal surfaces. These numbers are based on the contamination level determined for the *Viking* lander before sterilization (*Viking* pre-sterilization level) and apply to category IVa missions.
- For systems landing on the surface of Mars and carrying instruments for the investigation of extant Martian life, the contamination level shall be reduced by four orders of magnitude, compared to landing systems without life-detection experiments. The resulting "calculated" number is less than 30 bacterial spores per lander, which corresponds to the *Viking* lander post-sterilization levels. In cases where the nature and sensitivity of a particular life-detection experiment requires a more stringent bioburden reduction, then these shall be the driving requirements. These levels of bioburden reduction apply to category IVb missions.
- Landing missions to a "special region" of Mars, even if they do not include life-detection experiments, shall also comply

with *Viking* lander post-sterilization biological levels, according to the category IVc mission requirements. If the landing site is within the special region, the entire landed system shall be sterilized at least to *Viking* post-sterilization biological burden levels. If the special region is accessed though horizontal or vertical mobility, this high level of sterility applies not necessarily to the whole lander but to those parts or subsystems that will be in contact with the special region. If an off-nominal condition (such as a hard landing) would cause a high probability of inadvertent biological contamination of the special region by the spacecraft, the entire landed system must be sterilized to *Viking* post-sterilization biological burden levels.

- For any missions sent toward Jupiter's moon Europa, the probability of contaminating this body with terrestrial microorganisms shall be less than 10^{-4}.

Table 13.3 Current bioburden reduction requirements for Mars landers (*Viking* pre- and post-sterilization levels for surface and embedded bioburden)

Requirement	Category of mission		
	IVa: *Viking* Pre-sterilization	IVa: *Viking* Pre-sterilization with impacting hardware	IVb: *Viking* Post-sterilization
Surface spore density	300 spores m^{-2}	300 spores m^{-2}	No explicit requirement
Total surface spores	3×10^5	3×10^5	30[a]
Total spores (including surface, mated and embedded)[c]	[b]	5×10^5	[b]

Source: U. S. Space Studies Board Report, "Preventing the Forward Contamination of Mars."

a No surface spore assays required; number of spores established by the application of a certified bioburden reduction method (dry heat).

b No numerical requirements on embedded burden except for spacecraft with impacting hardware (e.g., heat shields). Embedded bioburden is assumed to remain embedded under nominal operations.

c Embedded bioburden refers to bioburden buried inside non-metallic spacecraft material.

All spacecraft integration operations must be performed in an ISO class 8 clean room (ISO 14644–1: Clean Rooms and Associated Controlled Environments, Part 1: Classification of Air Cleanliness) or better, with biological contamination monitoring. Periodic microbiological assessments must be performed on the flight hardware in order to check the contamination level. If the contamination is too high, surface biocleaning must be performed during integration. The traceability of all monitoring must be kept and the final assessments must demonstrate that the specification is met. After the last integration step, measures are taken in order to prevent any recontamination. As for orbiters, any unfortunate crash impact must be located, and organic materials brought to Mars must be known and recorded.

In order to comply with the contamination levels required for a certain category, the following cleaning or sterilization measures are applied during and after the assembly of the equipment and/or at the final integration:

- surface wiping with biocleaning agents,
- gamma radiation sterilization,
- dry-heat sterilization, and
- hydrogen peroxide gas plasma sterilization.

The selection of the appropriate cleaning and sterilization method (Table 13.4) is driven by two requirements: Firstly, the chosen method and associated procedures must be able to reach the specified decontamination level, and secondly, they should not affect the functional performance of the treated equipment.

Table 13.4 Bioburden reduction and sterilization methods used for the *Mars '96* lander.

Bioburden reduction or sterilization method	Spacecraft subsystem of concern
Biocleaning[a]	Finnish, U. S., Russian experiments on small stations; some subsystems and housekeeping systems of small surface stations and penetrators
Gamma-ray sterilization	Landers parachute and airbags systems; some subsystems of small stations and penetrators
Dry-heat sterilization	Finnish meteorological sensor of the small stations
H_2O_2 gas plasma sterilization	French payload on small stations; penetrators; some subsystems of the penetrators

a Alcohol, sporicide, UV, class 100 clean room.

However, below a certain level of cleanliness (<300 spores m^{-2}), the direct culturing and counting methodology is not sensitive enough for enumerating spacecraft bioburden. For this reason, a series of parametric measures were developed by NASA during the 1960 s. These can be used to estimate the bioload of spacecraft sterilized by dry heat (given as appendices in NASA document NPR8020.12C) based on parametric measurements and estimates. These include

- Time-temperature for sterility: A surface with a temperature exceeding 500 °C for more than 0.5 s during the mission may be considered sterile (viable microbial bioload statistically zero).

- D value, Z value: Dry-heat microbial reduction at a relative humidity less than 25 % (referenced to 0 °C and 1 000 hPa) for spores has a D value (the time for a 10-fold reduction of survival at a certain temperature) at 125 °C of 0.5 h for free surfaces and of 1 h for mated surfaces. The corresponding Z value (the change in temperature for a change in the D value by a factor of 10) is 21 °C for the temperature interval 104–125 °C.
- Hardy organisms: The maximum reduction factor that may be taken for a dry-heat microbial reduction is 10^4. This parameter specification is based on an assumed fraction of hardy organisms (these are of higher resistance to the process than the general microbial population) of 10^{-3} and a reduction of only 0.1 in a nominal sterilization cycle.
- Surface microbial density: Various parameter specifications were permitted for microbial bioload estimates where assays are not feasible; these may also eliminate the need for assay (Table 13.2).

13.3.2.1 Surface Wiping with Biocleaning Agents

Cleaning by wiping techniques in the spacecraft assembly facility shall be performed as often as necessary, depending on the bioburden assessed on surfaces (however, at least every day). A typical cleaning agent is isopropyl alcohol (IPA). This agent is not sporicidal, i.e., it does not kill spores; however, it adequately complies with the bioburden reduction specifications and is used to remove – but not kill – spores from the wiped surfaces. Sporicides may be used in order to kill the remaining microorganisms that are not removed from surfaces by the wiping technique; however, these may leave a residue, and such procedures need to be performed in conjunction with the contamination control staff for the spacecraft. Microbiological assessments must be continued according to a standard protocol in order to verify that the surface biocontamination level remains within acceptable limits.

13.3.2.2 Gamma Radiation Sterilization

Gamma radiation sterilization results from the interaction of photons in the range of 1.17–1.33 MeV (issued from high-activity radioactive ^{60}Co sources) with living systems. Such photons are able to penetrate several centimeters into steel. The typical dose is 25 kGy, which is sufficient to kill spores of the reference microorganism *Bacillus pumilus,* known as the most radiation-resistant representative of bacterial spores. Sterilization with gamma irradiation requires a few minutes to a few hours, depending on the activity of the source and the radiation dose required. Gamma sterilization must be conducted in a special certified blockhouse in order to protect the operators and the environment. Because of its limited compatibility with space hardware – in particular with electronics, optics, and some organic materials – this technique is suitable mainly for structural elements or for radiation-compatible subsystems, which may include, for example, airbags and/or parachutes.

13.3.2.3 Dry-heat Sterilization

Dry-heat thermal sterilization has been the technology most often used for spacecraft sterilization since the 1970 s, when it was applied as a terminal sterilization process on the *Viking* landers after their final integration. The fully integrated *Viking* landers encased in the bioshield were sterilized at 112 ± 1.7 °C for 30 h. This final dry-heat sterilization was applied after the lander was biocleaned to a level of 300 bacterial spores m^{-2} and after the life-detection instruments were sterilized at 120 °C for 54 h. Dry-heat cleaning is presently the only method qualified for space hardware sterilization by NASA. The optimal range recommended is 104 °C to 125 °C for 50 h and 5 h, respectively. The effects of thermal sterilization on microorganisms depend on the profile of temperature and time during the sterilization process. The typical parameters used for medical sterilization are given for spores of the reference microorganisms *Bacillus subtilis* or *Bacillus stearothermophilus*. The medical procedures generally give temperatures between 105 °C and 180 °C, for periods between 1.5 and 300 h. However, not all spacecraft material and equipment can withstand dry-heat sterilization; for example, modern spacecraft batteries would be destroyed by heat treatment at well below the minimum sterilization temperature.

13.3.2.4 Hydrogen Peroxide Vapor/Gas Plasma Sterilization

Hydrogen peroxide gas plasma is a recent sterilization method developed by industry for sterilization of medical instruments. The so-called Sterrad sterilization process is done in specific sterilization chambers at low temperature (45 °C) and low pressure (less than 1 hPa) where gaseous hydrogen peroxide sterilizes the equipment surface during a relatively short period (less than 1 h). Final plasma may be generated in order to increase the effect of sterilization and to break the remaining hydrogen peroxide molecules. The U. S. Food and Drug Administration successfully tested the effect of 50 sterilization cycles on different medical devices, including electronic and optical instruments. The medical procedures are qualified for spores of the reference microorganism *Bacillus subtilis* var. *niger*. Concerning spacecraft sterilization, this method was first tested (without the plasma phase) in 1992 on different materials and equipment of the *Mars* '96 mission (Table 13.4). It was found to be very compatible with various kinds of equipment. Later, elements of the *Beagle 2* spacecraft were sterilized using this technology. However, uncertainties about the compatibility of the process with complete spacecraft, together with the fact that this is a surface-only sterilization technology, have slowed its widespread adoption in the spacecraft community.

13.3.3
Prevention of Recontamination

One of the main problems during the spacecraft integration process is maintaining biological cleanliness after sterilization of the equipment. A special protocol is required to ensure that the spacecraft assembly facility is maintained in a biologically clean condition. This protocol must, on one hand, comply with required operations and access procedures for equipment and workers and, on the other hand, prevent recontamination of the sterilized equipment. To achieve this goal, various approaches can be adopted. Controlled access to the sterile clean room with operators being required to wear decontaminated overshoes, overalls, hoods, masks, and gloves is typical. Sterile tools need to be used, and continuous monitoring of the bioload level and frequent biocleaning operations must be included in the protocol. One example of this approach is the planetary protection protocol used for the *Mars '96* mission. In this case, two ISO class 5 clean rooms were used. To control and maintain the cleanliness of the clean rooms, all working surfaces (tables, chairs, etc.) in the integration area were cleaned daily with alcohol and a

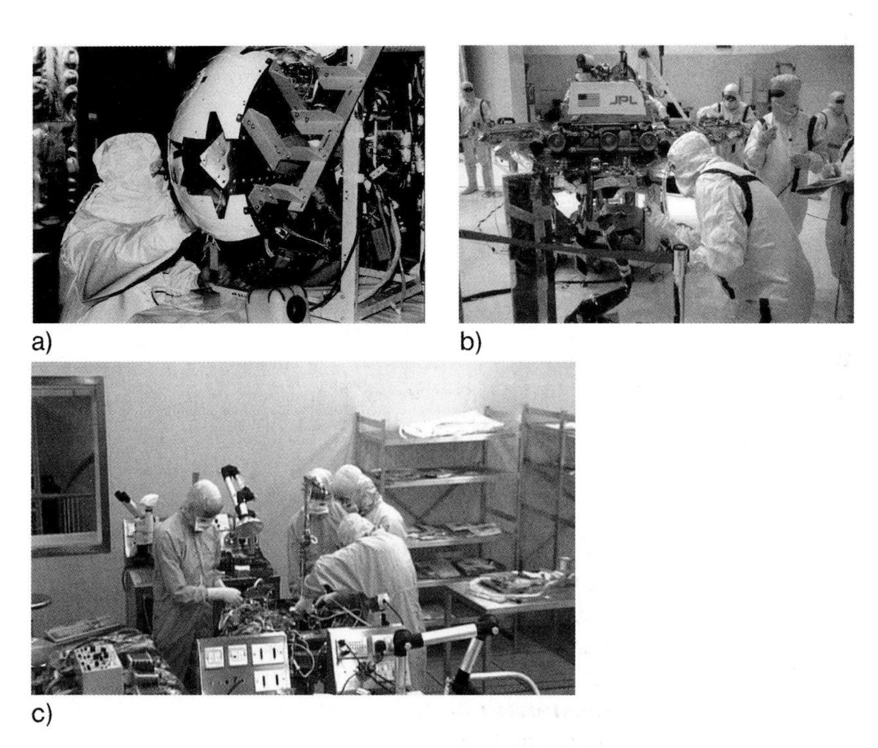

Fig. 13.3 Integration of Mars lander spacecraft. (a) *Mars '96* lander in a biologically controlled environment in NPO Lavochkin Moscow (Photo credit: CNES). (b) *Mars Pathfinder* at the Spacecraft Assembly Facility of the Jet Propulsion Laboratory (Photo courtesy of R. Koukol, JPL). (c) *Beagle 2* in the class 10 clean room of the Open University, U. K. (photo courtesy of Open University).

sporicide, and operators had to wear clean-room garments (Fig. 13.3). The clothes were washed in chlorinated water solutions, and one wash in 10 was followed by steam sterilization (20 min at 120 °C).

However, the prevention of contamination does not end with spacecraft assembly. Little or no recontamination should occur during launch. This can be achieved by encasing the spacecraft in a biobarrier or bioshield and removing this shield only after the system is launched into space.

13.4 Astrobiology Exploratory Missions of Concern to Planetary Protection

Within the ongoing and planned ventures to explore the planets, moons, and other bodies of our Solar System, both by orbiters and robotic landing missions, the search for signatures of life beyond the Earth is one of the major drivers. In the selection of candidate targets for this enterprise, the overarching argument is the putative habitability of the planet or moon under consideration (see Chapter 6). The planetary protection requirements are based on these considerations (Table 13.1).

13.4.1 Missions to the Moon

In preparation for the first landing mission to the Moon within the *Apollo* program, planetary protection regulations with regard to the possible contamination of a celestial body, in this case the Moon, were considered for the first time. However, because of the surface conditions, such as high vacuum and intense cosmic and solar UV radiation, the Moon's surface was regarded as highly unfavorable for the growth of terrestrial microorganisms. In contrast, the absence of oxygen, water, and other life-supporting substances was thought not to prohibit survival, especially of dormant anaerobic microorganisms. It was considered that the ubiquitous nature of bacterial spores and their durability under adverse conditions, including high vacuum and low temperatures (see Chapter 5), must be fully appreciated to prevent contamination. It was argued that a biological contamination of the Moon might affect studies on the early history of the Solar System, the origin of life on Earth, extraterrestrial life, and the chemical composition of matter in the remote past. Especially in order to avoid possible distortion of chemical evidence by microbial action and contamination of the Moon's deeper layers, it was recommended

- to minimize contamination to the extent technically feasible;
- to inventory all organic chemical constituents brought to the surface of the Moon; and
- to perform sterile drilling on the Moon in order to return an uncontaminated sample of the lunar subsoil for biological and geochemical investigations.

The latter recommendation shows that, at that time, it was not excluded that life might exist at some depth below the surface of the Moon where temperatures never exceed 100 °C and UV radiation does not penetrate.

Because minute traces of Earth-borne contaminants could lead to completely erroneous interpretations, the samples of rock and soil brought back from the Moon needed to be carefully protected. For this purpose, the Lunar Receiving Laboratory was set up at the Manned Spaceflight Center (now Johnson Spaceflight Center) in Houston, USA, where lunar samples were cataloged and subjected to preliminary examination. The *in vitro* and *in vivo* tests at the Lunar Receiving Laboratory provided data showing that no microorganisms were recovered from the lunar samples returned during the *Apollo* program. This was a result of both the efforts to limit contamination in the earlier unmanned launches and the efforts to take the samples under aseptic conditions in the *Apollo* manned landings. The contamination control plan developed and implemented eventually resulted in providing investigators with lunar samples containing less than 0.1 ppm total terrestrial organic contamination. It should be noted that this is as low as or even lower than the experimental blanks obtained in organic geochemistry research laboratories.

Another concern of back-contamination arose from the astronauts themselves who had landed on the lunar surface. Flight plans called for the astronauts to contain as much lunar dust as possible in the lunar module (LM) by vacuuming it off their space suits then removing the suits and sealing them in storage bags before returning to the command module (CM). During the transfer from the Moon to Earth, the LM pressure relief valve was opened and the oxygen flow in the CM was increased to maintain a current of gas flowing from the CM to the LM, thus minimizing the amount of dust carried into the CM. After return to the Earth, the crew of the first two *Apollo* landing missions was isolated in a Mobile Quarantine Facility aboard the recovery ship (Fig. 13.4), before being held in quarantine

Fig. 13.4 *Apollo 11* astronauts (from left to right) Neil A. Armstrong, commander; Michael Collins, command module pilot; and Edward E. Aldrin, Jr., lunar module pilot, as they are confined to the Mobile Quarantine Facility aboard the U. S. S. *Hornet*, prime recovery ship for the historic *Apollo 11* lunar landing mission (Photo NASA).

in the living quarters at the Lunar Receiving Laboratory for at least three weeks. In light of the results from the first two *Apollo* missions, the requirements for crew quarantine of *Apollo* astronauts were discontinued for the follow-on lunar missions. However, the biological containment practices in the Lunar Receiving Laboratory and the biological examination of the lunar samples were continued because, among other things, they assured the integrity of the samples.

There are reports that bacteria might have survived for 2.7 years on the surface of the Moon in the polyurethane foam insulation of the camera interior of the *Surveyor 3* spacecraft. This lander was visited by the crew of the *Apollo 12* LM, and about 10 kg of parts from the *Surveyor 3* spacecraft, including the TV camera, were removed for later examination back on Earth. Whereas most samples turned out to be sterile, the bacterium *Streptococcus mitis* was isolated from one part of the camera foam. Although it was first concluded that the bacterium was of terrestrial origin and had survived the lunar exposure and return trip, there are many indications that it came from contamination of the sample during the collection or return trip. The main reason is that *S. mitis* was identified as part of the microflora of the astronauts of *Apollo 12.*

The present view is that the Moon is part of a single Earth–Moon system; hence, from a planetary protection perspective the Moon does not need to be protected from terrestrial biological contamination. However, the Moon must be protected from contamination by extraterrestrial samples, defined as "restricted Earth return" of category V missions, in order to retain freedom from planetary protection requirements on future human lunar missions, including the establishment of a lunar base.

13.4.2 Missions to Mars

With the exception of the Earth, Mars is by far the most intensively studied of the planets of our Solar System (Table 13.5). In 1972, for the first time a spacecraft, *Mariner 9*, passed over the younger parts of Mars, revealing a wide variety of geological processes as indicated by volcanoes, canyons, and channels that resemble dry river beds and attest to a stable flow of water on Mars at some time in the past and even in more recent times (see Chapter 8). The COSPAR Planetary Protection Guidelines categorize missions to Mars within category III, IV, or V, depending on the kind of mission (whether they are orbiters, flyby missions, landers with *in situ* research, or landers with sample return tasks) (Table 13.1).

13.4.2.1 Orbiters or Flyby Missions to Mars

Currently four spacecraft are orbiting Mars: *Mars Global Surveyor, Mars Odyssey*, and *Mars Reconnaissance Orbiter* of NASA, and *Mars Express* of ESA (Table 13.5). They are categorized as category III missions. If they achieve a total bioburden of less than 500 000 spores per total spacecraft, then they do not need trajectory biasing.

Table 13.5 List of past and present missions to Mars.

Mission	Country/ Agency	Launch date m/d/y	Purpose	Results
Marsnik 1	USSR	10/10/60	Mars flyby	Did not reach Earth orbit
Marsnik 2	USSR	10/14/60	Mars flyby	Did not reach Earth orbit
Sputnik 22	USSR	10/24/62	Mars flyby	Achieved Earth orbit only
Mars 1	USSR	11/1/62	Mars flyby	Radio failed at 106 million km
Sputnik 24	USSR	11/4/62	Lander	Achieved Earth orbit only
Mariner 3	U.S.	11/5/64	Mars flyby	Shroud failed to jettison
Mariner 4	U.S.	11/28/64	First successful Mars flyby 7/14/65	Returned 21 photos
Zond 2	USSR	11/30/64	Mars flyby	Passed Mars but radio failed; returned no planetary data
Mariner 6	U.S.	2/24/69	Mars flyby 7/31/69	Returned 75 photos
Mariner 7	U.S.	3/27/69	Mars flyby 8/5/69	Returned 126 photos
Mariner 8	U.S.	5/8/71	Mars orbiter	Failed during launch
Cosmos 419	USSR	5/10/71	Mars lander	Achieved Earth orbit only
Mars 2	USSR	5/19/71	Mars orbiter/lander arrived 11/27/71	No useful data, lander destroyed
Mars 3	USSR	5/28/71	Mars orbiter/lander, arrived 12/3/71	Some data and a few photos
Mariner 9	U.S.	5/30/71	Mars orbiter, in orbit 11/13/71 to 10/27/72	Returned 7 329 photos
Mars 4	USSR	7/21/73	Mars orbiter	Flew past Mars 2/10/74
Mars 5	USSR	7/25/73	Mars orbiter, arrived 2/12/74	Lasted a few days
Mars 6	USSR	8/5/73	Mars orbiter/lander, arrived 3/12/74	Little data returned
Mars 7	USSR	8/9/73	Mars fly by /lander, arrived 3/9/74	Little data returned
Viking 1	U.S.	8/20/75	Mars orbiter/lander, orbit 6/19/76–1980, lander 7/20/76–1982	Combined, the *Viking* orbiters and landers returned more than 50 000 photos
Viking 2	U.S.	9/9/75	Mars orbiter/lander, orbit 8/7/76–1987, lander 9/3/76–1980	Combined, the Viking orbiters and landers returned >50 000 photos
Phobos 1	USSR	7/7/88	Mars/Phobos orbiter/lander	Lost 8/88 en route to Mars
Phobos 2	USSR	7/12/88	Mars/Phobos orbiter/lander	Lost 3/89 near Phobos
Mars Observer	U.S.	9/25/92	Orbiter	Lost just before Mars arrival 8/21/93
Mars Global Surveyor	U.S.	11/7/96	Orbiter, arrived 9/12/97	Science mapping
Mars '96	Russia	11/16/96	Orbiter and landers	Launch vehicle failed
Mars Pathfinder	U.S.	12/4/96	Mars lander and rover, landed 7/4/97	Last transmission 9/27/97
Nozomi (Planet-B)	Japan	7/4/98	Mars orbiter, currently in orbit around the Sun	Mars fly-by due to propulsion problem

Mission	Country/ Agency	Launch date m/d/y	Purpose	Results
Mars Climate Orbiter	U.S.	12/11/98	Orbiter	Lost on arrival at Mars 9/23/99
Mars Polar Lander/ Deep Space 2	U.S.	1/3/99	Lander/descent probes to explore Martian south pole	Lost on arrival 12/3/99
Mars Odyssey	U.S.	4/7/01	Orbiter	Science mapping
Mars Express with Beagle 2	ESA	6/2/03	Orbiter and lander	Science mapping by orbiter, lander lost on arrival
Mars Exploration Rovers MER A and B	U.S.	June/July 2003	Lander and rover	Landed on 1/4/04 (MER A); on 1/25/04 (MER B); extended science mission
Mars Reconnaissance Orbiter	U.S.	8/12/05	Orbiter	Entered orbit around Mars 3/10/06

For the historic mission set, *Mars Global Surveyor, Mars Odyssey*, and *Mars Express* adopted the trajectory biasing approach. *Mars Reconnaisance Orbiter* was required to adopt the bioburden control approach, as the mission design required a low science orbit that was too "risky" to meet the probability of impact requirements for the first 20 years of the mission.

13.4.2.2 Landers or Rovers with *in Situ* Measurements

All lander missions to Mars fall into category IV. Table 13.5 shows that there were at least five successful landing events on Mars, namely, *Viking 1* and *Viking 2, Mars Pathfinder*, and the two Mars Exploration Rovers *Spirit* and *Opportunity*. However, those lander missions that failed needed to follow the same stringent planetary protection requirements (Table 13.1).

The *Viking* landers themselves, with their overarching goal of searching for signatures in indigenous microbial life on the surface of Mars, were terminally sterilized. That is, at the end of the assembly phase, the landers were subjected to a high-temperature dry-heat sterilization process of $112 \pm 1.7\,°C$ for 30 h, rendering them effectively sterile (with a bioburden close to 0; statistically, 30 spore-forming microorganisms were accounted at launch).

However, when the scientific findings of the *Viking* landers showed that the Martian environment is inhospitable to then-known terrestrial microorganisms, COSPAR requirements for Martian landers carrying non-life-detection instrumentation (now category IVa) were relaxed to the level of contamination present on the *Viking* landers before sterilization, namely, 300 000 spore-forming microorganisms in total, at an average density of below 300 microorganisms per square meter for a single landing event. To achieve even this level of cleanliness is not straightforward and requires spacecraft assembly in a clean-room environment, with a set of monitoring and cleanliness criteria over and well above those normally adopted for spacecraft assembly, e.g., with missions into low Earth orbit. However, for landers carrying extant life-detection instrumentation (category IVb), the "gold standard" continues to be that of the contamination present on the *Viking* landers post-sterilization, namely 30 spores or fewer (Table 13.3).

Therefore, since *Viking*, all NASA (*Pathfinder, Mars Polar Lander*, Mars Exploration Rovers *Spirit* and *Opportunity*) and ESA (*Mars '96, Beagle 2*) missions to the surface of Mars (Table 13.5) have been classified as category IVa missions. (The Soviet *Mars 3* lander successfully landed on Mars in December 1971, but contact was lost 20 s after the landing.) NASA's 2007 *Phoenix* mission was categorized as IVa, but it meets category IVc requirements for its targeted landing site, which is a polar region rich in water ice. NASA's 2009 *Mars Science Laboratory* is assigned category IVc. ESA's 2013 *ExoMars* mission (see Chapter 12) has yet to be formally categorized.

For some missions, the scientific studies could potentially be affected or compromised by contaminants carried by the spacecraft. In these cases, the science team negotiates with the mission management to request additional requirements in terms of, e.g., chemical cleanliness over and above those of planetary protection. These are typically detailed in the Planetary Protection Implementation Plan. They can involve materials control in the lander and spacecraft assembly facility, together with monitoring of the organic compounds that would normally be "permitted" in the spacecraft. These scientific cleanliness requirements impact the environmental control of the spacecraft from initial assembly through launch. The *Beagle 2* spacecraft (Table 13.5) was one such mission. For the upcoming *Mars Science Laboratory* (MSL09) mission, an organic cleanliness level, measured as total organic carbon (TOC), of less than 1 ng cm^{-2} is required for all equipment that will come in contact with Mars samples.

As an example, the *Beagle 2* lander mission will be used to demonstrate how problem areas were identified during the design and construction of the lander and how final solutions were found (Table 13.6).

Once the scope of the planetary protection problem is understood, a sterilization strategy can be developed for a specific mission. This could rely on a single technology or on several and may require a single processing of the hardware or several. In the case of *Viking*, multiple dry-heat processings were used. For the Mars Exploration Rovers, typically only one dry-heat processing was used. In contrast, the *Beagle 2* program used several sterilization technologies, developing a "preference cascade" where gamma irradiation is preferred after dry heat, as it is also a penetrating technology. Hydrogen peroxide–based methodologies were next in preference because of their benign effect on the hardware. However, this technology sterilizes only surfaces. Finally, IPA (or other solvent) wiping is the last approach, as it removes the organism rather than killing it *in situ* and is used only as a last resort. Ideally, combinations could be used, although it is quite challenging to determine whether all the components of one system are compatible with a single technology.

The combination of parametric measures and culture-derived estimates of bioload is then summed to give the final bioburden at launch, and this is checked versus the requirement before launch is permitted. Table 13.7 gives the final bioload numbers from three Mars lander missions where very different planetary protection strategies were employed.

Table 13.6 *Beagle 2* sterilization challenges as highlighted in the Planetary Protection Plan.

Hardware item	Challenge identified	PPP[a] proposed solution	Evolution of solution by time of PPIP[b] report
Thermal insulation	Risk of shedding of organic debris from foams	Seal in Kapton foil and use UV to reduce microbial bioload	Seal in Kapton vented with Tyvek; bake to sterilize throughout (UV unsuitable as effective in line of sight only).
Batteries	Not suitable for heat sterilization	Wipe with alcohol wipes	Sterilize by H_2O_2 gas plasma to give better sterility assurance.
Solar array units	Phase change material in deployment mechanism not suitable for heat sterilization	None proposed (dependant on type of material to be used)	Frangibolt mechanisms to be alcohol wiped.
Parachutes	Sterilize/packing sequence may increase risk of debris ingress on deployment	Bake to reduce outgassing. Pack and rebake or use gamma radiation (effect on material strength?)	Pack "clean" and bake. Debris ingress prevented by airbag material.
Airbags	Sterilize/packing sequence may increase risk of debris ingress on deployment	Bake to reduce outgassing. Pack and rebake or use gamma radiation (effect on material strength?) Ensure insides are sterilized due to exposure on deflation. Material degradation?	Pack "clean" and bake. Debris ingress reduced by nature of airbag material. Small (tolerated) risk of fibers being present around the lander overcome by modification to scientific method.
EDLS/structure	Maintenance of sterility after assembly	Use bioseal and HEPA filter on the vent	Use bioseal and HEPA filter on the vent.
Front shield	Presence of organisms on the external surfaces	Bake to reduce bioload (is there a high-temperature bake during manufacturing that will reduce the internal bioload?) Maintenance through AIV and launch?	Sterilize by dry heat; rely on heating during atmospheric entry to maintain external sterility (>500°C predicted).
Back cover	Presence of organisms on the external surfaces	Bake to reduce bioload	Sterilize by dry heat; rely on heating during atmospheric entry to maintain external sterility (>500°C predicted for some areas, >200°C for all areas subsequently modeled to demonstrate sterility).

Hardware item	Challenge identified	PPP[a] proposed solution	Evolution of solution by time of PPIP[b] report
Solar panels	Large, delicate surface areas	Use sterile alcohol wipes	Requirement for no surface contact, as coatings are sensitive to solvent and abrasion; dry-heat sterilization acceptable. Post-sterilization, contamination prevention was the key. Witness plates used for contamination monitoring.
Pyro devices	Not suitable for heat sterilization	Wipe with alcohol wipes	Wipe with alcohol wipes.
Science instruments	Will they withstand baking?	None proposed	Sterilize by dry heat or H_2O_2 gas plasma where not compatible with heat.
Mole, Arm	Internal and external require sterilization	None proposed	Sterilize by dry heat.
Harness	Will require baking to reduce bioload	None proposed	Sterilize by dry heat or H_2O_2 gas plasma where associated with instruments not compatible with heat. (All wiring specified as Raychem 55 radiation hardened, which gives sterilization of effect on any embedded organisms.)
Mössbauer source	Because of the half-life of the radioactive source for the Mössbauer spectrometer, the source will be inserted at the last minute prior to delivery to the launch site	None proposed	Mössbauer source fitted as late as possible prior to closure (decay tolerated).
X-ray fluorescence spectrometer source	Because of the half-life of the radioactive source for the XRS, the source will be inserted at the last minute prior to delivery to the launch site	None proposed	XRS source fitted as late as possible prior to closure (decay tolerated).

a PPP: Planetary Protection Plan.
b PPIP: Planetary Protection Implementation Plan.

Table 13.7 Final bioload numbers from three lander missions to Mars where very different planetary protection strategies were employed.

Mission	*Viking* (NASA, 1975)	MER (NASA, 2003)[a]		*Beagle 2* (UK/ESA 2003)
		MER A	MER B	
Category	**IVb (equivalent)**	**IVa**	**IVa**	**IVa+**
Approach	Substantial qualification program, final system-level dry-heat sterilization	Mostly using either dry heat at subsystem level with wipe cleaning to maintain sterility	Mostly using either dry heat at subsystem level with wipe cleaning to maintain sterility	Science-driven program using sterilization cascade (heat > gamma > H_2O_2 > IPA wipe)
Spores accounted at launch (3sigma worst case)	Surfaces 30 (density <<1 m^{-2}) Embedded ND	Surfaces 1.2×10^5 (density $74/m^2$) Embedded[b] 3.5×10^4	Surfaces 1.3×10^5 (density 74 m^{-2}) Embedded[b] 3.5×10^4	Surfaces 2.3×10^4 (density $20.6/m^2$) Embedded 1.03×10^5
Assembly environment	Mostly class 100 000 clean rooms	Mostly class 100 000 clean rooms	Mostly class 100 000 clean rooms	Class 10 aseptic assembly following sterilization of subassemblies

a Data adapted from Newlin et al. Microbial cleanliness of the Mars Explorer Rover spacecraft, paper presented at the 35th COSPAR Scientific Assembly, Paris, 18–25 July 2004.

b Impacting hardware only (e.g., aeroshell) calculated from spore density specifications and adjusted for dry-heat microbial reduction. ND: not determined.

13.4.2.3 Landers or Rovers with Martian Samples Returned to the Earth

The long-term strategy of the European Space Exploration Program foresees a Mars Sample Return (MSR) mission as the most likely successor to *ExoMars* (see Chapter 12), and it is anticipated that in the future NASA and other space-faring nations will be interested in returning samples from Mars for in-depth scientific investigations.

A Mars sample return mission falls into category V as "restricted Earth return," which means that the samples brought from Mars to the Earth need special attention from the moment of collecting, through transport to Earth, and upon landing on Earth. After arriving on our surface, it has to be proven that there is no biohazard risk from the Martian samples. Even if there is only a remote possibility that the samples will contain specimens of microorganisms from Mars in a viable state, Martian soil, rocks, or atmosphere must be kept under strong containment according to the international treaty.

Although a joint team of American and French scientists, provided by NASA and the Centre National d'Etude Spatiale (CNES), working in the frame of a preliminary design of a common Mars sample return mission has already defined a preliminary biohazard assessment test protocol for samples returned from Mars, a similar protocol still needs to be developed and adopted by ESA.

In simple terms, the sample return container must meet the following requirements:

- preventing back-contamination, and
- preserving the integrity and the properties of the samples.

To avoid cross-contamination, the container must be hermetically sealed by means of a mechanical system. Otherwise, the samples may be recontaminated by terrestrial compounds during the Earth atmosphere entry phase. Different methods are envisaged to control the sealing before the release of the container system into Earth's atmosphere:

- Determination by gas chromatography–mass spectrometry (GC-MS) of the leak rate by outside detection of a characteristic gas, e.g., helium or CO_2 from the Martian atmosphere.
- If an atmospheric sample is collected, monitoring of the pressure inside the container – corrected by temperature variations – during cruise phase.

However, it must be recognized that each of these methods has its limits. A further requirement is bio-cleanliness and sterility of the container; otherwise, care must be taken that the outside surfaces of the container are never in contact with the extraterrestrial environment. Before release into Earth's atmosphere, three biological barriers are required in order to comply with World Health Organization recommendations for the transport of hazardous biological material. Even in the case of a non-nominal Earth return event, (in the worst case a crash or an unintentional landing in the sea), the container must maintain its hermetic seal and must be recovered.

As a next step, a Mars Sample Receiving Facility (MSRF) must be available to host the returned container and its content. Within this facility the material will be unpacked and analyzed accordingly to the defined scientific methods and standards in the biohazard assessment protocol. This MSRF, which must provide a very high degree of containment, must also prevent additional contamination of the samples from Earth-bound inorganic, organic, or biological material. To the authors' knowledge, no such facility exists today that meets all these requirements.

In order to formulate clear requirements for the MSRF, a functional project definition must be provided. This requires interactive involvement of several communities:

- the scientific community, in order to review and refine existing draft test protocols that define the requirements of the instruments and methods;
- experts in containment technologies, in order to work out a functional technical description of the high containment facility, resulting in a definition of space needs, technical installations, and cost estimation, as accurately as possible;
- a team providing a thorough risk assessment for the whole transport, including the way from Mars to Earth, the way from the landing site to the MSRF, and the work inside the facility;
- a group defining the research needs from the perspective of decontamination and sterilization of known Earth-bound and unknown Mars-bound life forms.

The starting point for the definition of a MSRF will be existing concepts of high containment and the necessary equipment, which exist on Earth mainly for biosafety reasons.

Existing Concepts of High Containment Organisms are currently classified into four risk groups: RG1 to RG4. Accordingly, there are also four biosafety levels: BSL1 to BSL4. (Note that there is no universally accepted standard for containment. BSL is the most widely used, so we have adopted it here.) For BSL4 laboratories, two distinct concepts exist, as follows:

- The cabinet laboratory, where all handling of the agent is performed in a class III biological safety cabinet (BSC), which is discussed later. The cabinet laboratory is arranged to ensure passage through a minimum of two doors prior to entering any room containing the class III BSC.
- The suit laboratory, where the personnel wear a protective suit. The use of class II BSC cabinets is common in the suit laboratory. (A suit laboratory was depicted in the movie *Outbreak*, starring Dustin Hoffman, who portrayed an Army medical expert sent to combat a deadly virus that destroyed a small town in California and threatened the world.) With this publicity event, suit laboratories became very popular and

many new ones are currently being built. The BSL4 containment zone should be located in a separate building or sealed room with independent air supply and exhaust systems. These air systems and components need to be sealed airtight and be accessible from outside the containment area. A ventilated airlock is required for the separation of higher and lower containment areas. The airlock is to have interlocking pneumatic or compressible sealed doors with manual overrides. The personnel working in the suit laboratory wear a one-piece, positive-pressure suit that is ventilated by a life-support system protected by HEPA filtration (see below). The life-support system includes redundant breathing air compressors, alarms, and emergency backup air tanks. A chemical shower is provided to decontaminate the surface of the suit before the worker leaves the area. An emergency power source is provided for the exhaust system, life-support systems, alarms, lighting, entry and exit controls, and BSCs. The air pressure within the personnel suit is positive to the surrounding laboratory. The air pressure within the suit area is lower than that of the adjacent area. The entry to the containment area is equipped with magnetic gauges or pressure-monitoring devices to prove directional flow. The containment area is equipped with audible alarms to detect depressurization. Supply air to the cabinet room or suit area and associated decontamination shower and airlock is protected by passage through a HEPA filter. The exhaust of the laboratory's suit area, decontamination shower, and decontamination airlock is treated by passage through two HEPA filters in a series prior to discharge to the outside. Redundant supply fans are recommended, and redundant exhaust fans are required.

HEPA Filtration The development of the HEPA filter during World War II was critical for providing the necessary containment in the modern biosafety laboratory. These filters are constructed of fiberglass "paper" that is pleated to maximize the surface area of the filter, and they have a minimum particulate removal of 99.97% for particles of 0.3 μm. The randomly oriented microfibers cause the particles in the air stream to move in a circuitous path, forcing even the smallest particles to collide with and adhere to the filter. The filter housings on contaminated exhaust streams should be constructed to facilitate decontamination or should be of the bag-in-bag-out design. Three glove sleeves are provided in each bag to facilitate the handling of the filter during exchange.

Biological Safety Cabinets The biological safety cabinet (BSC) design has come a long way since inception, with three main classifications: class I, class II, and class III.

Class I cabinets are defined as ventilated cabinets for personnel and environmental protection, with unrecirculated airflow away from the operator. Class I cabinets have a similar airflow pattern to a fume hood, except that the class I cabinet has a HEPA filter at the exhaust outlet, and it may or may not be connected to an exhaust duct system. Class I cabinets are suitable for work with agents that require BSL 1 or 2 containment.

Class II cabinets are ventilated cabinets having an open front with inward airflow for personnel protection, downward HEPA-filtered laminar flow for product protection, and HEPA-filtered exhaust airflow for environmental protection. Class II cabinets are suitable for work with agents that require BSL 1, 2, or 3 containment.

Class II cabinets are differentiated into various types based on their construction, air velocities, patterns, and exhaust system. For instance, class II type A cabinets may have contaminated plenums under positive pressure that are exposed to the room, while class II type B cabinets must surround all contaminated positive-pressure plenums with negative-pressure ductwork. Type A cabinets can be exhausted into the lab or outside by way of a canopy connection, whereas type B cabinets must have a dedicated, sealed exhaust system with remote blower and appropriate alarm system.

Class III cabinets are totally enclosed and ventilated of gas-tight construction. Operations are conducted through attached rubber gloves. The cabinet is maintained under negative air pressure of at least 125 Pa. Supply air is drawn into the cabinet through HEPA filters. The exhaust air is treated by double HEPA filtration. The cabinet also has a transfer chamber capable of sterilizing work materials before exiting the glove box containment system. Class III cabinets are suitable for work with agents that require BSL 1, 2, 3, or 4 containment.

A typical cabinet laboratory, as required for a MSRL, would be composed of a number of interconnected class III BSCs. All operations and handling would be performed within the modules, which could in addition be of the double-wall containment type. The double-wall containment module would be of the box-in-the-box type with an inter-space surrounding a standard single-wall containment. This space would be at negative pressure relative to that of the room and the inner cabinet in order to protect the personnel as well as to keep the sample pristine. If a Mars-like atmosphere is required inside the module, an inert gas, such as nitrogen, could be fed to the module or a Mars-analogue atmosphere. For this reason, gloves and glove material have to be evaluated very carefully. Material transfer in and out of the modules and the containment would need to maintain confinement and the pristine nature of the samples. Different methods are in use today in laboratories working with dangerous viruses.

Use of Robotics It might be necessary to include a robotics concept into the design of the MSRL in order to fulfill the requirements for non-contamination of the Mars samples by Earth-bound material. This could be for the whole spectrum of activities, including preparing, analyzing, and housing of samples, or it could be for single critical functions only, such as fine-scale manipulations to handle very small samples. It is not believed that existing robots from industrial applications could be used without further development. For obvious reasons, all the robots would have to be of the type suitable for clean-room operation, and ISO class 1–certified robots are now available. Aside from robot operations, class 1 certification is used in the semiconductor and pharmaceutical industries for achieving mini-containment environments.

In conclusion, from the above arguments, it is clear that for design and construction of a facility for Mars samples, new concepts and ideas are needed. Therefore, a program of research in high containment technology, in decontamination methods, and in material compatibility must be initiated well beforehand. It is estimated that on the order of 10 years preparation is required ahead of sample arrival.

13.4.2.4 Human Missions to Mars

A human mission to Mars has been a long-term goal in every space exploration program since the advent of space flight. Although any human endeavor to visit the red planet is still decades away, the implications of planetary protection requirements on system development and mission operation in combination with the long lead-time for the development of man-rated systems make this topic very timely. A detailed understanding of these aspects is also necessary in order to define the right robotic precursor missions to support the development of a human mission scenario for Mars.

It is not possible for all human-associated processes and mission operations to be conducted within entirely closed systems. Therefore, a human mission to Mars will contaminate the planet to a certain extent. However, this forward contamination can and must be minimized by using appropriate procedures. In order to define these procedures, we need to establish a better understanding of both the Martian environment and the fate of terrestrial contamination, particularly biological contamination, on this environment. Included in this context is the need for containment or sterilization of solid, liquid, and gaseous waste products during surface operations on Mars. Prior to leaving the surface, decontamination or stabilization of the surface elements with respect to biological contamination will be necessary.

The philosophy for backward contamination prevention is to break the chain of contact between Mars and Earth. Any contamination of the crew habitat on Mars by potential biological agents of Martian origin has to be avoided, as this contamination would be transferred to the crew and hence to Earth. Strict isolation of the crew upon return is not practicable and would be difficult to implement. Two specific issues are worth mentioning in this respect: the need for sophisticated dust management and the need for emergency decontamination procedures for the surface elements in case a contamination event takes place. The latter is simply a technology adaptation based on already-existing systems. Dust management, however, will be a challenging enterprise. The *Apollo* missions have already shown us that dust can be a major problem when not managed properly. It can effect mechanical systems and endanger the health of the crew, independent of its chemistry. In addition, dust is a potential carrier of reactive chemical agents and, potentially, biological entities. A well-developed dust management system required in order to implement effective backward planetary protection guidelines can therefore also reduce the threat to crew health and safety. One aspect that will need further reflection is the use of *in situ* resource utilization systems to produce consumables for the crew.

In summary, the planetary protection guidelines to limit contamination of Mars and to avoid contamination of the Earth by potential Martian life will not be relaxed to accommodate a human mission to Mars. There are, however, two issues that are very specific to a human mission to Mars. Regarding forward contamination, the biological contamination of the spacecraft will increase during the mission because of the presence of the crew. For robotic spacecraft the bioburden limit set forth in the planetary protection requirements has to be met at launch. The second, and certainly not less important, aspect is that general human factors have to be considered along with planetary protection issues for a human mission to Mars. Critical elements of forward and backward contamination control depending on procedures have to be designed in a way that makes them effective even in case of human error.

Although earlier human missions have a greater perceived risk, planetary protection requirements likely will not become more relaxed in subsequent missions. The main reason for this is that every mission will explore locations on Mars of high scientific (including astrobiological) interest that have not been visited before.

13.4.3
Missions to Venus

The current atmospheric environment on the surface of Venus, with an average surface temperature of 464 °C, is too hostile to support life, although the possibility of the production of organic compounds in its atmosphere, e.g., by lightning, has been suggested. Little is known about its history despite more than 20 probes sent to Venus since 1961 (see also Chapter 8). This lack of knowledge is due mainly to Venus's high surface temperature, high pressure (95 000 hPa), and permanent thick clouds – primarily composed of aqueous solutions of H_2SO_4 – which obstruct its exploration. It is assumed that in its early history Venus might have possessed a sizable inventory of water and a cooler climate (100–200 °C). Runaway greenhouse conditions that still exist today would then have led to evaporation of the water, its photolysis at higher altitudes, and escape of hydrogen. Today only traces of water remain in the atmosphere of Venus. Owing to the high temperature at the surface, the lack of water, and the presence of toxic chemicals such as sulfuric acid, indigenous biological activity on or below Venus's surface seems to be very unlikely. The same is valid for the clouds, where temperature and pressure are more moderate than on the surface but where drops of concentrated sulfuric acid prevail. However, Venus may have been considerably more Earth-like in its past, and thus it cannot be excluded that early environmental conditions on Venus might have been favorable for the emergence and early evolution of life.

Venus's present environment not only is inimical to life but also makes heavy demands on the robustness of the technical equipment landing on its surface. The first successful landing on the surface of Venus was achieved in 1970 with the Soviet *Venera 7* mission; the landing probe survived the heat and pressure for 23 min. This was followed in 1985 by the Soviet *Vega 1* and *Vega 2* twin missions, when the landers carried out surface studies for about 1 h. This environment would have killed any terrestrial microorganisms that might have traveled as blind passengers with the lander from Earth to Venus.

Venus is now being explored further by the European orbiter *VenusExpress*, which entered orbit around Venus on 11 April 2006. NASA's strategic exploration plans currently foresee the *Venus In Situ Explorer* and the *Venus Surface Explorer* to launch within the next decades.

The COSPAR Planetary Protection Guidelines categorize all missions towards Venus, whether orbiters, flybys, or landers, in category I. This infers that Venus is not of direct interest for understanding the processes of chemical evolution and the origin of life. Therefore, no planetary protection requirements are imposed. However, NASA and a recent study by the U. S. Space Studies Board of the National Academy of Sciences group all Venus missions, including sample return missions, into category II. This category applies to missions to celestial bodies that are of interest to astrobiology with regard to chemical evolution and the origin of life but pose only a remote risk of forward contamination by terrestrial microorganisms that would jeopardize future exploration. The Space Studies Board study even assumes that samples from the clouds or the surface of Venus would not raise a significant backward contamination risk to the inhabitants of Earth.

13.4.4
Missions to the Moons of the Giant Planets

Saturn's largest moon Titan (see Chapter 9), Jupiter's moon Europa (see Chapter 10), and probably also some of the other moons of the giant planets (see Chapter 9) are of special interest to astrobiology; therefore, missions to these moons need special consideration from a planetary protection point of view. Their relevance to planetary protection has already influenced the fate of the *Galileo* spacecraft, which explored Jupiter and its moons Io and Europa; at the end of its lifetime, in September 2003, it was deliberately steered into the gravitational pull of Jupiter to avoid accidental collision with – and possible contamination of – one of Jupiter's moons.

The surface of Titan was visited for the first time in 2005 by the European *Huygens* probe of the joint NASA–ESA *Cassini–Huygens* mission to the Saturn system (see Chapter 9).

Jupiter's moon Europa is considered to be potentially habitable for indigenous life (see Chapter 10). More than 95 % of the spectroscopically detectable material on its surface is H_2O. It has been established with high probability that this moon of Jupiter harbors an ocean of liquid water beneath a thick ice crust. In addition to liquid water, carbon and energy sources are required to support life as we know it on Earth. If carbon might have been delivered by impacts of various bodies (although crust resurfacing does not show very many impact craters), the question of energy sources is still open. The existence of liquid water beneath the ice crust may be the result of deep hydrothermal activity, radioactive decay, and/or tidal heating. In this case, conditions allowing prokaryotic-like life as we know it on Earth would have been gathered.

NASA's strategic exploration plan contains two missions to Jupiter's moon Europa: the *Europa Geophysical Explorer* in 2015 and the *Europa Astrobiological Lander* well after 2030. Although it is premature to conclude that either an ocean or biota exists on Europa, it is prudent to implement planetary protection procedures that assume the existence of both. Given the capabilities of terrestrial life to adapt to extreme conditions – such as heat, cold, pressure, salinity, acidity, dryness, and high radiation levels as well as combinations of these (see Chapter 5) – a conservative approach needs to be taken to protect the Europan environment. Properties of the Europan environment may present a target for contamination by terrestrial organisms carried on a spacecraft, which could colonize the entire moon via a subsurface ocean connection. Whereas the COSPAR Planetary Protection Guidelines classify orbiter and flyby missions to Europa as category III missions, the U. S. Space Studies Board, in the report "Preventing the Forward Contamination of Europa," as well as an International Expert Group at a COSPAR/IAU Workshop on Planetary Protection in 2002 recommended classifying all missions to Europa – including orbiters, landers, and penetrators – as category IV missions. This recommendation is based on the consideration that orbiting spacecraft have a significant probability of ultimately impacting on the surface. Any spacecraft component that is buried within the ice is effectively shielded from lethal levels of radiation, potentially allowing the survival of some contaminating organisms. Episodic resurfacing of the young Europan crust could then bring spacecraft components in contact with the ocean.

It was recommended that the bioload on spacecraft sent to Europa must be reduced to such a level that the probability of inadvertent contamination of its ocean is extremely low (10^{-4} per mission). Further studies are needed in order to determine

- the level of probability of survival and growth of terrestrial microorganisms in the Europan ice and ocean;
- methods of reducing the uncertainty in long-term survival of microorganisms in the Europan surface/near-surface environment and in the turnover time of Europa's icy crust; and
- the level of acceptable bioburden required for all parts of the spacecraft before launching.

13.4.5 Missions to Asteroids or Comets

Small bodies in the Solar System (asteroids and comets) represent a very large class of different types of objects. Imposing blanket forward-contamination controls on these missions seems not to be warranted. Currently, most such missions are classified as category I or II missions. Because such missions to comets and asteroids may become quite frequent in the future, further elaboration of the requirements may be needed.

Comets in particular are of interest to astrobiology, because among all celestial bodies, they contain the largest amount of organic molecules. They are considered the most pristine Solar System targets, bearing witness to the existence of a dynamic organic chemistry from the earliest stages of our Solar System (see Chapter 2). Therefore, comets have been the target of several recent missions, whether as a flyby (European *Giotto* mission to comet Halley), by collecting and returning samples of cometary dust from the tail (NASA *Stardust* mission to comet Wild 2), by creating a crater on its surface by sending a 350-kg copper projectile into a comet (NASA *Deep Impact* mission to comet 9P/Tempel 1), or by landing on a comet (ESA *Rosetta* mission to comet 67P/Churyumov-Gerasimenko).

The *Rosetta* mission, launched on 2 March 2004, includes an orbiter and a lander. After swing-by maneuvers at Earth (March 2005, November 2007, and November 2009) and Mars (February 2007) and asteroid flybys at Steins (2008) and Lutetia (2010), it will reach the target comet 67P/Churyumov-Gerasimenko in 2014. The COSPAR Planetary Protection Guidelines require different considerations at different mission phases:

- category III for the Mars gravity assist, which will bring the spacecraft as close as about 200 km to Mars, and
- category II for the comet exploration phase.

To meet the requirements of these categories, measures were taken to ensure very low probability of impact at Mars and control of organic contamination on the comet. Although the requirements of category II for the lander are less stringent, many measures and precautions were taken in order to control organic contamination and to avoid any confusion with regard to the results of onboard experiments searching for organic compounds and their precursors. This is an example where the requirements of the experiments are more stringent than those set by the COSPAR categorization. To meet the experiment requirements, the following measures were taken:

- The whole system was integrated in a class 100 000 clean room.
- All organic materials (list, nature, references, and amount) in the project Declared Material List handled by Product Assurance management were identified.
- At the design level, contamination sources were limited or controlled as far as possible, and the most possible contaminant coming from various sources (such as out-gassing, gas leaks, organic material pollution, or abrasion during lander or experiment operations, particularly drilling) or coming from the lander hardware was identified.
- Sensitive instrument parts were assembled in a class 100 or 1 000 clean room and were cleaned with specific detergents

and self-protection. Spectrometers were subject to purging and sealing, and some of them will perform an initial blank noise measurement in order to detect its internal contamination from terrestrial origin.

Japan's asteroid sample return mission *Hayabusa* (MUSES-C) aimed at collecting pieces from asteroid Itokawa and then bringing the samples back to Earth. It is planned that the sample return capsule will detach from the spacecraft and plunge through Earth's atmosphere for an intense reentry, with temperatures 30 times greater than those experienced by *Apollo* spacecraft. After reentry, the container should parachute to Earth, where it can be brought to a laboratory for study. Although the sampling of asteroid pieces has failed so far, this mission is of interest from the point of view of planetary protection. In view of the target comet type, the *Hayabusa* mission was classified as a category V mission with unrestricted Earth return.

13.5 Outlook: Future Tasks of Planetary Protection

After the first formulation of COSPAR Planetary Quarantine Objectives in 1969, which were targeted mainly towards missions to Mars and other planets, it was assumed that the question of indigenous life on the planets in consideration would be solved within 20 years, a period assumed to encompass about 100 exploratory-type missions. History has shown this to be an overoptimistic assessment, and it demonstrates that exploratory missions are influenced by a variety of parameters that are not in the hands of space scientists but are largely dependent on politics and economics.

Although the missions so far sent to Mars, Venus, Titan, and comets have provided a wealth of information on the habitability of the planets and their relevance for studies of chemical evolution and/or the conditions for origin and evolution of life, it will be the next generation of spacecraft that are targeted towards *in situ* search for signatures of extant or extinct life forms. The European lander *ExoMars*, to be sent to Mars in 2013, will be a prominent representative of those astrobiology missions (see Chapter 12). The COSPAR Planetary Protection Guidelines already sufficiently cover such missions of high astrobiology relevance looking for putative life forms on Mars. They are all classified as either category IVb or IVc missions.

Concerning Mars, future missions designed to bring planetary material back to Earth – such as Martian rocks, dust, or atmospheric samples – are classified as category V missions with restricted Earth return. For those missions, detailed protocols for sampling and sample handling and analysis to prevent both forward and backward contamination are under development and eventually need to be adopted by COSPAR.

A major task in view of planetary protection issues will be a mission sending humans to Mars, which is part of the strategic long-term planning of both ESA and

NASA. Activities associated with such human exploratory missions will interface with the terrestrial and extraterrestrial environment in three ways:

- The mission needs to be protected from the natural environmental elements of interplanetary space and the planet that can be harmful to human health, to the equipment, or to the operations. The following environmental elements need to be considered in order to protect humans and the equipment during their mission: cosmic ionizing radiation, solar particle events, solar ultraviolet radiation, reduced gravity, thin atmosphere, extremes in temperatures and their fluctuations, surface dust, and impacts by meteorites and micrometeorites.
- The specific natural environment of Mars should be protected so that it retains its value for scientific and other purposes. In order to protect the planetary environment, the requirements for planetary protection as adopted by COSPAR for lander missions need to be revised in view of human presence on the planet. Landers carrying equipment for astrobiological investigations require special consideration to reduce contamination by terrestrial microorganisms and organic matter to the greatest feasible extent. Records of human activities on the planet's surface should be maintained in sufficient detail that future scientific experimenters can determine whether environmental modifications have resulted from those explorations.
- The Earth and its biosphere need to be protected from potentially harmful agents brought back upon return of the explorers to their home planet Earth.

ESA, within its HUMEX study "Critical Issues in Connection with Human Exploratory Missions," has developed a roadmap identifying the research required to ensure astronauts' health and well-being during such a mission (Fig. 13.5).

Finally, the question arises whether the increasing robotic exploration of Mars and the eventual human exploration and settlement of that planet are likely to cause an environmental impact to scientifically important sites, regions of natural beauty, and historically important regions in the form of contamination with spacecraft parts and microbiota. The presence of crashed robots on Mars already raises important questions about the type of wilderness ethic one may apply to Mars and how this ethic is embodied within practical environmental policy.

The environmental ethics that underpin a system of preservation/conservation on the surface of other planets, such as Mars, have their origins in the "wilderness" debate, and therefore efforts for the protection of wilderness areas on Earth might be taken as an example. Anthropocentric reasons for declaring "natural parks" include the instrumental value of wilderness areas as sites for scientific study/preservation, human enjoyment, and the preservation of these regions for future

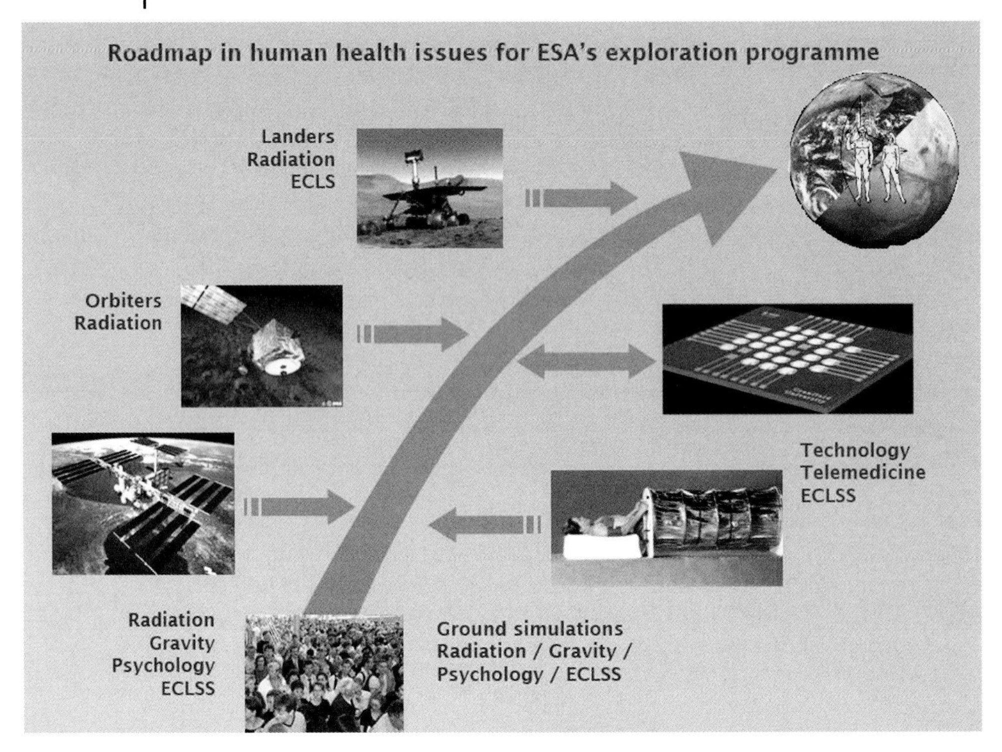

Fig. 13.5 Roadmap of steps regarding human-health issues required before a human mission to Mars takes place, as recommended in the HUMEX study in the frame of ESA's exploration program.

human enjoyment. However, even in the absence of humans, non-anthropocentric reasons – such as the intrinsic value of the Martian landscape and potential duties to indigenous microorganisms, if they are there – still exist and can drive policy even prior to human exploration. It is assumed that under the auspices of COSPAR (in addition to the current planetary protection policy, which is merely based on the scientific interpretation of the Outer Space Treaty), recommendations on how to deal with these ethical issues will be elaborated.

13.6 Further Reading

United Nations. *Treaty on Principles Governing the Activities of States in the Exploration and Use of Outer Space, including the Moon and Other Celestial Bodies* (the "Outer Space Treaty") referenced 610 UNTS 205 – resolution 2222(XXI) of December 1966.

13.6.1
Concerning COSPAR Planetary Protection Guidelines

COSPAR Planetary Protection Guidelines: http://www.cosparhq.org/scistr/PPPolicy.htm

DeVincenzi, D. L., Stabekis, P. D. Barengoltz, J B. Proposed new policy for planetary protection. *Adv. Space Res.* **1983**, *3,* 13–21.

DeVincenzi, D. L., Stabekis, P. D., Barengoltz, J. B. Refinement of planetary protection policy for Mars missions. *Adv. Space Res.* **1996**, *18, 311–316.*

Rummel, J. D. *Report of COSPAR/IAU Workshop on Planetary Protection,* Williamsburg, VA, USA, 2–4 April 2002. 2002, available at http://www.cosparhq.org/scistr/ppp.htm.

Rummel, J. D. Planetary exploration in the time of astrobiology: protecting against biological contamination. *Proc. Natl. Acad. Sci.* **2001**, *98,* 2128–2131.

Rummel, J. D., Stabekis, P. D., DeVincenzi, D. L., Barengoltz, J. B. COSPAR's planetary protection policy; a consolidated draft *Adv. Space Res.* **2002**, *30,* 1567–1571.

Rummel, J. D., Billings, L. Issues in planetary protection; history, policy and prospects, *COSPAR Information Bulletin* **2004**, *160,* 31.

Rummel, J. D., Billings, L. Issues in planetary protection: policy, protocol and implementation, *Space Policy* **2004**, *20,* 49–54.

13.6.2
Concerning Handbooks and Standards on Planetary Protection

CNES, *System Safety. Planetary Protection Requirements.* Référenciel Normatif, CNES *RNC–CNES-R-14* (Edition 4) December 2002.

Debus, A. and the working group on ECSS, European standard on planetary protection requirements, *Res. Microbiol.* **2006**, *157,* 13–18.

NASA, *NASA Standard Procedures for the Microbial Examination of Space Hardware,* NPR 5340.1B, NASA, Washington DC. 1980.

NASA; *Planetary Protection Provisions for Robotic Extraterrestrial Missions,* NPR 8020 12C; NASA, Washington DC. 2005

13.6.3
Concerning Bioload of Spacecraft

Debus, A., Auburtin, T., Darbord J. C., Hydrogen peroxide gas plasma sterilization process qualification for space application, *SAE publication ICES-128,* 2000.

La Duc, M. T., Nicholson, W., Kern, R., Venkateswaran, K. Microbial characterization of the Mars Odyssey spacecraft and its encapsulating facility. *Environ. Microbiol.* **2003**, *6,* 977–985.

La Duc, M. T., Satomi, M., Venkateswaran, K. *Bacillus odysseyi* sp. nov., a round-spore-forming bacillus isolated from the Mars Odyssey spacecraft. *Intern. J. System. Evolut. Microbiol.* **2004**, *54, 195–201.*

Link, L., Sawyer, J., Venkateswaran, K., Nicholson, W. Extreme spore UV resistance of *Bacillus pumilus* isolates obtained from an ultraclean spacecraft assembly facility, *Microbial Ecology* **2003**, DOI: 10.1007/s00248-003-1029-4 [online].

Nellen, J., Rettberg, P., Horneck, G., Streit, W. R. Planetary protection – Approaching uncultivable microorganisms, Adv. Space Res. **2006**, *38,* 1266–1270.

Newcombe, D. A., Schuerger, A. Benardini, J. N., Dickinson, D. and Venkateswaran, K. 2004. Survival of spacecraft associated microbes under simulated Martian UV irradiation. *Appl. Environ. Microbiol.* **2005**, *71,* 8147–8156.

Nicholson, W. L., Munakata, N., Horneck, G., Melosh, H. J., Setlow, P. Resistance of *Bacillus* endospores to extreme terrestrial and extraterrestrial environments. *Microb. Molecular Biol. Rev.* **2000**, *64,* 548–572.

Puleo, J. R., Fields, N. D., Bergstrom, S. L., Oxborrow, G. S., Stabekis, P. D., Koukol,

R. Microbiological profiles of the Viking spacecraft. *Appl. Environ. Microbiol.* **1977**, *33*, 379–384.

Rettberg, P., Horneck, G., Fritze, D., Verbarg, S., Stackebrandt, E., Kminek, G. Determination of the microbial diversity of spacecraft assembly facilities: first results of the project MiDiv, *Adv. Space Res.* **2006**, *38*, 1260–1265.

Schuerger, A. C., Mancinelli, R. L., Kern, R. G., Rothschild, L. J., McKay, C. P. Survival of endospores of *Bacillus subtilis* on spacecraft surfaces under simulated martian environments; implications for the forwards contamination of Mars, *Icarus* **2003**, *165*, 253–276.

Venkateswaran, K., Satomi, M., Chung, S., Kern, R., Koukol, R., Basic, C., White, D. Molecular microbial diversity of a spacecraft assembly facility. *Syst. Appl. Microbiol.* **2001**, *24*, 311–320.

Venkateswaran, K., Kempf, M., Chen, F., Satomi, M., Nicholson, W., and Kern. R. *Bacillus nealsonii* sp. nov., isolated from a spacecraft-assembly facility, whose spores are gamma-radiation resistant. *Int. J. Syst. Bacteriol.* **2003**, *53*, 165–172.

Venkateswaran, K., Hattori, N., La Duc, M. T., Kern, R. ATP as a biomarker of viable microorganisms in clean-room facilities. *J. Microbiol. Meth.* **2003**, *52, 367–377.*

13.6.4
Concerning Lunar Missions

Allton, J. H., Bagby, Jr. J. R., Stabekis, P. D. Lessons learned during Apollo lunar sample quarantine and sample curation. *Adv. Space Res.* **1998**, *22, 373–382.*

13.6.5
Concerning Missions to Terrestrial Planets

Cockell, C., Horneck, G. A planetary park system for Mars, *Space Policy* **2004**, *20*, 291–295.

Debus, A., Runavot, J., Rogovski, G., Bogomolov, V., Darbord J. C. *Mars '94/'96* mission planetary protection program, techniques for sterilization and decontamination level control, *IAF-95-I.5.10.* 1995.

Debus, A., Runavot, J., Rogovski, G., Bogomolov, V., Khamidullina, N., Trofimov, V. Landers sterile integration. Example of Mars 96 mission, *Acta Astronaut.* **2002**, *50*, 385–392.

Debus A., Planetary protection requirements for orbiter and netlander of the CNES/NASA Mars sample return mission, *Adv. Space Res.* **2002**, *30*, 1607–1616.

Debus, A. Planetary protection: organisation, requirements and needs for future planetary exploration missions, *Proceedings of the 37th ESLAB Symposium – Tools and Technologies for Future Planetary Exploration*, ESA SP-543, ESA-ESTEC, Noordwijk, The Netherlands, pp 103–114, 2003.

ESA web site: http://sci.esa.int/science-e/www/area/index.cfm?fareaid=9#

ESA, *Lessons Learned from the Planetary Protection Activities on the Beagle 2 Spacecraft*, Report Reference P3450026, ESA ESTEC, Noordwijk Holland, 2005.

Horneck, G., Facius, R., Reitz, G., Rettberg, P., Baumstark-Khan, C., Gerzer, R. Crticial issues in connection with human missions to Mars: protection of and from the Martian environment. *Adv Space Res.* **2003**, *31*, 87–95.

Horneck, G., Facius, R., Reichert, M., Rettberg, P., Seboldt, W., Manzey, D., Comet, B., Maillet, A., Preiss, H., Schauer, L., Dussap, C. G., Poughon, L., Belyavin, A., Heer, M., Reitz, G., Baumstark-Khan, C., Gerzer, R. *HUMEX, a Study on the Survivability and Adaptation of Humans to Long-Duration Exploratory Missions*, ESA SP 1264. ESA-ESTEC, Noordwijk, 2003.

Kminek, G., Rummel, J. D. (Eds.), *Planetary Protection – Human System Research and*

Technology, ESA-WPP-253, ESA-ESTEC, Noordwijk, 2006.

MEPAG Report: *Findings of the Special Regions Science Analysis Group* (Draft), MEPAG Special Regions Science Analysis Group, 2006; available at http://mepag.jpl.nasa.gov/reports/

NASA web site: http://mars.jpl.nasa.gov/overview/

Rummel, J. D., Race, M. S. *A Draft Test Protocol for Detecting possible Biohazards in Martian Samples Returned to Earth,* NASA/CP -2002–211842, NASA, Washington D. C. 2002.

Sims, M. R. (Ed.) *Beagle 2 Mission Report* University of Leicester UK, ISBN 1 898489 35 1, 2004.

Space Studies Board of the U. S. National Research Council, *Preventing the Forward Contamination of Mars,* The National Academies Press, Washington D. C. 2006, available online at http://fermat.nap.edu/catalog/11381.html

Space Studies Board of the US National Research Council Assessment of Planetary Protection Requirements for Venus Missions – Letter Report, 2006, available online at http://darwin.nap.edu/books/0309101506/html/3.html

Space Studies Board of the US National Research Council, *Safe on Mars, Precursor Measurements Necessary to Support Human Operations on the Martian Surface,* National Academy Press, Washington D. C., 2002.

Lessons Learned from the Viking Planetary Quarantine and Contamination Control Experience, Bionetics Corporation, contract NASW-4355. NASA, Washington D. C. 1990.

13.6.6
Concerning Missions to Jupiter's Moon Europa

Space Studies Board of the US National Research Council, *Preventing the Forward Contamination of Europa, 2000,* available online at http://www.nap.edu/openbook/NI000231/html/index.html

Planetary Protection and Contamination Control Technologies for Future Space Science Missions, Document D-31974, JPL/California Institute of Technology, Pasadena, 2005.

13.7
Questions for students

Question 13.1

Describe the COSPAR Planetary Protection Guidelines, the rationale and requirements for different missions within our Solar System.

Question 13.2

Develop a Planetary Protection Plan for a mission landing on a carbonaceous asteroid.

Question 13.3

Describe the planetary protection requirements and implementation plans for a robotic lander mission to Mars depending on the landing site and the purpose of the mission.

Question 13.3

Assess the impact of a human mission to Mars from the point of view of astrobiology and planetary protection (pros and cons).

Question 13.4

Give reasons, other than scientific ones, for requiring protection of the planets of our Solar System and justify them.

Index

Complete Course in Astrobiology. Edited by Gerda Horneck and Petra Rettberg

ISBN: 978-3-527-40660-9

d

f

g

q

r

s

Related Titles

Shaw, A.

Astrochemistry

From Astronomy to Astrobiology

256 pages

Hardcover

ISBN: 978-0-470-09136-4

Stahler, S. W., Palla, F.

The Formation of Stars

865 pages with 511 figures and 21 tables

2004

Softcover

ISBN: 978-3-527-40559-6

Roos, M.

Introduction to Cosmology

294 pages

Softcover

ISBN: 978-0-470-84910-1

Darling, D.

The Universal Book of Astronomy

From the Andromeda Galaxy to the Zone of Avoidance

576 pages

Hardcover

ISBN: 978-0-471-26569-6

Shore, S. N.

The Tapestry of Modern Astrophysics

888 pages

2003

Hardcover

ISBN: 978-0-471-16816-4

Shapiro, R

Planetary Dreams – The Quest to Discover Life Beyond Earth

320 pages

2001

Softcover

ISBN: 978-0-471-40735-5

Boss, A.

Looking for Earths

The Race to Find New Solar Systems

252 pages

2000

Softcover

ISBN: 978-0-471-37911-9